Renewable Energy and Resources

Renewable Energy and Resources

Edited by **George Thomson**

RCALLISTO REFERENCE

New York

Published by Callisto Reference,
106 Park Avenue, Suite 200,
New York, NY 10016, USA
www.callistoreference.com

Renewable Energy and Resources
Edited by George Thomson

International Standard Book Number: 978-1-63239-767-6 (Hardback)

Printed in the United States of America.

Contents

Preface

Renewable energy is derived from those resources that can be replenished in a shorter duration of time than conventional energy resources. Some of the common renewable resources are wind, sunlight, water, etc. Deployment of renewable resources for energy production may result in decelerated climatic degradation, economic benefits and energy security. This book contains some path-breaking studies in the fields of renewable energy and resources. The contents of this book will help the readers understand the modern approaches and applications of this subject. A number of latest researches have been included to keep the readers up-to-date with the global concepts in this area of study. Scientists and students actively engaged in this field will find this book full of crucial and unexplored concepts.

The researches compiled throughout the book are authentic and of high quality, combining several disciplines and from very diverse regions from around the world. Drawing on the contributions of many researchers from diverse countries, the book's objective is to provide the readers with the latest achievements in the area of research. This book will surely be a source of knowledge to all interested and researching the field.

In the end, I would like to express my deep sense of gratitude to all the authors for meeting the set deadlines in completing and submitting their research chapters. I would also like to thank the publisher for the support offered to us throughout the course of the book. Finally, I extend my sincere thanks to my family for being a constant source of inspiration and encouragement.

Editor

A Pedestrian Approach to Indoor Temperature Distribution Prediction of a Passive Solar Energy Efficient House

Golden Makaka

University of Fort Hare, Private Bag Box X1314, Alice 5700, South Africa

Correspondence should be addressed to Golden Makaka; gmakaka@ufh.ac.za

Academic Editor: Cheng-Xian Lin

With the increase in energy consumption by buildings in keeping the indoor environment within the comfort levels and the ever increase of energy price there is need to design buildings that require minimal energy to keep the indoor environment within the comfort levels. There is need to predict the indoor temperature during the design stage. In this paper a statistical indoor temperature prediction model was developed. A passive solar house was constructed; thermal behaviour was simulated using ECOTECT and DOE computer software. The thermal behaviour of the house was monitored for a year. The indoor temperature was observed to be in the comfort level for 85% of the total time monitored. The simulation results were compared with the measured results and those from the prediction model. The statistical prediction model was found to agree (95%) with the measured results. Simulation results were observed to agree (96%) with the statistical prediction model. Modeled indoor temperature was most sensitive to the outdoor temperatures variations. The daily mean peak ones were found to be more pronounced in summer (5%) than in winter (4%). The developed model can be used to predict the instantaneous indoor temperature for a specific house design.

1. Introduction

In the recent years indoor temperature distributions are gaining greater attention in building design and operations [1]. Due to high energy consumption by buildings and the ever energy price increase there is an urgent need to predict the indoor temperature distribution and energy consumption by building during the design stage [2]. Detailed and precise predictions of indoor thermal comfort and control of the indoor thermal conditions and fast dynamic model of the indoor temperature distribution are needed in the design stage of buildings so as to select the correct materials that will produce a thermally comfortable indoor environment [3]. This can result in the reduction of energy consumption in trying to keep the indoor environment within the temperature comfort levels (18°C to 28°C) and relative humidity (30% to 70%) [4]. Due to the diversity of building materials and climates there is need to pay particular attention to material properties and the availability of suitable materials in different local areas to minimize construction costs. Recently an amount of building computer design software has

been developed to simulate indoor temperature distributions including ECOTECT, DOE, ESP-r, and EnergyPlus [5].

ECOTECT is a complete building design and environmental analysis tool that covers the full range of simulation and analysis functions required to understand how a building design will operate and perform. It allows designers to work easily in 3D and apply all the tools necessary for an energy efficient and sustainable future.

The DOE provides the building construction and research communities with an up-to-date, unbiased, well-documented computer program for building energy analysis. Using DOE, designers can quickly determine the choice of building parameters which improve energy efficiency while maintaining thermal comfort. A user can provide a simple or increasingly detailed description of a building design or alternative design options and obtain an accurate estimate of the proposed building's energy consumption, interior environmental conditions, and energy operation cost.

The present predicting indoor temperature methods are cumbersome. However, fast and simpler predicting indoor temperature models need to be developed to help architects to design buildings with thermally comfortable indoor environment and thus advise building constructors to select better thermally performing building materials.

The mathematical description of thermal behaviour of building systems is complex [6] as most of the parameters involved are probabilistic in behaviour. The activities associated with each room (subsystem, e.g., kitchen and bathroom) complicate indoor temperature modeling process. The positioning of heat sources and their frequency of use do have a significant impact on the indoor temperature distribution. The modeling of the thermal behaviour involves the modeling of several interconnected subsystems, with each one containing long-time constants, nonlinearities, and uncertainties such as convection coefficients and material properties [7]. Moreover, external unpredicted perturbations such as temperature and relative humidity, soil temperature, solar radiation effects, and other sources of energy such as people illumination and equipment should be taken into account [8]. The generalized mathematical model for predicting indoor temperature of any building has to take into account the weather conditions to which the building is exposed and thermal properties of the building [9].

A number of indoor temperature predicting models have been developed. Givoni [10] developed very simple experimental predictive formula of the indoor temperatures of uninhabited houses. It was demonstrated that the indoor daily maximum and daily average temperatures (in degree Celsius) could be predicted on the basis of only the daily outdoors average temperatures. The formula developed by Givoni was as follows:

$$T_{\text{max-in}} = T_{\text{max-out}} - 0.3\left(T_{\text{max-out}} - T_{\text{min-out}}\right) + 1.6, \quad (1)$$

where $T_{\text{max-in}}$ is the indoor maximum temperature, $T_{\text{max-out}}$ is the outdoor maximum temperature, and $T_{\text{min-out}}$ is the outdoor minimum temperature.

Ogoli [11] also did a similar study and developed the following formula:

$$T_{\text{max-in}} = T_{\text{max-out}} - 0.4\left(T_{\text{max-out}} - T_{\text{min-out}}\right) + 2.44. \quad (2)$$

Construction materials also have a significant impact on the indoor temperature distribution. The properties of construction materials were also attributed to different leveling temperature constants in (1) and (2) which are 1.6°C and 2.44°C, respectively. The differences between the formulae of Givoni [12] and Ogoli [11] were also due to the different weather conditions and the fact that Givoni's study was carried out in open naturally ventilated test chambers, while Ogoli's study was for closed test chambers. Givoni and Ogoli only considered the outdoor temperature as the only influencing parameter to the indoor temperature. However, the indoor temperature is affected by a number of parameters, which include wind speed and direction, relative humidity, solar radiation, outdoor temperature, and landscape [13]. Some of these parameters are not independent, for example, relative humidity and temperature [14]. High wind speeds have the tendency to lower the indoor temperature and the effect depends on the wind direction with respect to

window orientations and their operation. High solar radiation heats the walls and roof which in turn transfer the thermal energy through conduction and convection to the indoor environment. In an unoccupied house one can have a complete control over the conditions of the house, whether to open or close, shade or unshade the windows, and so forth. In occupied buildings the situation is very different as the occupants have complete freedom to change the conditions according to their changing needs. The activities of the occupants have a significant influence on the indoor temperature distribution [15].

Hemmi [16] developed a model which divides the room air volume into several ideally mixed zones that were connected both in series and in parallel with respect to the mass air flow in the room. The precondition for the model of Hemmi was that the air mass flow between one zone and another could be prescribed but Hemmi never gave a theoretical method of how it could be prescribed. Peng and van Paassen [17] studied the prediction of indoor dynamic temperature distributions by using a fixed-flow-field obtained from (computational fluid dynamics) CFD calculation. The fundamental differential equations in CFD are the continuity, momentum conservation, and energy conservation equations. The basic idea of CFD is that the flow domain is first divided in thousands of finite volumes by setting a grid. For every finite volume of the grid, the conservation equations are solved iteratively until the solutions of all variables for all volumes have converged. However, CFD method is time-consuming [18] and faster easy methods are needed to predict indoor temperature distribution and assist architects to design buildings according to thermal performance.

Another very important approach is the statistical model approach. This approach allows estimating the probabilistic future thermal behaviour of buildings based on monitored statistical information, such as outdoor temperature and relative humidity. The thermal performance of buildings is probabilistic as it does depend on a number of stochastic parameters such as the occupants behaviour in operating the ventilation components and indoor human activities. Some can even choose to close doors to avoid pets getting inside thus compromising the thermal performance of the house. Indoor temperature depends not only on outdoor temperature but also on other parameters such as occupants and heat producing equipment (whether they are on or off). These parameters sometimes are uncertain and in some cases are difficult to find the exact information. This leads to the uncertainty in indoor temperature calculation result. Parameters which affect indoor temperature can be divided into two types, that is, uncontrollable and controllable parameters. Uncontrollable parameters such as ambient air temperature, wind direction and speed, and solar radiation affect the change in the probability of the indoor temperature distribution, while controllable parameters, such as type and shading coefficient of glass, do not affect the change in the probability of the indoor temperature variation.

The aim of this paper is to use the statistical method to predict the indoor temperature distribution in occupied low cost passive solar house constructed from fly ash bricks. Fly ash bricks were used because of their advantageous properties, low thermal conductivity, low water absorption, and high thermal compression [4]. The model house, passive solar house (PSH), was built in Somerset East (South Africa) which has the following properties: 32°42/S latitude, 25°33/E longitude, and an altitude of 790 m. It experiences a subtropical type of climate with long hot summer months and moderate sunny winter months with annual daily average of 7-8 sunshine hours [19].

During the monitoring period three people were staying in the house, a couple and their grandson. The two were pensioners and they were always at home, while the grandson attends school from 07:30 to 13:00.

2. Methodology

2.1. Simulation. The modeling approach in this study includes the following:

  (i) Selection of indoor temperature simulation software (ECOTECT and DOE were selected).

  (ii) Development of the passive solar house (PSH) model that could be deployed for the indoor temperature simulation.

  (iii) Construction of the PSH model.

  (iv) Monitoring the thermal behaviour of the PSH model.

  (v) Development of the statistical model for the prediction of the indoor temperature.

The simulation results from DOE and ECOTECT were compared with the developed statistical predicting model. ECOTECT was used in the simulation process because it has the ability to identify the input files of a building by drawing a model of the building and it has the ability to export the input data to other simulation modeling programs, such as DOE. For the purpose of simulation the following parameters were entered: temperature, relative humidity, diffuse solar radiation, global radiation, wind speed, cloud cover, rainfall, wall thickness, types of wall materials, types of roof materials, type of glass window, window closed or open, type of floor material, length of overhangs, and slab thickness.

2.2. Data Acquisition System. Monitoring sensors were installed indoors and outdoors. A weather station was installed to measure the comfort parameters, that is, wind speed and direction, solar radiation, temperature, and humidity. Figure 1 shows the weather station installed on the top of the roof and the automated data acquisition system. Sensors installed to monitor the thermal performance of the house included 26 Type K thermocouples for the measurement of outdoor and indoor air temperatures, 2 HMP50 temperature-humidity probe, model 03001 wind sentry anemometer and vane, and a pyranometer.

Thermocouples were placed at three height levels, thus creating three planes: the lower plane level at a height of $z = 1$ m, the middle plane level at a height of $z = 1.5$ m, and the upper plane level at a height of $z = 2.6$ m. The thermocouples for lower plane level are labeled: 26, 19, 20, 21, 22, 23, 18, and 25; for the middle plane level

they are labeled: 17, 10, 11, 12, 13, 14, 6, and 16; and for the upper plane level they are labeled: 8, 1, 2, 3, 4, 9, 5, 15, 24, and 7. Figure 2 shows how thermocouples were distributed to measure indoor and outdoor temperatures. A HMP50 temperature-humidity sensor was placed in the centre of the building at a height of about 2 m above the floor, while the second HMP50 temperature-humidity sensor was placed on the top of the roof. All sensors were connected to a CR1000 data logger and readings were recorded at 30-minute interval.

2.3. Design of the Model House (PSH). The house design was made to minimize space-conditioning loads by using fly ash brick walls and a concrete floor as the thermal mass. The house floor plan measured 6880 mm by 6580 mm. An open plan layout was adopted in order to optimize natural ventilation since mechanical ventilation systems were avoided to keep the running cost of the house low. The roof was split into two, the lower and upper roof. The lower roof faces north, while the higher roof faces south. This was done to insert clerestory windows making it possible to direct solar radiation to the desired rear zone (floor and southern wall) and to maximize day lighting, thus minimizing the use of electricity during the day. A door and two windows are on the north wall, a door and a small window are on the west wall, two windows are on the south wall, and one small window is on the east wall. Table 1 shows the performance parameters of materials used to construct the PSH. Thermal admittance is a measure of a material's ability to absorb heat from and release heat to a space over time and this gives an indication of the thermal storage capacity of a material. Thermal admittance (Y) is calculated as $Y = (Q/A)\Delta T$, where Q is the heat transfer, A is the surface area, and ΔT is the temperature differences between the surfaces of the material. The thicker and more resistive the material is the longer it will take for heat waves to pass through. The time delay due to the thermal mass is known as a time lag. The reduction in temperature on the inside surface compared to the outside surface is known as decrement factor. The decrement factor is calculated as $\Gamma = (T_{out} - T_{in})/T_{out}$, where $T_{out} > T_{in}$.

2.4. Operation of the House. In winter (May to August) the sun rises almost at the northeast but following a low northern path in the sky and then sets at the northwest. From May to August the maximum sun's angle of each day ranges from 34° to 45° and these maximum angles occur at around 12:15 pm local time with 21st June having the smallest angle. So the north facing windows allow solar radiation to enter the house and the clerestory windows allow the south wall and the far south floor to receive solar radiation. The thermal masses of high heat capacity (i.e., concrete floor of 100 mm in thickness and the wall made from fly ash bricks) absorb the short wave solar radiation during the day. Figure 3 shows how the lower winter sun enters indoors and heats the thermal mass and how the summer sun is prevented from entering indoors by the overhangs.

Since the thermal masses have high heat capacity, they will absorb large amount of solar energy but with minimal temperature variation. This minimal temperature variation

FIGURE 1: Experimental setup. (a) External sensors. (b) Automated data acquisition system.

TABLE 1: Properties of the basic material elements used in the construction of the PSH.

Element	U value (W/m^{2}K)	Admittance (W/m^{2}K)	Thermal decrement	Thermal lag (hours)
Floor	2.56	4.20	0.70	4
Roof	7.14	7.10	0.00	1
Wall	1.78	4.59	0.37	8
Window	6.00	6.00	—	—

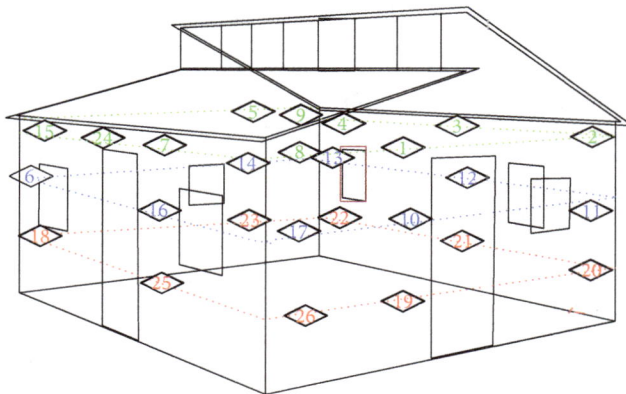

Red: lower level   Green: upper level
Blue: middle level

FIGURE 2: Distribution of thermocouples.

avoids the indoor overheat, thus keeping the indoor temperature within the comfort levels. At night, as the outdoor temperature falls, the thermal mass slowly radiates long-wave radiation heating the indoor air therefore keeping the indoor air temperature within the thermal comfort levels. The window glazes are opaque to the long-wave thermal radiation, so the thermal radiation emitted by the thermal mass is trapped indoors and this minimizes the heat losses.

In summer the sun almost rises from the east and sets on the west and the overhangs on the windows eliminate the possibility of the sunrays to penetrate indoors. With reference to the clerestory windows, the upper roof was extended out by 200 mm, while the lower roof was extended in by 100 mm and this arrangement eliminates the possible entrance of solar radiation in summer while allowing maximum entrance in winter. Somerset East experiences westerly prevailing winds in summer, so the small windows on the west and east make it possible to control the ventilation rate. The clerestory windows and the south windows enhance controllable natural ventilation rate and help to control the temperature to remain within the comfort zone (18°C to 28°C).

3. Results

3.1. Outdoor and Indoor Temperatures. The behaviour of occupants was found to have a significant impact on the indoor thermal environment, that is, the management of the ventilation components (closing and opening of windows and doors). The indoor temperature within the building was found to be the result of heating and heat transfer between the building and the environment. Figure 4 shows the fluctuations of the indoor temperature in response to the outdoor temperature fluctuations.

A common feature observed was the tendency of similarity between the peak indoor temperatures and the peak outdoor temperatures. Givoni also saw this observation in several studies in California [10, 12] and Israel [20]. The maximum indoor temperature was found to lag behind the maximum outdoor temperature by an average of two hours. From Figure 4 it can be observed that there is distinct general variation of the indoor temperature in response to the outdoor temperature for summer and winter. With reference

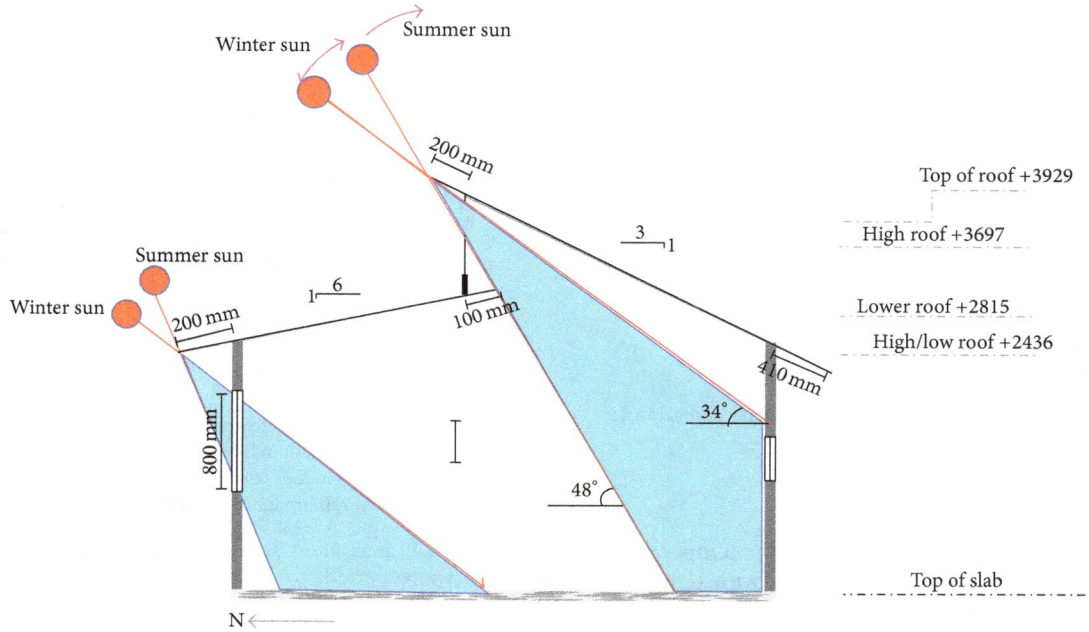

FIGURE 3: Operation of the passive solar house.

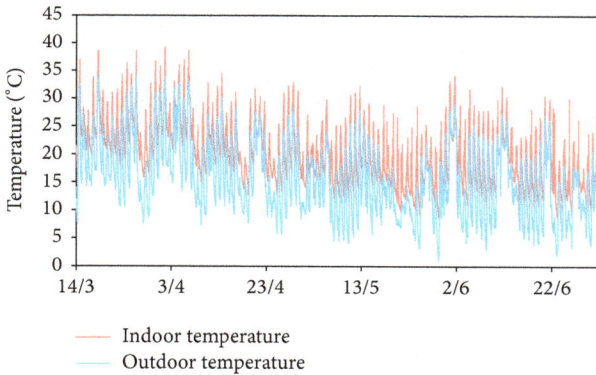

FIGURE 4: The response of the indoor and outdoor temperature variations.

to Figure 4, the average temperature and the daily mean temperature amplitude for the whole period were calculated. Using these parameters the best-fit equations for the daily temperature variations were established. Equations (3) and (4) are the best-fit equations for the daily general variation of the indoor and outdoor temperatures for the summer period:

$$T_{\text{in}} = 21.3 + 5.2 \cos\left(\frac{\pi}{12}t - \frac{7\pi}{6}\right), \tag{3}$$

with $R^2 = 0.974$ and standard error of $\pm 2\%$;

$$T_{\text{out}} = 23.4 + 7.7 \cos\left(\frac{\pi}{12}t - \frac{4\pi}{3}\right), \tag{4}$$

with $R^2 = 0.988$ and standard error of $\pm 1.8\%$.

Equations (5) and (6) are the best-fit equations for the variation of the indoor and outdoor temperatures for the winter period:

$$T_{\text{in}} = 19.5 + 4.5 \cos\left(\frac{\pi}{12}t - \frac{17}{12}\right), \tag{5}$$

with $R^2 = 0.967$ and standard error of $\pm 2.2\%$;

$$T_{\text{out}} = 16.7 + 9.2 \cos\left(\frac{\pi}{12}t - \frac{5\pi}{4}\right) \tag{6}$$

with $R^2 = 0.979$ and standard error of $\mp 2.1\%$, where T_{out} and T_{in} are outdoor and indoor temperature, respectively. For the summer period the maximum outdoor temperatures were observed to occur at around 14:00 and the maximum indoor temperature occurred at around 16:00, while for the winter period the maximum outdoor temperature occurred at around 15:00 and the maximum indoor temperature occurred at around 17:00. With reference to (3) to (6) it was observed that in summer the average indoor temperature swing was about 10.4°C, while in winter the average indoor temperature swing was about 9°C.

The time delay (2 hours) of the indoor temperature to attain the peak value in relation to the outdoor temperature indicates that the wall materials used have high heat retention capacity with minimal temperature variation. This makes the indoor environment warm in winter, and in summer the sun shaded off thus keeping the indoor temperature within the comfort levels. When the sun sets at around 17:00 in winter the indoor air will be at highest temperatures thus keeping the indoor environment within the comfort levels for a considerable time eliminating the need of using heating devices.

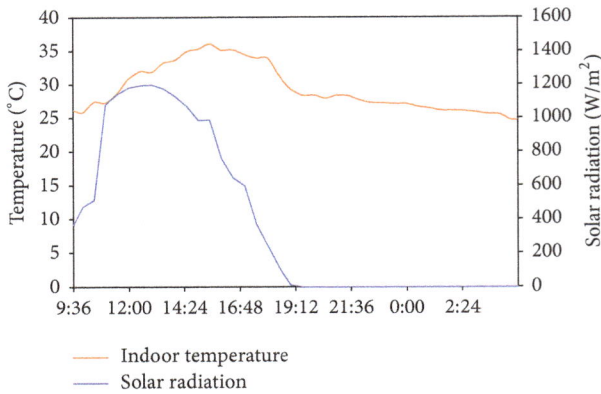

FIGURE 5: Variation of indoor temperature with solar radiation (5 November).

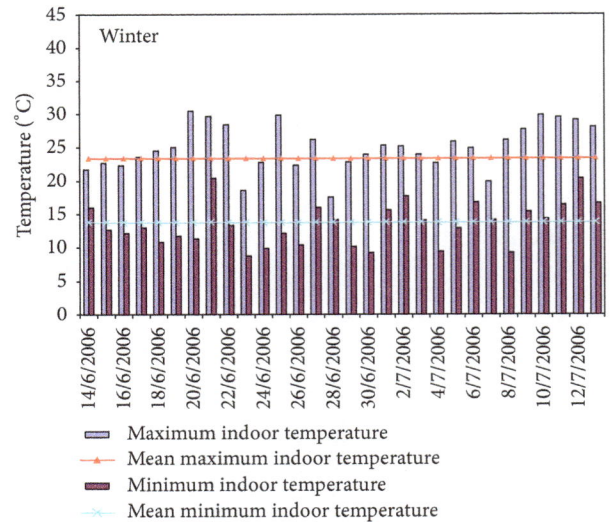

FIGURE 6: Daily maximum and minimum temperatures indoors during the winter period.

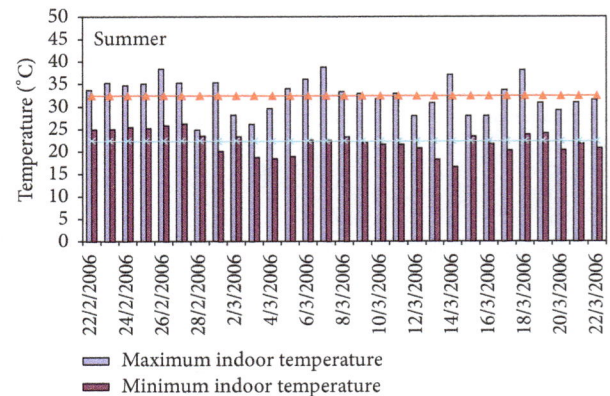

FIGURE 7: Daily maximum and minimum temperatures indoors during the summer period.

In South Africa the recommended indoor temperature comfort levels range from 18°C to 28°C and with reference to Figure 4 the indoor temperature was observed to be within the comfort levels for 85% of the time. This means that during this period there will be no need of heating/cooling the indoor environment resulting in energy saving. The indoor temperatures were always lower than the outdoor temperatures during the daytime but the situation was reversed during night time. As the sun sets the outdoor temperature falls much faster than the indoor temperature and heat flows from the inner walls to the outer walls. With reference to Figure 4 the mean indoor temperature was lower than the mean outdoor temperature by about 3.1°C, while the mean indoor minimum temperatures were greater than the mean minimum outdoor temperature by 4.5°C.

3.2. Variation of the Indoor Temperature with Solar Radiation.

The solar radiation was seen to have a significant impact on the indoor temperature distribution. Figure 5 shows the variation of the indoor temperature with the solar radiation for a particular day (5 November). From Figure 5 it can be observed that the indoor temperature generally follows the fluctuations of the solar radiation but with a thermal time delay of about 2 h 30 min.

The thermal time delay is due to the thermal inertia of the building materials used. The thermal time delay makes it possible for the indoor temperature to remain within the comfort levels during the winter nights.

Figures 6 and 7 show the daily maxima and minima temperature variations for winter and summer seasons, respectively. From the two graphs it can be seen that the summer period experienced higher daily indoor temperatures swings. In winter the mean maximum indoor temperature was found to be within the temperature comfort levels, while in summer it was above the upper bound comfort level by 5°C. However, in winter, the mean minimum indoor temperature was lower than the lower bound comfort levels by 3°C, while in summer the mean indoor temperature was within the comfort levels. During winter and night, the zinc roof loses thermal energy fast as the ambient temperature falls rapidly. However, this fast fall of temperature can be mitigated by the installation of ceiling thus maintaining the indoor temperature within the comfort levels.

3.3. Correlating Indoor and Outdoor Temperatures.

Figure 8 shows the best-fit correlation of the indoor to the outdoor temperature. The correlation function was found to be as follows:

$$T_{in} = 1.0699 T_{out} - 6.1731: \quad \text{for } p = 0.01;$$

$$R^2 = 0.8594: \text{standard error of } \pm 2.312°C. \tag{7}$$

From Figure 8 it was observed that the outdoor temperature has a significant impact on the indoor temperature. The data dispersion indicates that there are other outdoor weather parameters which affect the indoor temperature distribution. It must be noted that human activities significantly affect the indoor temperatures and this gives data dispersion. In the

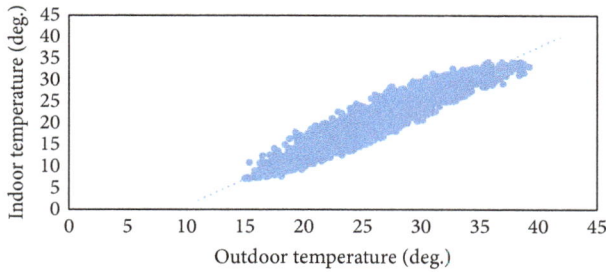

FIGURE 8: Variation of indoor and outdoor temperatures.

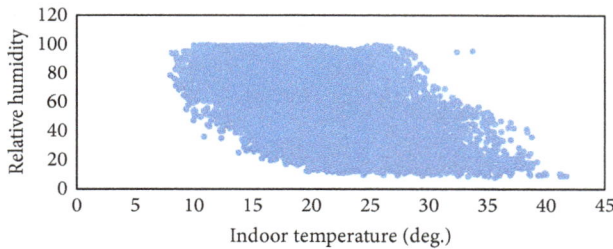

FIGURE 9: Variation of the indoor temperature with outdoor relative humidity.

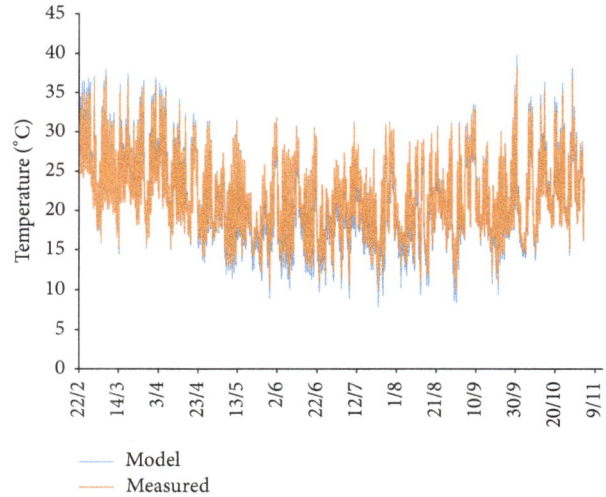

FIGURE 10: Comparison of the measured and the modeled temperatures.

morning the indoor temperature is generally higher than the outdoor temperature; however, as the sun rises, the outdoor temperature increases faster than the indoor temperature. The time lag of the indoor temperature in response to the outdoor temperature is contributed mainly by the thermal mass (walls), thus resulting in outdoor temperatures higher than the indoor temperatures.

3.4. Correlating T_{in} and Outdoor Relative Humidity. Figure 9 shows the correlation of indoor temperature to outdoor relative humidity and high data dispersion was observed. The wide data dispersion means that the outdoor temperature is not the only outdoor weather parameter which affects the indoor temperature. The correlation function was found to be as follows:

$$T_{in} = 0.1184\,(RH)_{out} + 33.389:$$

$$R^2 = 0.3099; \text{standard error of} \pm 7.6°C.$$

(8)

3.5. Indoor Temperature Modeling for the Passive Solar House. The performance of the house was observed to depend on how the occupants operate the house and it was assumed that there was no heating or cooling appliance used to modify the indoor thermal environment. Based on the monitoring results the first stage in the development of the experimental predictive model was to analyze the patterns of how the indoor temperature varies with outdoor weather parameters. Taking into account this possible dependence, a predictive model for the indoor temperature was developed. A linear dependence was proposed. Consider

$$T_{in}\,(T_{out}, I, V_{ws}, RH) = 0.$$

(9)

The above formula can be expressed in the following form:

$$T_{in} = a_1 * T_{out} + a_2 * (RH)_{out} + a_3 * I + a_4 * V_w + a_5,$$

(10)

where T_{in} is indoor temperature, T_{out} is outdoor temperature, V_w is wind speed, I is solar radiation, RH_{out} is outdoor relative humidity, and a_i are regression coefficients, where $i = 1, 2, 3, 4, 5$.

The above proposed model is a simplification of a complex dependence. It must be noted that all the parameters involved are not independent. For example, the outdoor temperature greatly depends on solar radiation and wind speed. Using regression analysis the coefficients in (10) were obtained. The obtained indoor temperature predicting model is

$$T_{in} = 0.81820 T_{out} + 0.013562\,(RH)_{out} - 0.12907 V_w + 0.00038 I + 8.18$$

(11)

with $R^2 = 0.950$ and standard error of ±5.

Figure 10 shows the comparison of the measured indoor temperature (February to July) and the predicted indoor temperature using the prediction model (11). The predicted indoor temperature using the developed model is in good agreement (95%) with the measured temperature values.

From (11) it can be noted that the outdoor temperature, solar radiation, and outdoor relative humidity have a positive contribution to the indoor temperature; that is, when the outdoor temperature, solar radiation, and relative humidity increase the indoor temperature also increases. But when the wind speed increases the tendency is to lower the indoor temperature, that is, negative contribution. From the models it was seen that the outdoor weather parameters do have varying impact on the indoor temperature but with highest sensitivity to the outdoor temperature variation. The determination of the sensitivity of the models revealed that

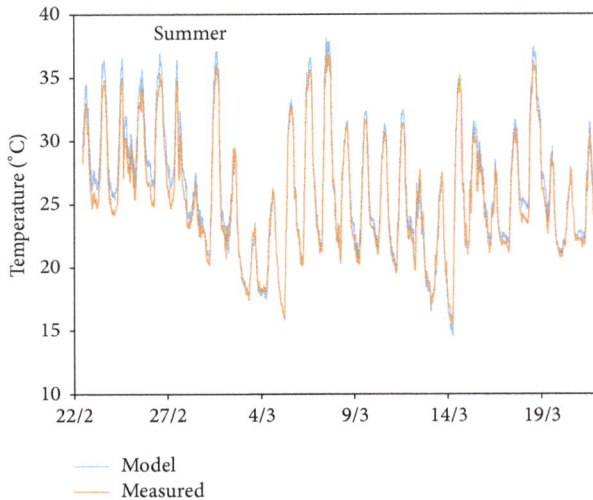

FIGURE 11: Summer comparison of the measured and the modeled temperatures.

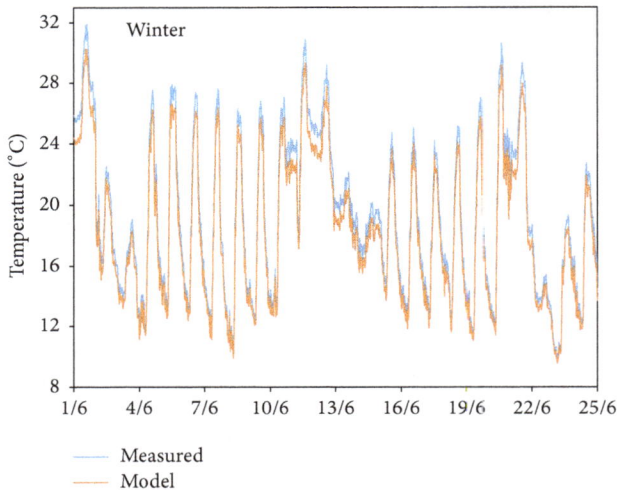

FIGURE 12: Winter comparison of the measured and the modeled temperatures.

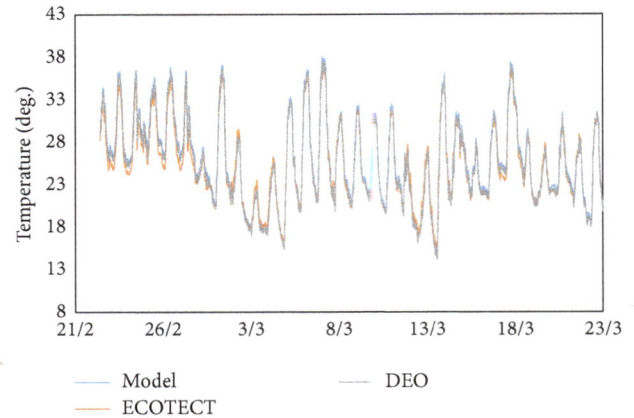

FIGURE 13: Comparison of simulation results of ECOTECT, DOE, and the model.

much higher measured indoor temperature than the model can predict. However, measured and predicted minimum temperatures were found to be in good agreement with small differences of about 0.5% for summer and 0.6% for winter. The model was found to predict lower than the minimum measured temperatures. It is worth mentioning that the activities of the occupancies play an important role in indoor temperature distribution. The results suggest that the new developed model equation can be used to predict the indoor temperature for a high thermal building (houses constructed from fly ash bricks) with a passive solar design in the South African type of climate.

3.6. The Comparison of the Model with ECOTECT and DOE Simulation Results. Figure 13 shows a comparison of the simulation results using ECOTECT and DOE and the developed predictive model. It was observed that the simulation results agree very well with the developed model. At minimum temperatures ECOTECT was observed to record the lowest values followed by DOE, but at peak temperatures the three were observed to agree very well. This means the model can be used to predict the instantaneous indoor temperature if the instantaneous outdoor weather parameters are known thus allowing architects to predict the indoor temperature.

4. Conclusions

Indoor temperatures within buildings are dynamic as there are a number of stochastic activities which take place indoors. The indoor temperature prediction formula was modeled. It was found that the instantaneous indoor temperature could be predicted if the outdoor weather variables are known for a given house design constructed using specific building materials. Outdoor temperature was found to be the most influential parameter on the indoor temperature distribution. Relative humidity, temperature, and solar radiation were found to have an incremental effect on the indoor temperature; that is, when these parameters increase the tendency is to increase the indoor temperature. Wind speed was found to have lowering effect on the indoor temperature; that is, when

the indoor temperature is 89% dependent on the outdoor temperature, 2% on relative humidity, 3.5% on wind speed and direction, 4.6% on solar radiation, and 0.9% on other undefined parameters. From Figure 10 it can be seen that the predicted temperatures agree well with the measured data; however, at high temperatures there are significant differences with the measured temperature being higher than the predicted. The difference between the measured and the predicted temperatures is mostly noticed at peak temperatures. The daily mean peak temperature differences of the measured and the predicted temperatures are more pronounced in summer (5%) than in winter (4%). Figures 11 and 12 show the comparison of the measured and modeled indoor temperatures for summer and winter, respectively.

Significant differences were noted during cooking periods when a lot of heat was generated indoors, resulting in the wind speed increases the effect is to decrease the indoor temperature. The model was found to be in good agreement with simulations results using ECOTECT building design computer software and DOE building computer simulation software. Results also suggest that the developed model can be used to predict the instantaneous indoor temperature distribution for a specific house design resulting in thermally comfortable indoor environment with minimal energy consumption.

Conflict of Interests

The author declares that there is no conflict of interests regarding the publication of this paper.

References

[1] T. Olofsson, J.-U. Sjögren, and S. Andersson, "Energy performance of buildings evaluated with multivariate analysis," in *Proceedings of the Building Simulation Conference*, pp. 891–898, Montreal, Canada, 2005.

[2] C. Federspiel, Q. Zhang, and E. Arens, "Model-based benchmarking with application to laboratory buildings," *Energy and Buildings*, vol. 34, no. 3, pp. 203–214, 2002.

[3] H. Li, B. Lin, X. Zhou, Y. Zhu, G. Han, and F. Meng, "The design of a comfortable thermal environment by simulation: a case study," in *Proceedings of the Building Simulation Conference*, pp. 1820–1825, Beijing, China, September 2007.

[4] G. Makaka and E. Meyer, "Temperature stability of traditional and low-cost modern housing in the Eastern Cape, South Africa," *Journal of Building Physics*, vol. 30, no. 1, pp. 71–86, 2006.

[5] W. Abanomi and P. Jones, "Passive cooling and energy conservation design strategies of school buildings in hot, arid region, Riyadh, Saudi Arabia," in *Proceedings of the International Conference "Passive and Low Energy Cooling for the Built Environment"*, Santorini, Greece, 2005.

[6] T. A. Reddy and D. E. Claridge, "Using synthetic data to evaluate multiple regression and principal component analyses for statistical modeling of daily building energy consumption," *Energy and Buildings*, vol. 21, no. 1, pp. 35–44, 1994.

[7] B. W. Olesen and G. S. Brager, "A better way to predict comfort: The New ASHRAE standard 55-2004," August 2004, http://escholarship.org/uc/item/2m34683k.

[8] B. Givoni and F. Vecchia, "Predicting thermal performance of occupied houses," in *Proceedings of the 18th International Conference on Passive and Low Energy Architecture (PLEA '01)*, pp. 701–706, Florianoplis, Brazil, 2001.

[9] ISO, "Thermal environments-specifics relating to appliance and methods for measuring physical characteristics of the environment," ISO Standard 7726, International Standards Organization, Geneva, Switzerland, 1985.

[10] B. Givoni, *Climate Considerations in Building and Urban Design*, Van Nostrand Reinhold, New York, NY, USA, 1998.

[11] D. M. Ogoli, "Predicting indoor temperatures in closed buildings with high thermal mass," *Energy and Buildings*, vol. 35, no. 9, pp. 851–862, 2003.

[12] B. Givoni, "Effectiveness of mass and night ventilation in lowering the indoor daytime temperatures. Part I: 1993 experimental periods," *Energy and Buildings*, vol. 28, no. 1, pp. 25–32, 1998.

[13] E. L. Kruger and E. M. Dumke, "Thermal performance evaluation of the technological village in Curitiba-Brazil," in *Proceedings of the 18th International Conference on Passive and Low Energy Architecture (PLEA '01)*, pp. 707–711, Florianopolis, Brazil, 2001.

[14] H. M. Künzel, A. Holm, D. Zirkelbach, and A. N. Karagiozis, "Simulation of indoor temperature and humidity conditions including hygrothermal interactions with the building envelope," *Solar Energy*, vol. 78, no. 4, pp. 554–561, 2005.

[15] C. Inard, H. Bouia, and P. Dalicieux, "Prediction of air temperature distribution in buildings with a zonal model," *Energy and Buildings*, vol. 24, no. 2, pp. 125–132, 1996.

[16] P. Hemmi, *Room temperature distribution [Ph.D. dissertation]*, ETH Zürich, Zürich, Switzerland, 1967.

[17] X. Peng and A. H. C. van Paassen, "A type of calculation of indoor dynamic temperature distributions," in *Proceedings of the 4th International Conference on System Simulation in Buildings (SSB '94)*, pp. 57–68, Liege, Belgium, 1994.

[18] Q. Chen, X. Peng, and A. H. C. van Paassen, "Prediction of room thermal response by CFD technique with conjugate heat transfer and radiation models," *Ashrae Transactions*, vol. 101, part 2, 11 pages, 1995.

[19] South Africa Weather Service, 2006, http://www.weathersa.co.za/.

[20] B. Givoni, *Climate and Architecture*, Applied Science Published, London, UK, 1976.

Biodiesel Production Process Optimization from Sugar Apple Seed Oil (*Annona squamosa*) and Its Characterization

Siddalingappa R. Hotti[1] and Omprakash D. Hebbal[2]

[1]*Department of Automobile Engineering, PDA College of Engineering, Gulbarga Karnataka 585102, India*
[2]*Department of Mechanical Engineering, PDA College of Engineering, Gulbarga Karnataka 585102, India*

Correspondence should be addressed to Siddalingappa R. Hotti; hottisr@gmail.com

Academic Editor: Abdurrahman Saydut

This paper presents the production of biodiesel from nonedible, renewable sugar apple seed oil and its characterization. The studies were carried out on transesterification of oil with methanol and sodium hydroxide as catalyst for the production of biodiesel. The process parameters such as catalyst concentration, reaction time, and reaction temperature were optimized for the production of sugar apple biodiesel (SABD). The biodiesel yield of 95.15% was noticed at optimal process parameters. The fuel properties of biodiesel produced were found to be close to that of diesel fuel and also they meet the specifications of ASTM standards.

1. Introduction

Energy is the critical input factor for the socioeconomic development and welfare of human being of any country. Fossil fuels are the major sources for the energy demand since their exploration. India is highly dependent on crude oil; net import of crude oil during 2011-12 is 171.73 million metric tons [1]. Due to limited reserves of fossil fuels, environmental degradation, and volatility in fuel prices, there is a growing need for energy security and protection of the environment. Country like India with an agricultural background has wasteland of about 55.27 million hectares [2], which can be utilized for growing plants/crops, which produce nonedible oil in appreciable quantity. Thus indigenously produced biodiesel, which is defined as the mono-alkyl esters of vegetable oils or animal fats, obtained by transesterifying oil or fat with an alcohol [3], is considered one of the options to substitute the petroleum fuels.

Many researchers have produced the biodiesel from nonedible oil, which include Jatropha (*Jatropha curcas*) oil [4–8], Karanja or Honge (*Pongamia pinnata/glabra*) seed oil [7, 9–12], Polanga (*Calophyllum inophyllum*) seed oil [13], rubber (*Hevea brasiliensis*) seed oil [14, 15], mahua (*Madhuca indica*) oil [16], tobacco (*Nicotiana tabacum*) seed oil [17], bitter almond (*Prunus dulcis*) oil [18], castor (*Ricinus communis*) seed oil [19, 20], okra (*Hibiscus esculentus*) seed oil [21], Kusum (*Schleichera trijuga*) oil [22], Simarouba (*Simarouba glauca*) [23], milo (*Thespesia populnea*) seed oil [24, 25], milk thistle (*Silybum marianum*) seed oil [26], and wild safflower (*Carthamus oxyacantha* Bieb) seed oil [27]. In the production process of biodiesel, the effects of process parameters such as alcohol to oil molar ratio, catalyst concentration, reaction time, and reaction temperature have been studied and optimized. the fuel properties of produced biodiesel have been investigated and compared with the standard specifications for assessing their feasibility to substitute the petroleum fuels. However there are many other nonedible oils for which process parameters are not being optimized. One among them is sugar apple (*Annona squamosa*) seed oil.

Annona squamosa, the sugar apple, sweetsop, or sugar-pineapple, is a species in the Annonaceae family. Sugar apple is grown in lowland tropical climates worldwide, including southern Mexico, the Antilles, and Central and South America, Tropical Africa, Australia, Indonesia, Polynesia,

TABLE 1: Fatty acid composition of sugar apple seed oil.

Fatty acid composition	Weight %
Lauric acid (C12:0)	0.08
Palmitic acid (C16:0)	17.79
Stearic acid (C18:0)	4.29
Oleic acid (C18:1)	39.72
Linoleic acid (C18:2)	29.13
Linolenic acid (C18:3)	1.37
Arachidonic acid (C20:4)	1.06
Behenic acid (C22:0)	2.01

Source: [29].

FIGURE 1: Sugar apple (*Annona squamosa*) fruits and seeds.

and USA, Hawaii and Florida. It was introduced to India and the Philippines by the Spanish and Portuguese in the 16th century and has been cultivated there ever since [28].

The sugar apple tree ranges from 3 to 6 m in height with open crown of irregular branches and somewhat zigzag twigs. Deciduous leaves, alternately arranged on short, hairy petioles, are lanceolate or oblong, blunt tipped, 5 to 15 cm long, and 2 to 5 cm wide with dull-green on the upper side, and pale, with a bloom on the lower side. The leaves are slightly hairy when young and aromatic when crushed. Along the branch tips, opposite the leaves, the fragrant flowers are borne singly or in groups of 2 to 4. They are oblong, 2.5 to 3.8 cm long, never fully open with 2.5 cm long, drooping stalks and 3 fleshy outer petals, yellow-green on the outside, and pale-yellow from the inside with a purple or dark-red spot at the base. The 3 inner petals are merely tiny scales. The compound fruit is nearly round, ovoid, or conical and it is 6 to 10 cm long. The fruit is thick rind composed of knobby segments, pale-green, gray-green, bluish-green, or, in one form, dull, deep-pink externally (nearly always with a bloom), separating when the fruit is ripe and revealing the mass of conically segmented, creamy-white, glistening, delightfully fragrant, juicy, sweet, and delicious flesh. Many of the segments enclose a single oblong-cylindrical, black or dark-brown seed about 1.25 cm long. There may be a total of 20 to 38, or perhaps more, seeds in the average fruit [30]. The sugar apple, fruits, and seeds are shown in Figure 1.

The emphasis of present work is to produce biodiesel from *Annona squamosa* seed oil by transesterification process

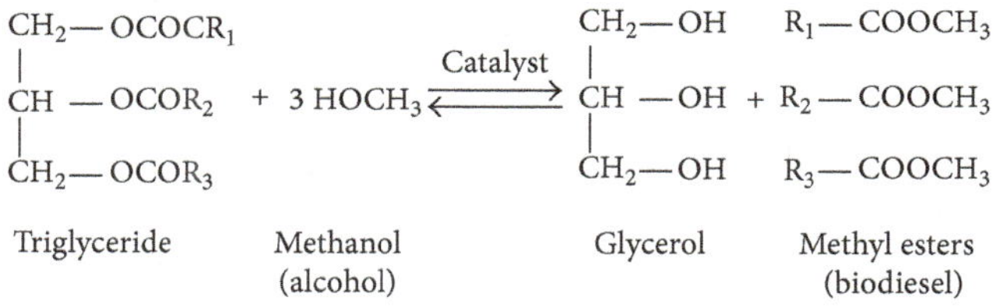

FIGURE 2: Transesterification reaction mechanism (R_1, R_2, and R_3 are long chain hydrocarbons, sometimes called fatty acid chains).

using methanol and sodium hydroxide (NaOH) as catalyst and to study the effect of process parameters such as alcohol to oil molar ratio, reaction time, and reaction temperature on the yield of biodiesel. Further the physicochemical properties of produced biodiesel are investigated and compared with diesel and standard specifications of biodiesel to assess its feasibility to replace the petroleum fuel.

2. Materials and Method

2.1. Oil Extraction. The seeds were collected from the different households as one discards the seeds after consuming the fruit. The collected seeds were dried and crushed in a mechanical expeller. For complete extraction of oil the seeds were passed four times through the expeller. The neat oil is allowed to settle for 48 hours and after that oil is stored in an airtight container to avoid oxidation.

2.2. Fatty Acid Composition of Sugar Apple Seed Oil. The vegetable oil extracted from a plant is composed of triglyceride, which is an ester derived from three fatty acids and one glycerol. The fatty acid composition of sugar apple seed oil (SASO) is given in Table 1.

2.3. Transesterification Reaction. The transesterification reaction was carried out in a laboratory scale batch reactor equipped with thermometer and condenser; the heating and stirring were done with a hot plate magnetic stirrer system. In each set of experiment 50 g of oil was heated to the predefined temperature and after attainment of predefined temperature the mixture of catalyst and methanol was transferred to reactor and all the predefined sets of transesterification reaction conditions were measured from this point for each set of experiment. The transesterification reaction mechanism is as shown in Figure 2. Stoichiometrically 3 : 1 molar ratio of alcohol to oil is needed for completion of transesterification reaction, but many researchers reported that biodiesel yield is maximum with excess molar ratio of alcohol to oil. Hence in the present investigation, in each set of experiment, 6 : 1 molar ratio of alcohol to oil and constant stirrer speed were maintained.

After the completion of predefined set of transesterification reaction conditions the reaction mixture was transferred into a separating funnel left for 60 minutes to separate into biodiesel and glycerol. The lower layer of glycerol was removed and the upper layer of crude biodiesel is washed several times with hot water at 50°C to remove the impurities,

TABLE 2: ASTM standard test methods used for determination of fuel properties.

Property	Test method	Unit
Flash point	D93-10	°C
Density	D1298-99	kg/m³
Kinematic viscosity	D445-09	mm²/s
Calorific value	D240-09	MJ/kg
Calculated cetane index (cetane number)	D976-06	—
Acid number	D5555-95	mg KOH/g
Copper strip test	D130-04	—
Sulfated ash	D874-07	w/w (%)
Conradson carbon residue	D189-06	w/w (%)
Distillation	D86-09	°C

such as residual catalyst, methanol, soap, and glycerol. The removal of impurities was confirmed by measuring the pH of water. The biodiesel was dried by heating it to a temperature of 110°C and allowed overnight for evaporation and cooling. The final product was weighed to determine the biodiesel yield.

2.4. *Analytical and Test Methods.* The mean molecular weight, saponification number (SN), iodine value (IV), and cetane number (CN) were determined from the fatty acid composition of oil using (1), (2), (3), and (4); respectively [18, 31]:

$$MW_{oil} = 3 \times \sum (MW_i \times x_i) + 38, \qquad (1)$$

where MW_{oil} stands for molecular weight of SASO and MW_i and x_i stand for molecular weight and mass fraction of ith fatty acid; respectively,

$$SN = \sum \left(560 \times \frac{A_i}{MW_i} \right), \qquad (2)$$

$$IV = \sum \left(254 \times D \times \frac{A_i}{MW_i} \right), \qquad (3)$$

$$CN = 46.3 + \frac{5458}{SN} - 0.225 \times IV, \qquad (4)$$

where A_i is the percentage, D is the number of double bonds, and MW_i is the molecular mass of each component.

The fuel properties of sugar apple seed oil (SASO) and sugar apple biodiesel (SABD) were determined as per the ASTM standards as given in Table 2.

In this paper the biodiesel yield was calculated using

$$\text{Biodiesel yield} = \frac{m_{biodiesel}}{m_{oil}} \times 100, \qquad (5)$$

where $m_{biodiesel}$ is the weight of SABD after purification and m_{oil} is the weight of SASO.

3. Results and Discussion

3.1. *Oil Content of Sugar Apple Seeds.* The oil extracted from the mechanical expeller was weighed after filtering; it was found that the sugar apple seeds contain moderate quantity of oil, 24.5 w/w % oil, and thus it can be suitable feedstock for the production of biodiesel.

3.2. *Fatty Acid Composition Analysis.* Generally three main types of fatty acids are present in triglyceride and they are saturated, monounsaturated, and polyunsaturated. The quality of biodiesel will be affected by the fatty acid composition of oil, preferably the vegetable oil should have low saturated and low polyunsaturated fatty acid composition, and the composition of monounsaturated fatty acid should be high. Table 1 shows the fatty acid composition of SASO and it contains predominant amount of monounsaturated fatty acid composition, that is, 39.72%, followed by 31.56% of polyunsaturated fatty acid and 24.07% of saturated fatty acids. The major fatty acids present in the SASO were oleic acid, 39.72%, linoleic acid, 29.13%, palmitic acid, 17.79%, and stearic acid, 4.29%. Thus the SASO can be classified as oleic-linoleic oil.

3.3. *Physicochemical Properties of SASO and SABD.* The physical and chemical properties of SASO and SABD were determined as per the ASTM standard test procedures and tabulated in Table 3. The iodine value of SASO is 118 mgI₂/g, as the iodine value is higher, which indicates the unsaturation of SASO. The heating of these higher fatty acids results in polymerization of glycerides, which necessitates the limitation of unsaturated fatty acids; otherwise, it leads to formation of deposits and deterioration of lubricating oil. The saponification of SASO was 192, which indicates that the SASO is normal triglyceride and useful in production of soaps. The FFA (free fatty acid) content of oil is 0.965%, as it is less than the biodiesel that can be produced using single stage transesterification process, that is, by base-catalyzed transesterification process. The flash points of SASO, SABD, and diesel were 235, 161, and 54 degrees Celsius, respectively. The flash points of SASO and SABD are found to be much higher in comparison with diesel, which helps in safe storage and transportation. The densities of SASO and SABD were higher than that of diesel, which may be due to the presence of higher molecular weight triglycerides. The kinematic viscosity of SASO was found to be 42.53 mm²/s, which is much higher than that of diesel; hence the direct use of SASO may lead to poor combustion, untimely wear of fuel pumps, and injector. The viscosity of SASO was reduced by converting it to biodiesel and it was found to be 5.90 mm²/s, which is within the limits of standard specification for biodiesel fuel. Calorific values of SASO, SABD, and diesel were found to be 37.95, 40.48, and 43.00 MJ/kg, respectively. The calorific values of studied oil and biodiesel are found to be lower than that of diesel, which may be due to the difference in chemical composition or presence of oxygen molecule in molecular structure of oil. The other parameters, that is, cetane number, calculated cetane index, sulfated ash, carbon residue, copper strip corrosion, and distillation temperature, were found to be within the limits of standard specifications for biodiesel fuel.

TABLE 3: Physiochemical properties of SASO and SBD in comparison with commercially available diesel and ASTM standard specifications for biodiesel fuel.

Property	SASO	SABD	Commercially available diesel	Standards specifications for biodiesel fuel, ASTM D6751-09a
Iodine value (mg I$_2$/g)	118	—	—	
Saponification number	192	—	—	
Acid number (mg KOH/g)	1.93	0.34	—	0.50 max
Flash point (°C)	235	161	54	130 min
Density (kg/m^3)	910	865	830	870–900
Kinematic viscosity (mm^2/s) at 40°C	42.63	5.90	2.4	1.9–6.0
Calorific value (MJ/kg)	37.95	40.48	43.00	
Cetane number	47.93	—	—	47 min
Calculated cetane index	—	53.57	50.98	47 min
Sulfated ash (w/w, %)	—	0.0015	—	0.020 max
Carbon residue (w/w, %)	0.697	0.033	—	0.050 max
Copper strip corrosion	—	3 h, 50°C/1a	—	No. 3 max
Distillation temperature, 90% recovered (°C)				
IBP		338	161	360 max
10%		340	202	
20%		342	215	
50%		348	254	
90%		356	348	
Molecular weight (g/mol)	872.50	—	—	—

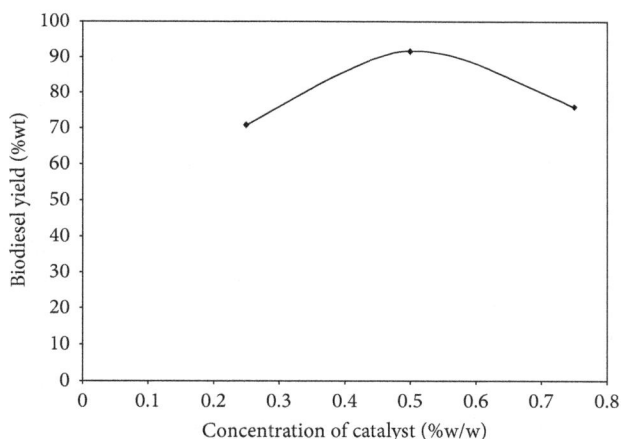

FIGURE 3: Effect of catalyst concentration on biodiesel yield (molar ratio of 6 : 1, reaction temperature of 60°C, and reaction time of 60 minutes).

3.4. Effect of Catalyst Concentration on Biodiesel Production.
The catalyst concentration is one of the most significant parameters which affects the biodiesel yield. Figure 3 shows the effect of NaOH catalyst concentration on the yield of biodiesel; during the process the other parameters are kept constant. It is observed that the biodiesel yield increases when catalyst concentration is increased from 0.25 w/w % to 0.50 w/w % and decreases when increased from 0.50 w/w % to 0.75 w/w %. The decrease in yield at 0.25 w/w % of catalyst concentration may be due to incomplete reaction and at 0.75 w/w % of catalyst concentration may be due to the fact that the higher amount of catalyst concentration favors the saponification reaction; thus further increase in catalyst concentration was not studied. From the present investigation, optimum amount of catalyst concentration was found to be 0.50 w/w % with 90.69% biodiesel yield.

3.5. Effect of Reaction Time on Biodiesel Production.
In order to study the effect of reaction time on production of biodiesel, the transesterification reaction was carried out by varying the reaction time as 45, 60, 75, and 90 minutes and by keeping other process parameters constant. Figure 4 shows the effect of reaction time on the yield of biodiesel. Varying the reaction time from 45 to 60 minutes and 60 to 75 minutes, the percentage yield of biodiesel is increased. The percentage yield of biodiesel is decreased with further increase in reaction time from 75 to 90 minutes; this may be due to formation of more amount of soap. The highest percentage yield of biodiesel was noticed when the reaction time was 75 minutes and it is found to be 95.15%.

3.6. Effect of Reaction Temperature on Biodiesel Production.
Further experiments were conducted by keeping the reaction temperature at 40, 50, 60, and 75°C in order to study the effect of reaction temperature. The catalyst concentration of 0.5 w/w % and reaction time of 75 minutes were maintained. Figure 5 shows the effect of reaction temperature on biodiesel

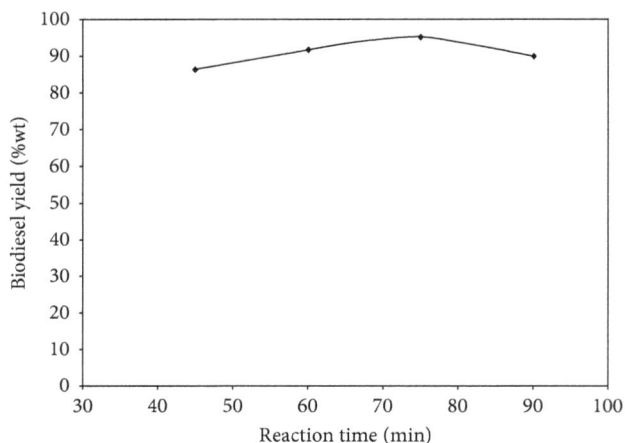

FIGURE 4: Effect of reaction time on biodiesel yield (molar ratio of 6 : 1, reaction temperature of 60°C, and catalyst concentration of 0.5 w/w %).

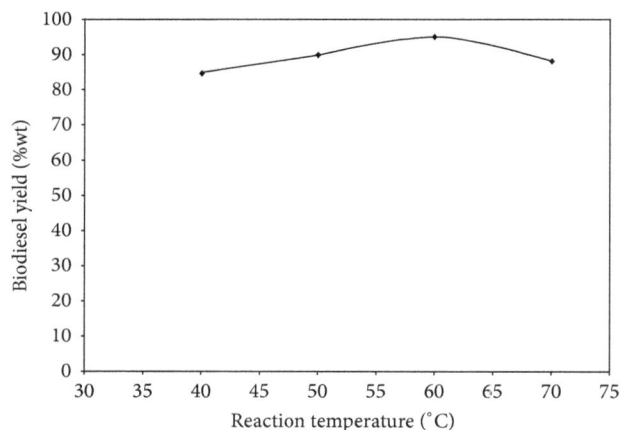

FIGURE 5: Effect of reaction temperature on biodiesel yield (molar ratio of 6 : 1, reaction time of 75 minutes, and catalyst concentration of 0.5 w/w %).

yield. As the methanol boils at 65°C, varying the reaction temperature from 40 to 60°C the biodiesel yield is increased and biodiesel yield was decreased at temperature of 70°C; this may be attributed to occurrence of saponification reaction at higher temperature. The highest biodiesel yield is found to be 95.15% at 60°C.

4. Conclusions

The purpose of the present study was to evaluate the sugar apple seed oil as a potential raw material for the production of biodiesel and to assess its feasibility for the replacement of petroleum fuel. The sugar apple seed oil was converted into biodiesel successfully by transesterification process and following conclusions were drawn.

(1) The sugar apple oil was converted to biodiesel by single stage base-catalyzed transesterification process

without any pretreatment as the FFA content is found to be less than 1%.

(2) The optimized process parameters are catalyst concentration of 0.5 w/w %, reaction time of 75 minutes and reaction temperature of 60°C with alcohol to oil molar ratio of 6 : 1, and constant stirrer speed. The biodiesel yield was found to be 95.15% at the optimized process parameters.

(3) The physical and chemical properties of biodiesel produced were found to be close to those of diesel fuel and also they meet the ASTM standard specifications for biodiesel.

Conflict of Interests

The authors declare that there is no conflict of interests regarding the publication of this paper.

Acknowledgments

Authors wish to thank Professor G. R. Naik, the Convener of Biodiesel Information and Demonstration Centre, Gulbarga University, Gulbarga, for permitting using the facilities available at the center for crushing of seeds for oil extraction and other test facilities and authors also thank supporting staff of centre, Mr. Pramod Kulkarni, Girish, and Ramesh for their support.

References

[1] "Energy statistics 2013," Central statistics office, National statistics organization, Ministry of statistics and programme implementation government of India, 2013, http://mospi.nic.in/mospi_new/upload/Energy_Statistics_2013.pdf.

[2] C. P. Reddy, "Bio-diesel production through jatropha (ratan jyot) plantations in wastelands," Soil & Water Conservation Today, vol. 2, no. 1, pp. 4–5, 2007.

[3] G. Knothe, "Biodiesel and renewable diesel: a comparison," Progress in Energy and Combustion Science, vol. 36, no. 3, pp. 364–373, 2010.

[4] H. J. Berchmans and S. Hirata, "Biodiesel production from crude Jatropha curcas L. seed oil with a high content of free fatty acids," Bioresource Technology, vol. 99, no. 6, pp. 1716–1721, 2008.

[5] P. Nakpong and S. Wootthikanokkhan, "Optimization of biodiesel production from Jatropha curcas L. oil via alkali-catalyzed methanolysis," Journal of Sustainable Energy & Environment, vol. 1, no. 3, pp. 105–109, 2013.

[6] P. D. Patil, V. G. Gude, and S. Deng, "Biodiesel production from Jatropha curcas, waste cooking, and camelina sativa oils," Industrial & Engineering Chemistry Research, vol. 48, no. 24, pp. 10850–10856, 2009.

[7] S. R. Kalbande and S. D. Vikhe, "Jatropha and Karanj bio-fuel: an alternate fuel for diesel engine," ARPN Journal of Engineering and Applied Sciences, vol. 3, no. 1, pp. 7–13, 2008.

[8] H. C. Ong, A. S. Silitonga, H. H. Masjuki, T. M. I. Mahlia, W. T. Chong, and M. H. Boosroh, "Production and comparative fuel properties of biodiesel from non-edible oils: Jatropha curcas, Sterculia foetida and Ceiba pentandra," Energy Conversion and Management, vol. 73, pp. 245–255, 2013.

[9] L. C. Meher, V. S. S. Dharmagadda, and S. N. Naik, "Optimization of alkali-catalyzed transesterification of *Pongamia pinnata* oil for production of biodiesel," *Bioresource Technology*, vol. 97, no. 12, pp. 1392–1397, 2006.

[10] M. Naik, L. C. Meher, S. N. Naik, and L. M. Das, "Production of biodiesel from high free fatty acid Karanja (*Pongamia pinnata*) oil," *Biomass and Bioenergy*, vol. 32, no. 4, pp. 354–357, 2008.

[11] L. C. Meher, S. N. Naik, and L. M. Das, "Methanolysis of *Pongamia pinnata* (karanja) oil for production of biodiesel," *Journal of Scientific and Industrial Research*, vol. 63, no. 11, pp. 913–918, 2004.

[12] M. N. Nabi, S. M. N. Hoque, and M. S. Akhter, "Karanja (*Pongamia Pinnata*) biodiesel production in Bangladesh, characterization of karanja biodiesel and its effect on diesel emissions," *Fuel Processing Technology*, vol. 90, no. 9, pp. 1080–1086, 2009.

[13] P. K. Sahoo, L. M. Das, M. K. G. Babu, and S. N. Naik, "Biodiesel development from high acid value polanga seed oil and performance evaluation in a CI engine," *Fuel*, vol. 86, no. 3, pp. 448–454, 2007.

[14] A. S. Ramadhas, S. Jayaraj, and C. Muraleedharan, "Biodiesel production from high FFA rubber seed oil," *Fuel*, vol. 84, no. 4, pp. 335–340, 2005.

[15] O. E. Ikwuagwu, I. C. Ononogbu, and O. U. Njoku, "Production of biodiesel using rubber [*Hevea brasiliensis* (Kunth. Muell.)] seed oil," *Industrial Crops and Products*, vol. 12, no. 1, pp. 57–62, 2000.

[16] S. V. Ghadge and H. Raheman, "Process optimization for biodiesel production from mahua (*Madhuca indica*) oil using response surface methodology," *Bioresource Technology*, vol. 97, no. 3, pp. 379–384, 2006.

[17] V. B. Veljković, S. H. Lakićević, O. S. Stamenković, Z. B. Todorović, and M. L. Lazić, "Biodiesel production from tobacco (*Nicotiana tabacum* L.) seed oil with a high content of free fatty acids," *Fuel*, vol. 85, no. 17-18, pp. 2671–2675, 2006.

[18] M. Atapour and H.-R. Kariminia, "Characterization and transesterification of Iranian bitter almond oil for biodiesel production," *Applied Energy*, vol. 88, no. 7, pp. 2377–2381, 2011.

[19] S. M. P. Meneghetti, M. R. Meneghetti, C. R. Wolf et al., "Ethanolysis of castor and cottonseed oil: a systematic study using classical catalysts," *Journal of the American Oil Chemists' Society*, vol. 83, no. 9, pp. 819–822, 2006.

[20] M. H. Chakrabarti and R. Ahmad, "Transesterification studies on castor oil as a first step towards its use in biodiesel production," *Pakistan Journal of Botany*, vol. 40, no. 3, pp. 1153–1157, 2008.

[21] F. Anwar, U. Rashid, M. Ashraf, and M. Nadeem, "Okra (*Hibiscus esculentus*) seed oil for biodiesel production," *Applied Energy*, vol. 87, no. 3, pp. 779–785, 2010.

[22] Y. C. Sharma and B. Singh, "An ideal feedstock, kusum (*Schleichera triguga*) for preparation of biodiesel: optimization of parameters," *Fuel*, vol. 89, no. 7, pp. 1470–1474, 2010.

[23] H. Manjunath, H. Omprakash, and R. K. Hemachandra, "Process optimization for biodiesel production from Simarouba, Mahua, and waste cooking oils," *International Journal of Green Energy*, vol. 12, no. 4, pp. 424–430, 2014.

[24] U. Rashid, F. Anwar, R. Yunus, and A. H. Al-Muhtaseb, "Transesterification for biodiesel production using *Thespesia populnea* seed oil: an optimization study," *International Journal of Green Energy*, vol. 12, no. 5, pp. 479–484, 2014.

[25] B. Panchal, S. Dhoot, S. Deshmukh, and M. Sharma, "Optimization of extraction of oil and biodiesel from *Thespesia populnea*

seed oil by alkali-catalyst in India," *International Journal of Green Energy*, 2012.

[26] M. Ahmad, M. Zafar, S. Sultana, A. Azam, and M. A. Khan, "The optimization of biodiesel production from a novel source of wild non-edible oil yielding plant *Silybum marianum*," *International Journal of Green Energy*, vol. 11, no. 6, pp. 589–594, 2014.

[27] H. Sadia, M. Ahmad, M. Zafar, S. Sultana, A. Azam, and M. A. Khan, "Variables effecting the optimization of non edible wild safflower oil biodiesel using alkali catalyzed transesterification," *International Journal of Green Energy*, vol. 10, no. 1, pp. 53–62, 2013.

[28] http://eol.org/pages/1054831/overview.

[29] K. V. Yathish, B. R. Omkaresh, and R. Suresh, "Biodiesel production from custard apple seed (Annona Squamosa) oil and its characteristics study," *International Journal of Engineering*, vol. 2, no. 5, pp. 31–36, 2013.

[30] https://www.hort.purdue.edu/newcrop/morton/sugar_apple.html.

[31] M. M. Azam, A. Waris, and N. M. Nahar, "Prospects and potential of fatty acid methyl esters of some non-traditional seed oils for use as biodiesel in India," *Biomass & Bioenergy*, vol. 29, no. 4, pp. 293–302, 2005.

Effect of Operating Conditions on Pollutants Concentration Emitted from a Spark Ignition Engine Fueled with Gasoline Bioethanol Blends

Haroun A. K. Shahad and Saad K. Wabdan

College of Engineering, University of Babylon, P.O. Box 4, Hilla, Babylon, Iraq

Correspondence should be addressed to Haroun A. K. Shahad; hakshahad@yahoo.com

Academic Editor: Onder Ozgener

This study is an experimental investigation of the effect of bioethanol gasoline blending on exhaust emissions in terms of carbon dioxide CO_2, carbon monoxide CO, unburnt hydrocarbons UHC, and nitric oxide NO_x of a spark ignition engine. Tests are conducted at controlled throttle and variable speed condition over the range of 1200 to 2000 rpm with intervals 400 rpm. Different compression ratios are tested for each speed, namely (7,8,10, and 11). Pure gasoline and bioethanol gasoline blends are used. The bioethanol used is produced from Iraqi date crop (Zehdi). Blending is done on energy replacement bases. Ethanol energy ratio (EER) used is 5%, 10%, and 15%. At each of the three designated engine speeds, the torque is set as 0, 3, 7, 10, and 14 N·m. It is found that ethanol blending reduces CO and UHC concentration in the exhaust gases by about 45% and 40.15%, respectively, and increases NO_x and CO_2 concentrations in the exhaust gases by about 16.18% and 7.5%, respectively. It is found also that load and speed increase causes an increase in CO_2 and NO_x concentrations and reduces CO and UHC concentrations. It is also found that increasing the compression ratio causes the emissions of CO_2 and NO_x to decrease and those of CO and UHC to increase.

1. Introduction

In recent years, given the dramatically increasing demand of energy, public concern has steadily increased regarding a possible shortage of fossil fuel resources, energy safety policies, and environmental pollution regulations. The degradation of the global environment and the foreseeable future depletion of worldwide fossil fuel reserves have been the driving force to searching for alternative fuels that are sustainable and environmentally friendly. Ethanol fuel is one of the renewable fuels for addressing these issues. The potential of ethanol fuel in improving the performance of internal combustion engines has been recently the focus of many investigations [1, 2]. In 2005, the Australian Government's Biofuels Taskforce reported that the environmental and human health impact of using ethanol as a biofuel was a major issue requiring resolution in order to guide national policy measures aimed at reducing greenhouse gas emissions. Because of the excellent

miscibility of bioethanol with common gasoline, it can be used as an additive to partially replace the gasoline as an automotive fuel [3]. Such mixtures are normally named after the amount or percentage of ethanol contained in the blended fuel [3].

Bioethanol is a renewable, biodegradable, and environmentally friendly alternative fuel, because it can be produced from agricultural products and scrapped resources. The road transport network using conventional fuels accounts for 23% of total greenhouse gas. These emissions can be reduced by using bioethanol fuel. Because of these benefits, bioethanol and ethanol-gasoline blends are widely investigated and used as alternative fuels in automotive vehicle [4–6]. The effects of ethanol addition to gasoline on engine performance and exhaust emissions were investigated experimentally and theoretically. It was found that the ethanol addition to gasoline has caused leaner operation and improved the combustion process. The potential of ethanol fuel in reducing the emission pollution of internal combustion engines has been extensively investigated. Liao et al., 2005, [7] performed an experimental study in a closed combustion chamber to investigate combustion characteristics and pollutants emission of ethanol-gasoline blends at low temperature, which is related to the cold-start operation of engines fueled with ethanol-gasoline. The exhaust emissions were purposely measured in terms of unburned hydrocarbon UHC, CO, and NO_x. It was confirmed that the emissions of UHC during rich combustion at relatively low temperature increased with increasing the addition of ethanol.

Najafi et al. 2009 [8] analyzed experimentally the pollutant emissions of a four-stroke SI engine operated with ethanol-gasoline blends of 0%, 5%, 10%, 15%, and 20% with the aid of artificial neural network (ANN) theoretically. The concentrations of CO and UHC emissions in the exhaust pipe decreased when ethanol blends were increased. This was due to the high oxygen percentage in the ethanol. In contrast, the concentrations of CO_2 and NO_x were found to be increased when ethanol is introduced. Yusaf et al. 2009 [9] evaluated the use of potato waste bioethanol as an alternative fuel for gasoline engines. The pollutant emissions of a four-stroke SI engine operating on ethanol-gasoline blends have been investigated experimentally and theoretically. Experiments were performed with the blends containing 5%, 10%, 15%, and 20% by volume of ethanol. Exhaust gas emissions were measured and analyzed for UHC, CO_2, CO, O_2, and NO_x at engine speed ranging from 1000 to 5000 rpm. The concentrations of CO and UHC emissions in the exhaust pipe were decreased and the concentrations of CO_2 and NO_x were increased when ethanol was introduced. Results obtained from both theoretical and experimental studies were compared. The simulation results have been validated against data from experiments and a good agreement was noticed between the trends in the predicted and measured results. Seshaiah, 2010, [10] performed tests on a variable compression ratio spark ignition designed to run on pure gasoline, LPG (Isobutene), and gasoline blended with ethanol 10%, 15%, 25%, and 35% by volume. In addition, the gasoline was mixed with kerosene at 15%, 25%, and 35% by volume without any engine modifications. The CO and CO_2 emissions had been also compared for all tested fuels. It was observed that the LPG is a promising fuel at all loads, which produced lesser carbon monoxide emission compared with other fuels tested. Ethanol was used as a fuel additive to the mineral gasoline; (up to 30% by volume) without any engine modification and with no efficiency loss. Ozsezen and Canakci, 2011, [11] studied the exhaust emissions of a vehicle fueled with low content alcohol (ethanol and methanol) blends and pure gasoline. The vehicle tests were performed at wide-open throttle using an eddy current chassis dynamometer with vehicle speeds of 40, 60, 80, and 100 km/hr. The test results obtained with the use of alcohol gasoline blends, 5% and 10% alcohol by volume, were compared with the pure gasoline results.

In general, alcohol gasoline blends provided higher combustion efficiency compared to pure gasoline. In exhaust emission results, a stable trend was not seen, especially for

CO emission, but, on average, alcohol gasoline blends exhibited decreasing UHC emissions. In the 100 km/hr vehicle speed test, the alcohol gasoline blends provided lower NO_x emission values compared to pure gasoline. At all vehicle speeds, minimum CO_2 emission was obtained when 5% methanol was added in gasoline. Sales and Sodré, 2012, [12] presented the exhaust emission levels from a flexible fuel engine with heated intake air and fuel during cold start operation. Electric resistances provided heating of intake air and fuel. The exhaust emissions from the engine equipped with heated intake air and ethanol injector were compared with the levels obtained from the conventional cold start system that uses gasoline as auxiliary fuel. The use of heated air and ethanol in substitution to the conventional system, that introduces gasoline in the intake pipe, to help cold start of a flexible fuel engine fuelled with hydrous ethanol (ethanol with 6.8% water mass content) produced significant reductions on raw exhaust UHC and CO emissions, especially in the first 150 s. Raw exhaust NO_x emissions were slightly reduced after 200 s from cold start. Yang et al. 2012 [13] studied the effects of ethanol-blended gasoline on emissions of regulated air pollutant and carbonyls from motorcycles. In addition, durability testing was performed on two brand-new motorcycles of the same model, using E3 in one and E0 in the other, to assess the effects of E3 usage on motorcycle emissions. The results show that average emission factors of CO and UHC decreased by 20% and 5.27%, respectively, using E3 fuel. However, NO_x and CO_2 emissions increased by 5.22% and 2.57%, respectively.

2. Experimental Apparatus and Procedure

The aim of work is to study the effect of bioethanol blending on the exhaust gas pollutants concentrations of a spark ignition engine. The bioethanol blending is done on energy replacement basis. Different blending ratios are to be tested. No engine modification is made. The test engine, the instrumentations, and the experimental program are described briefly in the following sections.

2.1. Test Engine and Instrumentation. The experiments are performed on a research engine, which is a variable compression ratio (varicomp), single cylinder, water cooled, dual fuel (gasoline/diesel) manufactured by prodit company, see Figure 1. The specifications of this engine are shown in Table 1. The exhaust gases are analyzed using MEG001 gas analysis and T156D gas analysis units while temperature measurement is done using thermocouple type K.

2.2. Test Fuel. The pollutants concentrations of bio-ethanol-blended gasoline (E5, E10, and E15) are to be evaluated and compared with that of neat gasoline fuel (E0). The purity ratio of bioethanol is 99.9% [14]. The fuel blends are prepared just before starting the experiment to provide homogenous fuel mixture. The ethanol-supplementation ratio by energy replacement is defined by the following equation:

$$EER = \frac{EE}{GE + EE} \times 100\%, \tag{1}$$

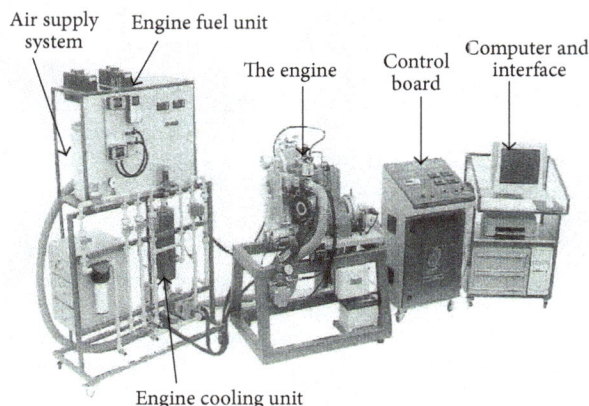

FIGURE 1: Test rig [14].

TABLE 1: Engine specifications.

Manufacture	Prodites.a.s.
Cycle	Otto or Diesel four stroke
Diameter	90 mm
Stroke	85 mm
Swept volume	541 mm³
Compression ratio	4–17.5
Max. power output	4 kW at 2800 rpm
Max. torque	28 N·m at 1600 rpm
Cooling type	Water cooled
No load speed range	500–3600 rpm
Load speed range	1200–3600 rpm

where EER is ethanol energy ratio, EE is ethanol energy content, and GE is gasoline energy content:

$$EE = (mxLCV)_{eth},$$
$$GE = (mxLCV)_{Gas}. \tag{2}$$

The properties of the two blended fuels are shown in Table 2.

2.3. Experimental Procedures.

Tests are carried out at three different engine speeds ranging from 1200 rpm to 2000 rpm, by 400 rpm increments at various loads starting from no load to 14 N·m and at four different compression ratios (7 : 1, 8 : 1, 10 : 1, and 11 : 1). At each of these engine speeds, four different fuels are used which are neat unleaded gasoline (E0) and three bio-ethanol-blended gasoline, namely, E5, E10, and E15. The letter E refers to bioethanol while the followed number refers to the percentage of bioethanol in the blended fuel. For each experiment, the engine is allowed to reach a stable condition and then the measurements are recorded. The full experimental program is shown in Table 3.

TABLE 2: Test fuel properties.

Properties	Gasoline	Ethanol
Chemical formula	$C_{8.23}H_{15.39}$	C_2H_5OH
Molecular weight (kg/kmol)	114.15	46.07
Density (kg/m³ at 20°C)	732	792
Oxygen (% wt.)	0	35
Octane number (RON)	86–94	105–108
Boiling point (°C)	25–230	78.5
Latent heat of vaporization (kJ/kg)	289	854
Autoignition temperature (°C)	257	423
A/F ratio (by mass)	14.7	9
Lower heating value (MJ/kg)	43.8	26.7
Flash point	−43	9

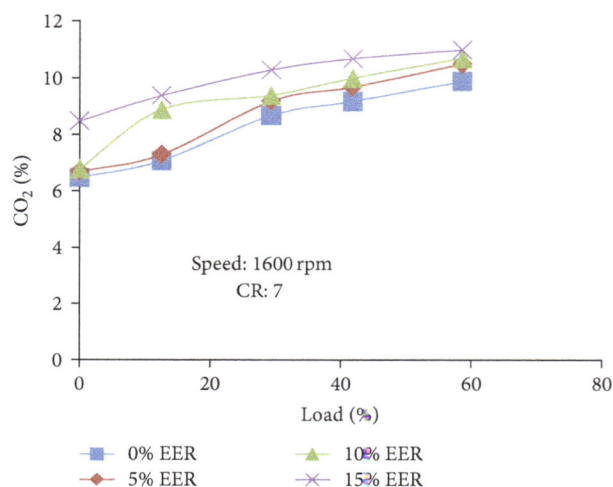

FIGURE 2: Effect of load on CO_2 concentration.

3. Results and Discussion

In this section, the experimental results of the effect of bioethanol addition to gasoline fuel on the pollutants emissions of a spark ignition engine have been presented and discussed. It must be mentioned here that the ethanol blending is based on energy replacement basis; see (1). The experimental program is limited to a bioethanol blending ratio ranging from 0% to 15% since at higher ratios the engine does not run smoothly.

3.1. The Effect of Load.

Carbon dioxide is product of complete combustion of fuel. Normally, CO_2 emission increases with increase in load due to enhancement in combustion process as seen from Figure 2. Further, the presence of alcohol provides more oxygen for burning of fuel thus the emission of CO_2 increases with increasing the alcohol blending ratio. The stoichiometric air-fuel ratio of ethanol is about 2/3 that of gasoline; hence, the required amount of air for complete combustion of the blended fuel is reduced and the mixture becomes leaner. When the engine condition goes leaner, the

TABLE 3: The experimental program.

Torque (N·m)	CR = 7, 8, 10, 11											
	1200 rpm				1600 rpm				2000 rpm			
	EER				EER				EER			
	0%	5%	10%	15%	0%	5%	10%	15%	0%	5%	10%	15%
0												
3												
7												
10												
14												

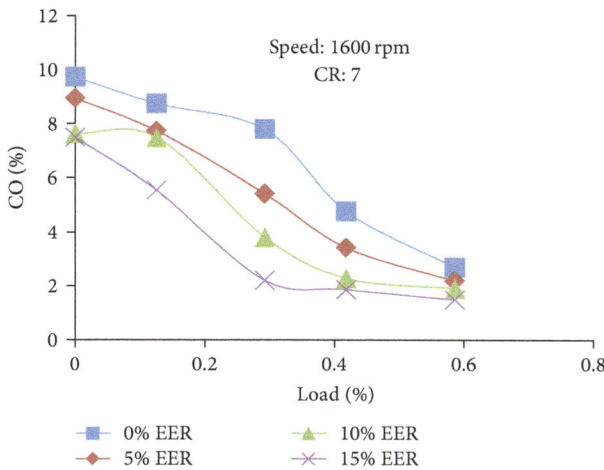

FIGURE 3: Effect of load on CO concentration.

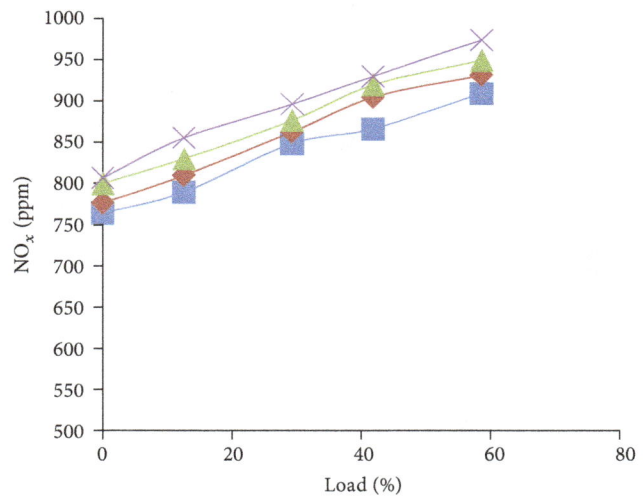

FIGURE 5: Effect of load on NO_x concentration.

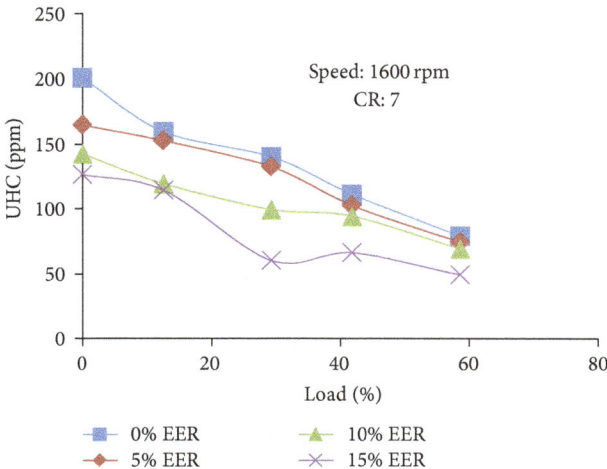

FIGURE 4: Effect of load on UHC concentration.

combustion process is more complete and the concentration of CO_2 emission gets higher.

The carbon monoxide concentration shows opposite behavior as compared with carbon dioxide as shown in Figure 3. The carbon monoxide concentration decreases as EER increases. This is because of the fact that addition of ethanol makes the mixture leaner, which gives better combustion and less CO production. The formation of carbon monoxide indicates loss of power because of oxygen deficiency in combustion chamber and hence incomplete combustion. The UHC emission decreases with increasing load and EER as shown in Figure 4, because increasing load results in stable combustion processes and faster flame speed. This is further improved by the addition of oxygenated alcohol. It provides more oxygen for the combustion process and leads to the so-called "leaning effect." Its final result is that better combustion is achieved therefore the concentration of UHC emission decreases as the ethanol content increases. The NO_x concentration results are very complicated. It depends on combustion temperature, availability of oxygen, and time for combustion process. The NO_x increases as the EER increases and as the load increases as shown in Figure 5. This is due to better combustion process, leading to higher combustion temperature, which favors NO_x formation. As load on engine was increased, the NO_x emissions for all blending ratios are also increased gradually. This is due to higher combustion temperature.

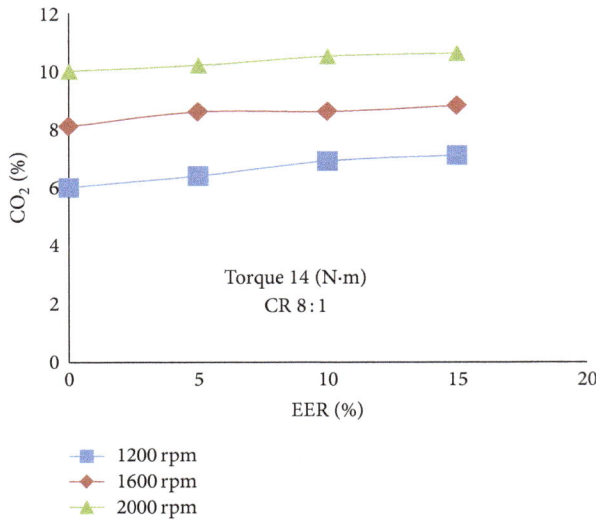

FIGURE 6: Effect of speed on CO_2 concentration.

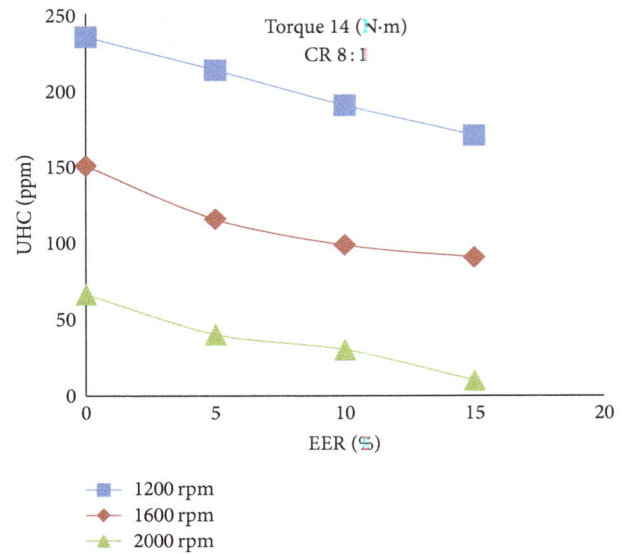

FIGURE 7: Effect of speed on CO concentration.

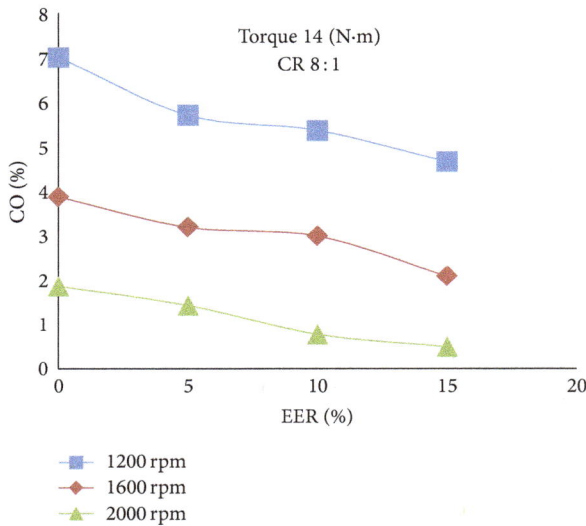

FIGURE 8: Effect of speed on UHC concentration.

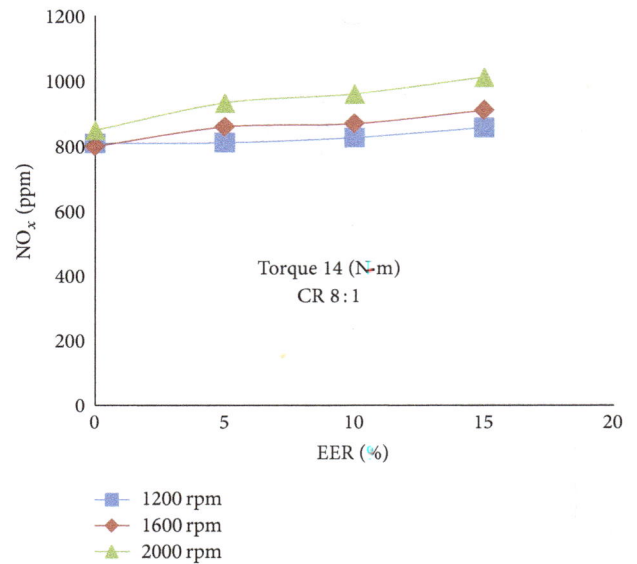

FIGURE 9: Effect of load on NO_x concentration.

3.2. The Effect of Engine Speed. The carbon dioxide concentrations increase with increasing engine speed and EER while the CO decreases. This is due to larger oxidation rate of fuel carbon to CO_2 which is caused by presence of extra oxygen when using ethanol blending. The increase in engine speed improves engine volumetric efficiency and mixing process, leading to better combustion process. This leads to increasing CO_2 emissions and reducing CO emissions as shown in Figures 6 and 7. However, very high engine speed reduces volumetric efficiency which deteriorates combustion process. The unburnt hydrocarbon emission shows the same trend as CO since both are products of incomplete combustion of fuel; see Figure 8.

Figure 9 shows that the concentration of NO_x increases with increasing engine speed and EER at constant load due to the increase in the cylinder temperature. This is due to

higher temperature caused by better combustion process. The maximum level of NO_x emission is obtained at maximum speed and maximum EER which is about 1010 ppm.

3.3. Effect of Engine Compression Ratio. The results show that the concentration of CO_2 decreases, Figure 10, while the concentrations of CO and UHC increase, Figures 11 and 12, respectively, with increasing compression ratio for all EER values. The decreasing in CO_2 concentration and the increasing in CO concentration may be due to the dissociation of CO_2 at high combustion temperature caused by increasing compression ratio and the presence of ethanol. The increasing in UHC concentration may be caused by

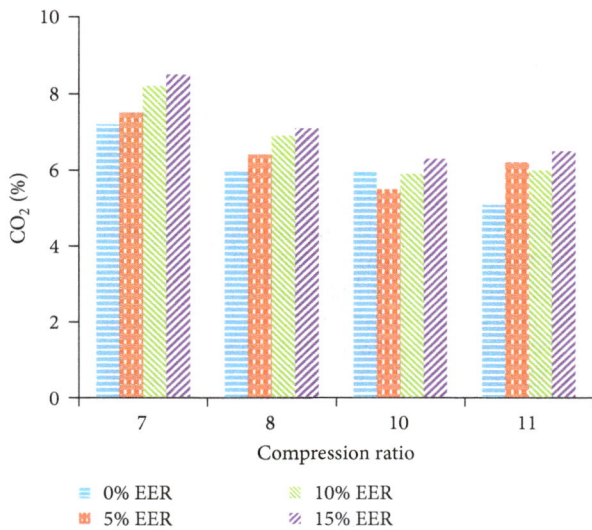

FIGURE 10: Effect of compression ratio on CO_2 concentration for different EER at 1200 rpm and load 14 N·m.

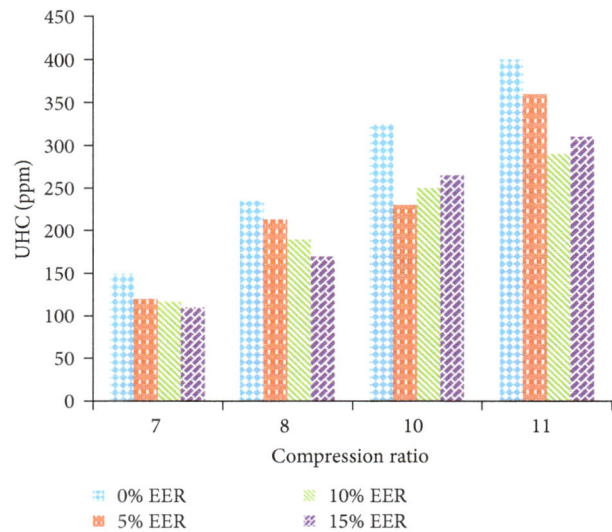

FIGURE 12: Effect of compression ratio on UHC concentration for different EER at 1200 rpm and load 14 N·m.

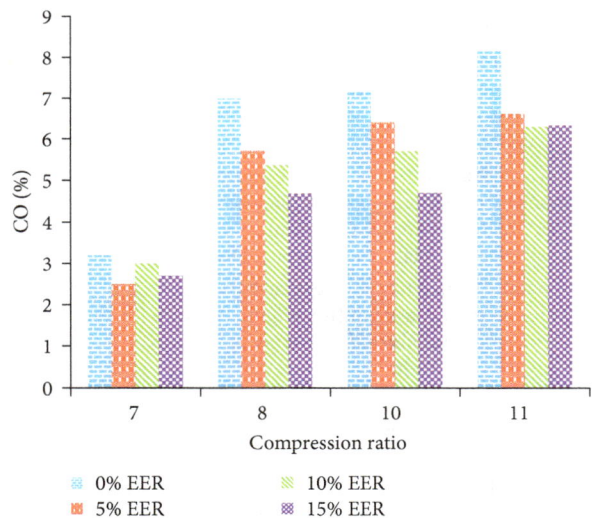

FIGURE 11: Effect of compression ratio on CO concentration for different EER at 1200 rpm and load 14 N·m.

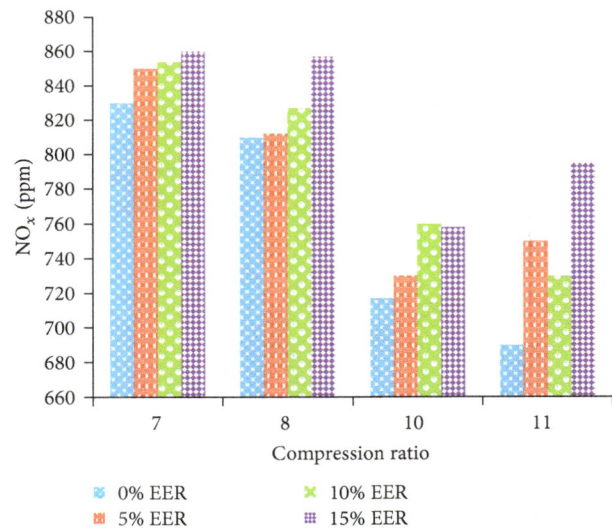

FIGURE 13: Effect of compression ratio on NO_x concentration for different EER at 1200 rpm and load 14 N·m.

the increasing of crevice volume ratio caused by increasing compression ratio.

The variation of NO_x concentration is shown in Figure 13. The figure shows that NO_x concentration decreases slightly at low compression ratios (7 and 8) for all values of EER while the decrease is more noticeable at higher compression ratios as shown in Figure 13. This may be due to longer expansion stroke which gives lower temperature at later stages of expansion stroke.

Figures 14 and 15 show a comparison of results of present work with results of [15] for CO_2 and CO emissions. Comparison shows acceptable agreement in trends.

4. Conclusions

The main conclusions that can be drawn from the results and discussions in the previous section are as follows.

(1) The concentrations of CO_2 and NO_x increase while the concentrations of CO and UHC decrease as EER increases.

(2) Increasing engine load causes an increase in the CO_2 and NO_x emissions and a decrease in CO and UHC emissions.

(3) It was observed that the emission values of CO_2 and NO_x increase while those of CO and UHC decrease with increased speed of engine.

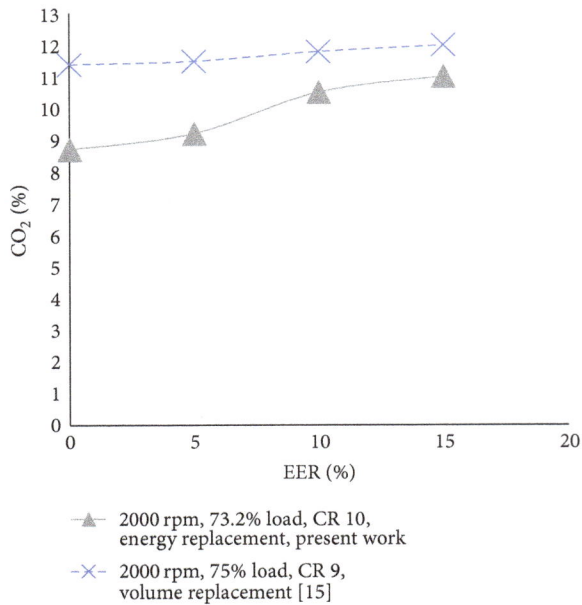

FIGURE 14: Comparison of present results with results of [15].

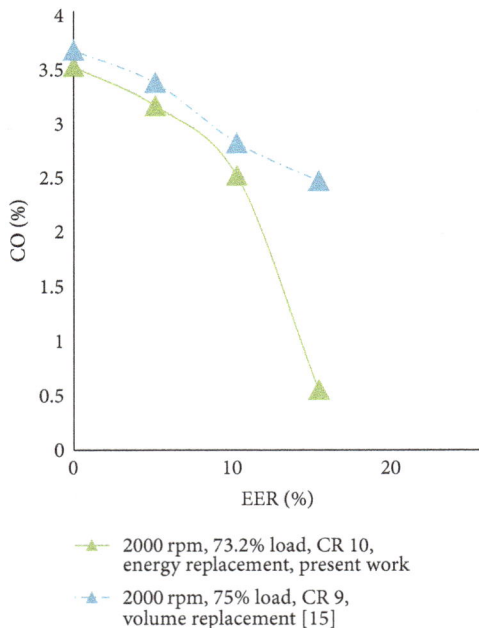

FIGURE 15: Comparison of present results with results of [15].

(4) With increasing the compression ratio the concentration values of the CO_2 and NO_x decrease while the concentration values of the CO and UHC increase.

Conflict of Interests

The authors declare that there is no conflict of interests regarding the publication of this paper.

References

[1] S. H. Yoon and C. S. Lee, "Lean combustion and emission characteristics of bioethanol and its blends in a spark ignition (SI) engine," *Energy & Fuels*, vol. 25, no. 8, pp. 3484–3492, 2011.

[2] Y. Zhuang and G. Hong, "Primary investigation to leveraging effect of using ethanol fuel on reducing gasoline fuel consumption," *Fuel*, vol. 105, pp. 425–431, 2013.

[3] T. Beer and J. Carras, "The health impacts of ethanol blend petrol," *Energies*, no. 4, pp. 352–367, 2011.

[4] A. M. Francisco and R. G. Ahmad, "Performance and exhaust emissions of a single cylinder utility engine using ethanol fuel," SAE Technical Paper 2006-32-0078, 2006.

[5] C. D. Marriott, M. A. Wiles, J. M. Gwidt, and S. E. Parrish, "Development of a naturally aspirated spark ignition direct-injection flex-fuel engine," SAE Technical Paper 2008-01-0319, 2008.

[6] S. H. Yoon, S. H. Park, and C. S. Lee, "Experimental investigation on the fuel properties of biodiesel and its blends at various temperatures," *Energy & Fuels*, vol. 22, no. 1, pp. 652–656, 2008.

[7] S. Y. Liao, D. M. Jiang, Q. Cheng, Z. H. Huang, and Q. Wei, "Investigation of the cold-start combustion characteristics of ethanol-gasoline blends in a constant-volume chamber," *Energy and Fuels*, vol. 19, no. 3, pp. 813–819, 2005.

[8] G. Najafi, B. Ghobadian, T. Tavakoli, D. R. Buttsworth, T. F. Yusaf, and M. Faizollahnejad, "Performance and exhaust emissions of a gasoline engine with ethanol blended gasoline fuels using artificial neural network," *Applied Energy*, vol. 86, no. 5, pp. 630–639, 2009.

[9] T. Yusaf, D. Buttsworth, and G. Najafi, "Theoretical and experimental investigation of SI engine performance and exhaust emissions using ethanol-gasoline blended fuels," in *Proceedings of ICEE 3rd International Conference on Energy and Environment*, pp. 195–201, Malacca, Malaysia, December 2009.

[10] N. Seshaiah, "Efficiency and exhaust gas analysis of variable compression ratio spark ignition engine fuelled with alternative fuels," *International Journal of Energy and Environment*, vol. 1, no. 5, pp. 861–870, 2010.

[11] A. N. Ozsezen and M. Canakci, "Performance and combustion characteristics of alcohol-gasoline blends at wide-open throttle," *Energy*, vol. 36, no. 5, pp. 2747–2752, 2011.

[12] L. C. M. Sales and J. R. Sodré, "Cold start emissions of an ethanol-fuelled engine with heated intake air and fuel," *Fuel*, vol. 95, pp. 122–125, 2012.

[13] H.-H. Yang, T.-C. Liu, C.-F. Chang, and E. Lee, "Effects of ethanol-blended gasoline on emissions of regulated air pollutants and carbonyls from motorcycles," *Applied Energy*, vol. 89, no. 1, pp. 281–286, 2012.

[14] A. K. Ebraheem and A. K. S. Haroun, *Production and analysis of bio-ethanol from dates syrup [M.Sc. thesis]*, University of Babylon, Hillah, Iraq, 2013.

[15] Y. Çaya, A. Çiçekb, F. Karac, and S. Sağiroğlua, "Prediction of engine performance for an alternative fuel using artificial neural network," *Applied Thermal Engineering*, vol. 37, pp. 217–225, 2012.

4

Computational Actuator Disc Models for Wind and Tidal Applications

B. Johnson,[1] **J. Francis,**[1] **J. Howe,**[1,2] **and J. Whitty**[1]

[1] *School of Computing Engineering and Physical Sciences, University of Central Lancashire, Preston PR1 2HE, UK*
[2] *Thornton Science Park, University of Chester, Parkgate Road, Chester, Cheshire CH1 4BJ, UK*

Correspondence should be addressed to B. Johnson; bmcjohnson@uclan.ac.uk

Academic Editor: Tarek Ahmed-Ali

This paper details a computational fluid dynamic (CFD) study of a constantly loaded actuator disc model featuring different boundary conditions; these boundary conditions were defined to represent a channel and a duct flow. The simulations were carried out using the commercially available CFD software ANSYS-CFX. The data produced were compared to the one-dimensional (1D) momentum equation as well as previous numerical and experimental studies featuring porous discs in a channel flow. The actuator disc was modelled as a momentum loss using a resistance coefficient related to the thrust coefficient (C_T). The model showed good agreement with the 1D momentum theory in terms of the velocity and pressure profiles. Less agreement was demonstrated when compared to previous numerical and empirical data in terms of velocity and turbulence characteristics in the far field. These models predicted a far larger velocity deficit and a turbulence peak further downstream. This study therefore demonstrates the usefulness of the duct boundary condition (for computational ease) for representing open channel flow when simulating far field effects as well as the importance of turbulence definition at the inlet.

1. Introduction

The actuator disc method has been used together, with the Reynolds averaged Navier-Stokes (RANS) equations, for many years and for many applications including helicopter rotors [1], horizontal axis wind turbines [2], and horizontal axis tidal turbines [3, 4] alike. The actuator disc method represents a turbine as a simple disc of similar dimensions to the rotor and is used to approximate the forces applied to the flow. The forces are implemented as body loads or as negative momentum source terms on the flow as it passes through the disc.

The actuator disc approximation has a number of benefits over modelling the full rotor geometry. The most significant benefit amongst these is the reduction in computational expense especially for multiple rotor simulations. Full rotor simulations require a fine mesh to capture the boundary layer and separation along the blade surface, as well as the solution of the unsteady compressible Navier-Stokes equations. A full transient rotor simulation is needed, allowing the rotor blades to rotate in order to capture the wake. The actuator disc method allows for coarser meshes to be used and the incompressible Navier-Stokes equations to be solved as long as the Mach number is below 0.3, for this study the mach number is below 0.00021. Additionally, steady-state solutions can be obtained, vastly reducing the computational expense.

The actuator disc method has been a key tool of the renewable energy industry and has been used in a large number of studies [2, 5]. Even though there are more complex models such as the actuator line and full rotor models the low computational expense of the actuator disc method means it is still widely used [6, 7] and can be used to model multiple turbine interactions and wind farm simulations [8]. Although the actuator disc method has been used for many years, the majority of studies have used in-house code, as opposed to commercially available software, to conduct their studies.

The work described in this paper is essentially a benchmarking study, that is, a comparison of modelling data of previous theoretical, numerical, and experimental studies. Here the actuator disc method was implemented using the commercially available computational simulation suite ANSYS-Workbench with ANSYS-CFX v13 [9] and compared to the one-dimensional (1D) momentum theory as described in [10], as well as previous studies featuring both numerical [3, 4] and experimental data [11, 12].

2. Benchmarked Studies

Three previous studies were chosen for benchmarking featuring one theoretical [10], one numerical [4], and one experimental study [11]. The 1D momentum theory [10] also known as the simple actuator theory is an application of the 1D momentum equation applied to an idealized turbine. It uses control volume analysis to consider an infinitely thin frictionless disc with a constant momentum sink within an inviscid and incompressible fluid. The experimental data to be compared with is detailed in [11] and features three different porous discs to simulate different turbines. The experiment was conducted in a water channel measuring 21 m by 1.37 m with a depth of 0.3 m. Three 0.1 m diameter discs of various porosities were placed into the channel. The various porosities were used to represent different thrust coefficients (C_T), which were measured using a pivot arm mounted onto a load cell. The water velocities were measured at various locations using an acoustic Doppler velocimeter (ADV) at a sample rate of 50 Hz and the data was averaged over 3 minutes. The previous numerical studies chosen for comparison are described in [3, 4] and used ANSYS-CFX to reproduce analogous experimental data [11]. These studies were chosen, because they were conducted using the same software as in the work presented in this paper, hence providing a benchmark to verify modelling methods.

3. Numerical Method

The numerical simulations used ANSYS-Workbench specifically ANSYS-CFX [9] and the steady-state solution of the Reynolds averaged Navier-Stokes (RANS) equations [13] together with the k-ω SST turbulence model [14]. This model was chosen over the k-ϵ model based on the literature and on some preliminary simulations which showed that the k-ω SST model performs better in flows featuring adverse pressure gradients [15] in terms of accuracy to predict the flow properties. The k-ω SST model was also used in the benchmark studies [3, 4]. The model domain was defined to the dimensions of the experimental channel setup [11]. It featured a 2 m long inlet, a 3 m outlet, and a 0.3 m deep-water column along with a 0.1 m diameter disc with a thickness of 0.001 m at the centre. The flow was assumed to be symmetrical, allowing a symmetry plane to be setup through the centre of the disc, dividing the domain in half creating a width of 0.685 m as opposed to the 1.37 m width of the experimental channel; this therefore reduces the computational expense. All simulations were carried out using water at 25 degrees centigrade corresponding to a density of 997 kg/m^3 and dynamic viscosity of 8.899×10^{-4} kg/ms. As part of this study three discs were simulated with two different sets boundary conditions to represent a channel and a duct each with two different inlets totaling 12 simulations.

3.1. Boundary Conditions. The inlet velocity was defined in the same manner as the numerical study and based on the empirical data [11]. The equation used to define the inlet velocity was given in [4] as

$$U_{\text{in}} = 2.5U^* \ln\left(\frac{y_w U^*}{\nu}\right) + A, \tag{1}$$

where U_{in} is the inlet velocity across the width of the domain, U^* is the friction velocity, y_w is the depth of the water, ν is the kinematic viscosity, and A is a constant. Curve fitting methods were used to define U^* and A. The numerical paper [4] used values of $U^* = 0.00787$ m/s and $A = 0.197$ m/s were also used in this study. Figure 1 shows the inlet velocity used in this work and the experimental study [11], normalized with a free stream velocity of 0.331 m/s for the experimental study [4] and 0.33 m/s in this work. The vertical height was also normalized with the diameter of the disc $2R$.

The turbulence intensity, which is defined by (2), was described in two different ways to define two simulations referred to in this paper as inlet 1 and inlet 2. Both inlets were set with a turbulence intensity of 5% at the inlet to produce agreement with the experimental data [11] for $y/2R > 0.5$. The difference between the inlets is that inlet 2 was also defined with a length scale of 0.3 (height of the domain). Both these approaches are different to [4] which defined the turbulent kinetic energy and eddy dissipation

$$I \equiv \frac{u'}{\overline{U}}, \tag{2}$$

$$u' \equiv \sqrt{\frac{2}{3}k}, \tag{3}$$

$$\overline{U} \equiv \sqrt{U_x^2 + U_y^2 + U_z^2}. \tag{4}$$

In (2)–(4) I is the turbulence intensity, u' is the root-mean-square of the turbulent velocity fluctuations, $\overline{U}$ is the mean velocity, and k is the turbulent kinetic energy and $U_i = [x, y, z]$ is the velocity in the x, y, z directions.

The outlet was defined as a static pressure outlet with a relative pressure of zero. The floor and far side of the domain were defined as a nonslip wall. In this study two separate models were produced featuring different boundary conditions at the top or roof of the domain; the first featuring an opening creating a channel and the second featuring a nonslip wall creating a duct. These boundary condition sets were analogues of those in [3, 4]. Although a free-surface approach may be considered more suitable, as the experiment was carried out in a channel featuring water and air interactions, it was shown to only produce a 0.2% depth change at the disc [4].

3.2. Disc Definition. The disc was defined with a diameter of 0.1 m and a thickness of 0.001 m as a subdomain with a uniform momentum loss across the disc in the longitudinal (z-) direction. The momentum loss was defined using a directional loss model, which added a momentum source term (S) to the flow, which was defined as

$$S = K\frac{\rho}{2}U\,|U|, \tag{5}$$

where K is the resistance coefficient, ρ is the density, and U is the velocity. The resistance is applied as the loss across the disc thickness and so was specified by the user as K/d, where d is the thickness of the disc. The work described in this paper and previous numerical studies [3, 4] used identical resistance coefficient of 1, 2, and 2.5 in separate simulations to represent the three different porous discs used in the experimental study [11]. The resistance coefficient was derived in [3, 4] based on the thrust coefficient observed in the experimental data and was estimated using (6) which is a theoretical relationship between C_T and K [3, 4]. Here the value of b is obtained from $U_0/KU_r - 1/K$ to render the required momentum deficit of 80%, 66%, and 61%, in this study:

$$C_T = \frac{K}{(1 + bK)^2}. \tag{6}$$

3.2.1. Thrust Coefficient C_T. The thrust coefficient (C_T) is a nondimensional variable used to describe rotor's characteristics. The greater the C_T value the greater the wake expansion and turbulence levels within the wake. The C_T value of the porous discs in the experimental study [11] was measured using a pivot arm attached to a load cell. The thrust coefficient can be described numerically using (7). It requires the thrust (T) to be estimated which can be achieved in a number of ways. In [4] the thrust coefficient was estimated from the results using (8) to define the thrust. However in this work, as in [3], the thrust was calculated using (9):

$$C_T = \frac{T}{(1/2)\,\rho U_0^2 A_r}, \tag{7}$$

$$T = \frac{\rho}{2}KU_r\,|U_r|\,A_r, \tag{8}$$

$$T = \Delta p A_r. \tag{9}$$

In (7)–(9) C_T is the thrust coefficient, T is the thrust, ρ is the density, U_0 the free stream velocity, A_r is the disc area, K is the resistance coefficient, U_r is the velocity at the rotor, and Δp is the change in pressure over the disc. The free stream velocity was calculated as the average velocity between $0.5 < y/2R < 2.5$ at the inlet and was 0.331 m/s in the experimental study [11], 0.337 m/s in [4], and 0.33 m/s in this work.

3.2.2. Mesh. In this work an unstructured hybrid mesh was constructed consisting of various mesh densities ranging from 2.4×10^6 to 6.2×10^6 cells. The majority of the domain was constructed out of tetrahedral cells with an inflated zone of wedge cells at the boundary of the floor and symmetry plane. The data presented in this work corresponds to a mesh density of approximately 6.2×10^6 cells unless otherwise stated. Figure 2 shows the velocity profiles of various mesh

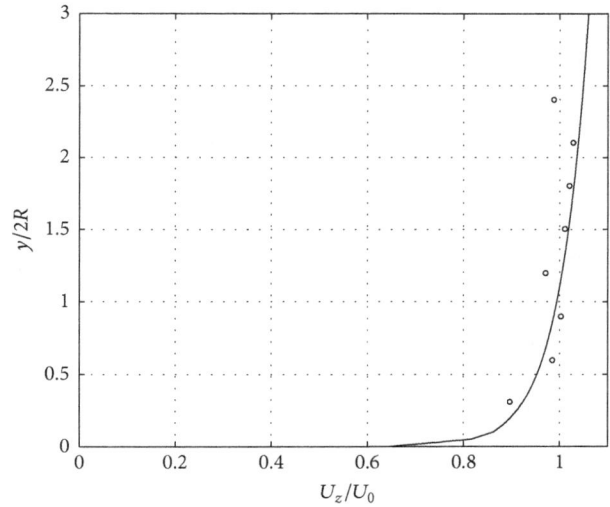

FIGURE 1: Normalized velocity at the inlet of this study (solid line) and experimental data (∘) [11].

densities along the centre line behind the disc and at 14 radii (14R) downstream of the disc. There was very little difference between the predictions of the four different mesh densities showing little advantage in refining the mesh. Figure 2(a) demonstrates a realistic velocity recovery beyond the peak velocity drop just before $x/2R = 5$. Figure 2(b) shows that the main differences between the different mesh densities are within the floor boundary layer and at the peak velocity deficit.

4. Results

All the data presented in this paper were produced using ANSYS-CFX and calculated with a root-mean-square residual of 1×10^{-5} which was in line with [4]. The velocities which are compared to both previous numerical and experimental results were normalized using the free stream velocity of the flow described between $0.5 < y/2R < 2.5$ at the inlet which was 0.331 m/s in the experimental study [11], 0.337 m/s in [4], and 0.33 m/s in the work described in this paper. This range is being consistent with the empirical data. The inlet velocity profile, as shown in Figure 1, shows good agreement with the experimental data [11].

4.1. Influence of the Boundary Conditions. Before detailing the results a comparison of the effects of the boundary types is needed. To do this the same domain and mesh were set up excluding the momentum loss to observe how the velocity profiles develop without the influence of the discs. Figure 3 shows the differences between the channel (solid lines) and duct (dashed lines) velocity profiles as they develop through the domain. The figure shows how the channel flow is almost unchanged as the inlet was defined with a channel velocity profile. The duct profile changes significantly as expected with the additional wall boundary causing a sharp decrease in

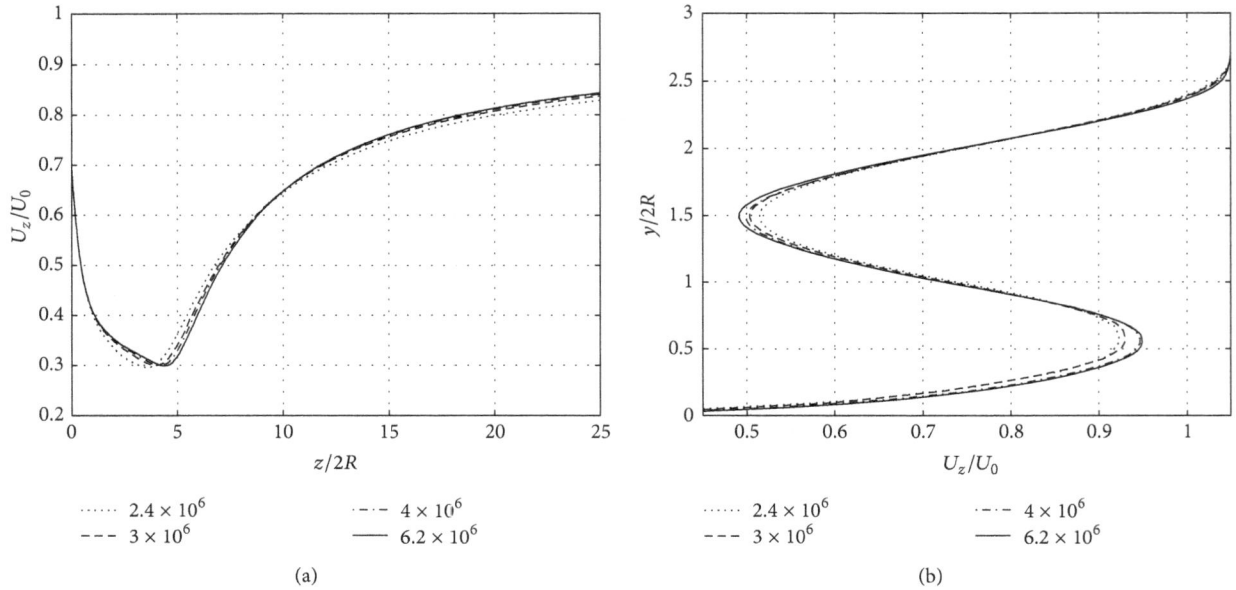

FIGURE 2: Normalized velocity profiles showing different mesh densities (a) along the centre line and (b) $14R$ downstream of the disc.

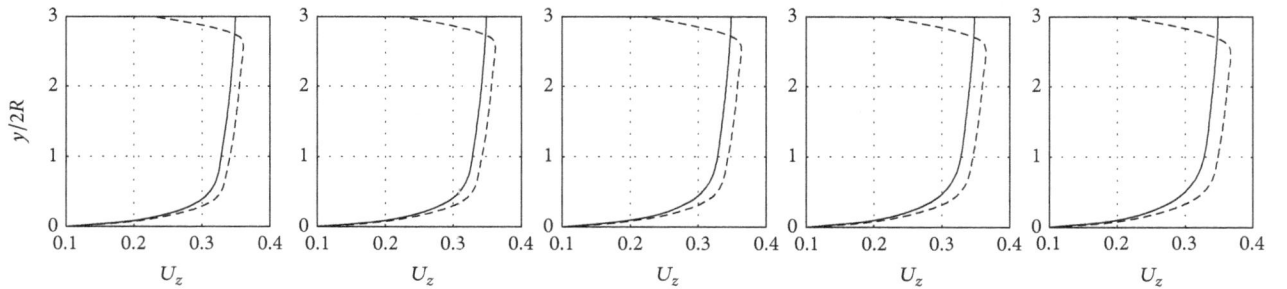

FIGURE 3: Velocity profile with no disc of the channel flow (solid line) and the duct flow (dashed line) at 24 m, 27 m, 31 m, 35 m, and 40 m from the inlet.

velocity at the top of the domain which forces the central velocity to increase to maintain the same mass flow rate.

4.2. Wake Predictions. The model in this work was compared to the 1D momentum theory as described in [10], specifically the pressure and velocity profiles along the centre line of the domain. Figure 4(a) shows the pressure and velocity profiles given by the 1D momentum equation. Figure 4(b) shows the pressure and velocity profiles produced by the model in this work. The overall profiles are in good agreement with only the magnitudes of the graphs changing depending on the C_T value and the characteristics of the disc.

The experimental study [11] conducted experiments using discs with different porosity measurements to represent different values of C_T. C_T was measured in the experimental study [11] to be 0.61, 0.86, and 0.97, respectively for each experiment. The simulations in the numerical paper [4] produced C_T values of 0.65, 0.91, and 0.98, respectively, for each disc using (8). The work presented in this paper

calculated C_T values of 0.60, 0.86, and 0.93 for the channel simulation and 0.64, 0.93, and 1.00 for the duct simulation, respectively, for each disc using (9).

Figure 5 shows the velocity profiles along the centre line of the domain and shows good agreement for all simulations in terms of the velocity characteristics, although the velocity magnitude is underpredicted compared to the numerical [4] and experimental data [11] for inlet 1 which has a delayed velocity recovery and appears to be offset from the other data sets. Inlet 2 shows a much better prediction of the experimental data [11] and both inlets show the duct has a quicker velocity recovery.

Figure 6 shows the turbulence intensity along the centre line of the domain and the difference between inlets 1 and 2 (Section 3.1) with inlet 1 having a lower starting turbulence intensity and subsequent peak. Both inlets show very little change in turbulent intensity just behind the rotor and then an almost linear increase up to the maximum intensity. This increase is most likely due to the presence of the wake edge shear layer with the maximum turbulence intensity indicating

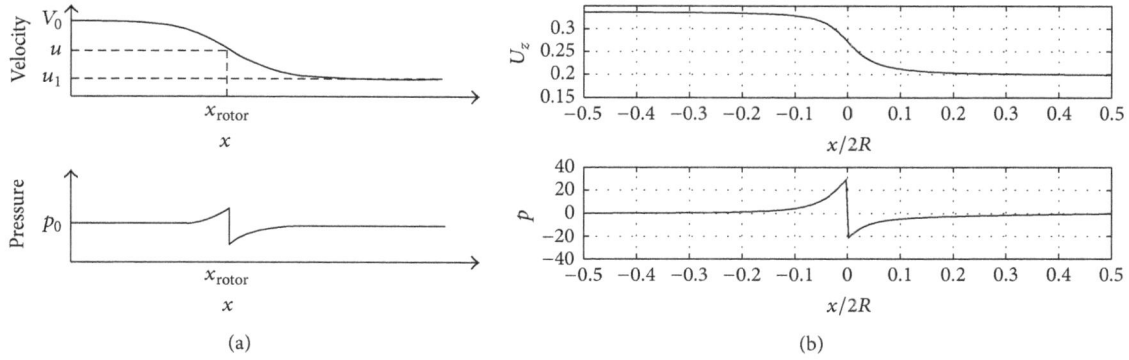

FIGURE 4: Pressure and velocity profiles along the centre line given by (a) the 1D momentum theory [10] and (b) the developed model.

the merger of the layers and subsequent end of the near wake region. The near wake region of the flow which is defined behind the disc up until the wake edge shear layers meet at the centre line of the wake. The near wake region varies in distance generally from about $4R$ to around $10R$ downstream depending on the disc geometry and flow conditions. Figure 6 shows that the model is able to predict the intensity accurately far down stream of the disc, although it is unable to predict the peak in the turbulence intensity behind the rotor both in terms of magnitude and location.

Figures 7 and 8 show the velocity and turbulence intensity profiles at various distances downstream of the disc and for the three different discs. The distances downstream correspond to $8R$, $14R$, $22R$, $30R$, and $40R$ downstream of the disc. These locations where chosen as they were the locations where the experimental data [11] was measured.

For all profiles in Figure 7 agreement was achieved (at least from a qualitative viewpoint) for the majority of the profile characteristics, such as the locations of highest and lowest velocities with the main numerical discrepancy at the maximum velocity deficit for all simulations. While the initial velocity drop is overpredicted at the centre, the free stream and floor boundary layer features are predicted well. Figure 7 shows that the duct simulations, displayed as the dashed line, predicted a smaller velocity deficit than the channel simulations at the centre. However, the duct model predicts higher velocity values towards the boundaries. Figure 7 shows quite well how the velocity deficit of the experimental data recovers quicker than the numerical data with inlet 1 simulations recovering the slowest. The experimental data seems to have almost recovered by $22R$ and completely recovered by $30R$ downstream whereas all numerical simulations still show some velocity deficit at $40R$.

Figure 8 shows the turbulence intensity of the models in comparison with the experimental data. There is little difference between the solid and dashed lines representing the channel and duct flows, respectively, and all models predicted intensities below that of the experiment data. The figure shows how the experimental data peaks earlier and higher than the modelled numerical data. Beyond approximately $22R$ downstream of the rotor the modelled and experimental data are very close.

5. Discussion

Two types of simulations were carried out in the work described in this paper to represent a channel and duct flow. It was observed that the duct had a higher central velocity magnitude and marginal lower turbulence intensity than the channel flow with two different turbulent inlets. This is to be expected due to the presence of the additional wall at the top of the domain. The wall creates an additional boundary layer which restricts and slows the flow near the wall. This deceleration along the wall focuses the flow increasing the central velocity magnitude in order to maintain the same mass flow rate which is clearly visible in Figure 3. However, the influence in the region $0.5 < y/2R < 2.5$ is minimal, meaning that representing the open channel flow as a duct incurs minor error, whilst reducing computational expense.

The difference between inlet 1 and inlet 2, through defining the turbulence length scales, had a significant effect on the simulation results. Inlet 2 predicted a more realistic velocity and turbulence intensity profile when compared to the experimental data [11]. This is due to a reduction of the turbulent dissipation throughout the domain prolonging the turbulence generated at the inlet and disc which was overly dissipated using inlet 1.

The models detailed in this work seem to have an inherent weakness in the definition of the momentum source as a predefined constant unidirectional loss. Allowing this, they performed well and predicted some characteristics of the velocity profile and turbulence levels. The model generally predicted both the velocity and turbulence intensity magnitudes lower than the experimental data [11] in the near wake region, with the discrepancy reducing as the flow moves downstream.

All the simulations carried out in this work predicted the turbulence intensity peak at far lower magnitude and further downstream of the disc than the experimental data [11]. This is due to the definition of the discs within the model as opposed to the physical discs. The discs within the experimental study [11] were porous discs with different C_T values created through different porosities. These porous discs extracted momentum from the flow by converting the velocity into small scale turbulence and, thus, creating a

(a)

(b)

(c)

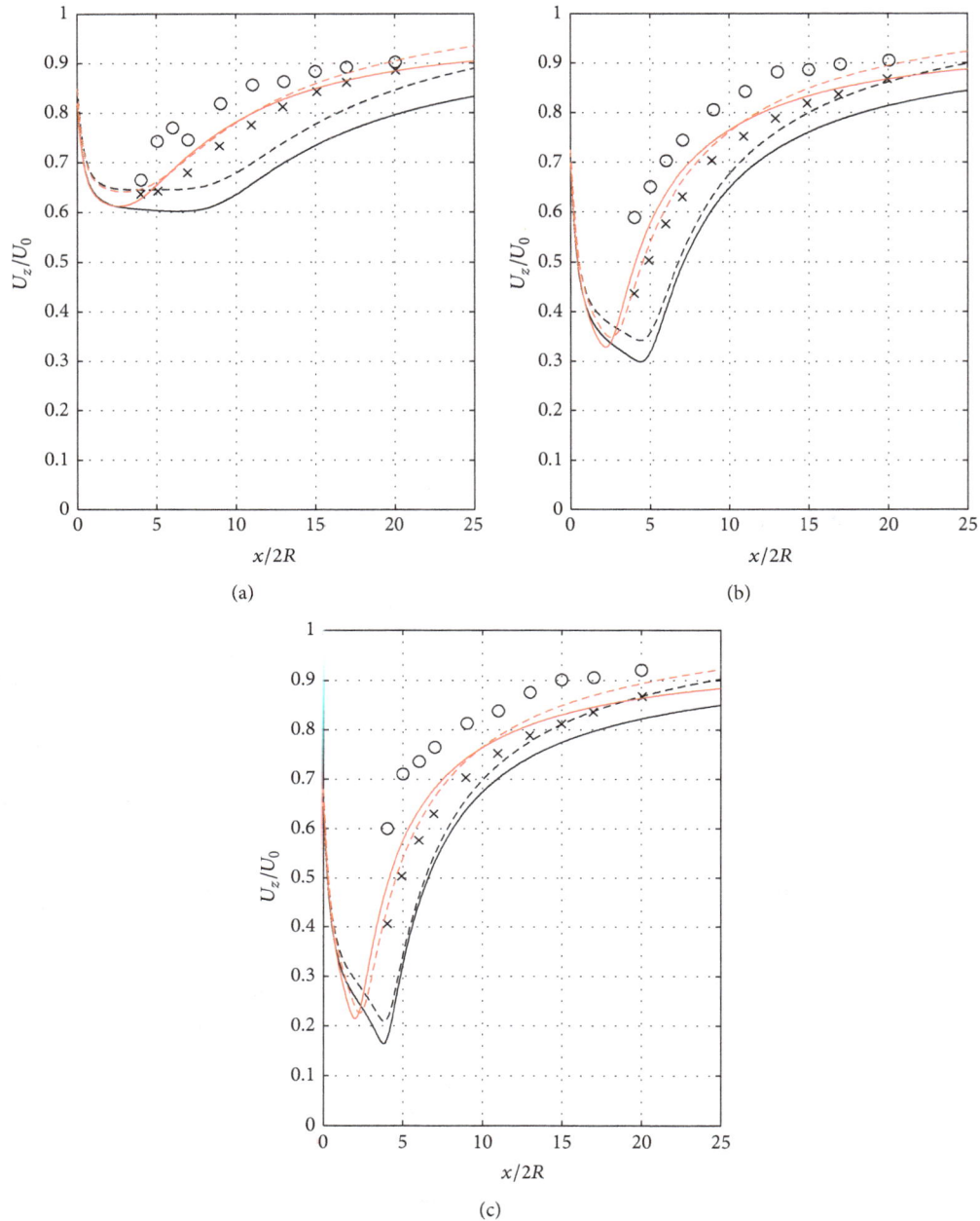

FIGURE 5: Velocity along the centre line showing the channel (solid line), duct (dashed line), numerical (×) [4], and experimental data (○) [11] for the C_T values of (a) 0.61, (b) 0.86, and (c) 0.97 for the experimental study [11]. The inlet 1 simulations are shown in black and inlet 2 simulations are in red.

high level of turbulence behind the disc. However the discs within the model extract momentum from the flow explicitly, reducing the velocity with no added turbulence. This explains the very high levels of turbulence behind the rotor for the experimental data and the lack of this peak in the modelled results. The turbulence intensity of the model peaked further downstream than observed experimentally (Figure 6); this is due to the merger of the boundary layers created by the velocity deficit. The porous discs within the experimental study [11] produced a variable 3D momentum loss. Although

the discs in this work were defined with an isotropic 1D momentum loss, this reduced the amount of mixing and therefore produced a longer wake, implying the presence of anisotropic momentum losses.

The differences between the C_T values calculated from the channel and duct flow can be attributed to the added boundary layer and subsequent small velocity increase. The C_T values were calculated using a free stream velocity based on the velocity inlet which is perfectly reasonable for the channel flow as this velocity profile is fully developed.

(a)

(b)

(c)

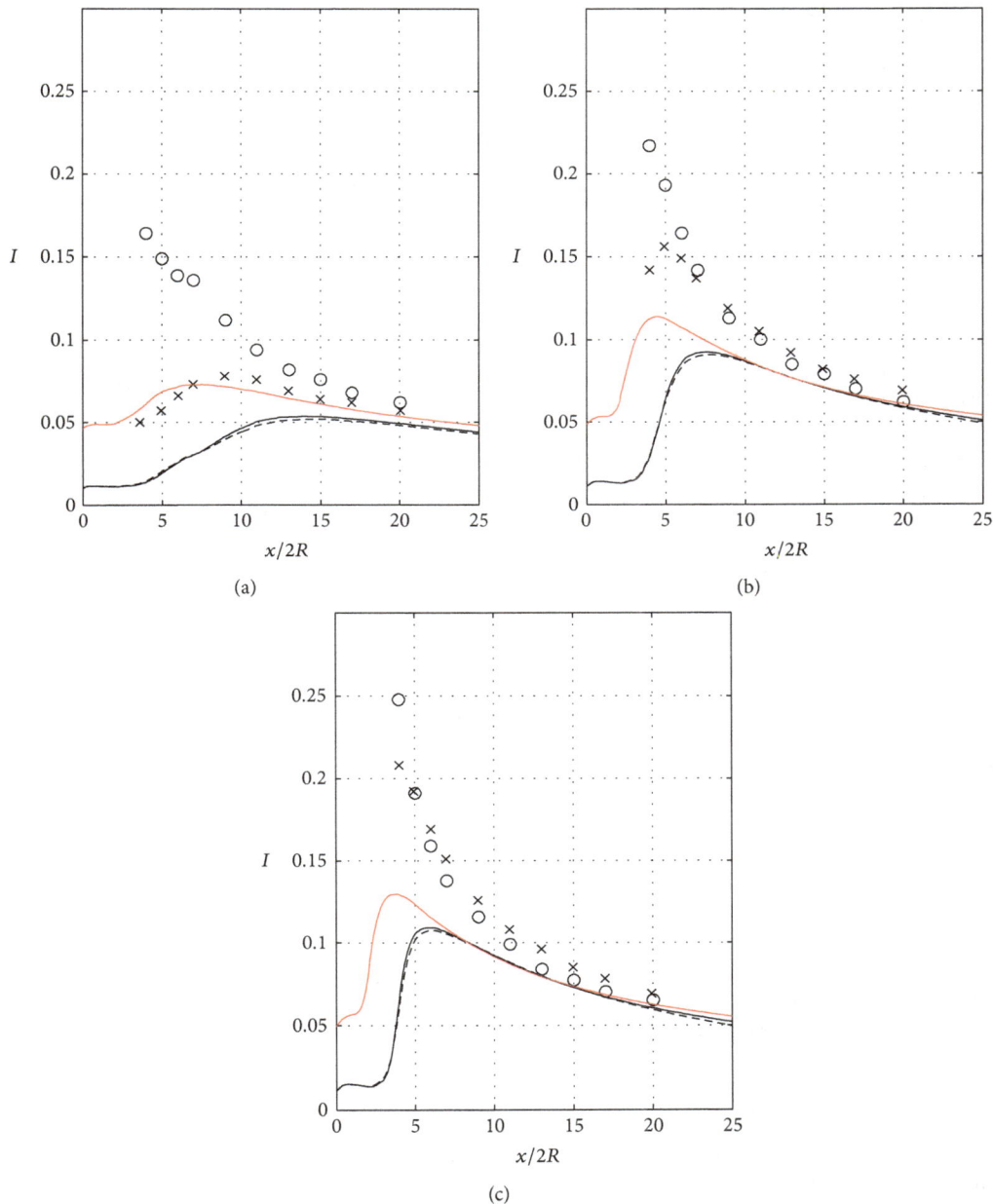

FIGURE 6: Turbulence intensity along the centre line showing the channel (solid line), duct (dashed line), numerical (×) [4], and experimental data (○) [11] for the C_T values of (a) 0.61, (b) 0.86, and (c) 0.97 for the experimental study [11]. The inlet 1 simulations are shown in black and inlet 2 simulations are in red.

However, this is not the case for the duct flow. Taking this into account the C_T values were recalculated, using a new free stream velocity of 0.34 m/s obtained at the domain origin in the absence of the disc. Application of (9) produced new C_T values of 0.61, 0.88, and 0.96 for the three discs, respectively.

Use of the k-ω SST model was based on the literature and Figure 4 implies that the adverse pressure gradient before the disc is well predicted as well as the floor effects shown in Figures 3 and 7. The k-ω SST model is most appropriate in situations with adverse pressure gradient and 3D flow phenomena featuring strong swirl but the work described

here only considered a 1D momentum source. The ability of the k-ω SST model to prevent the overprediction of eddy viscosity may have inadvertently reduced turbulence in the wake and led to the longer wake seen in Figure 7 when compared with the experimental data [11] that had higher turbulence levels and 3D effect from the porous discs.

6. Conclusion

The work described in this paper has used the steady-state RANS solution method resident within the commercially

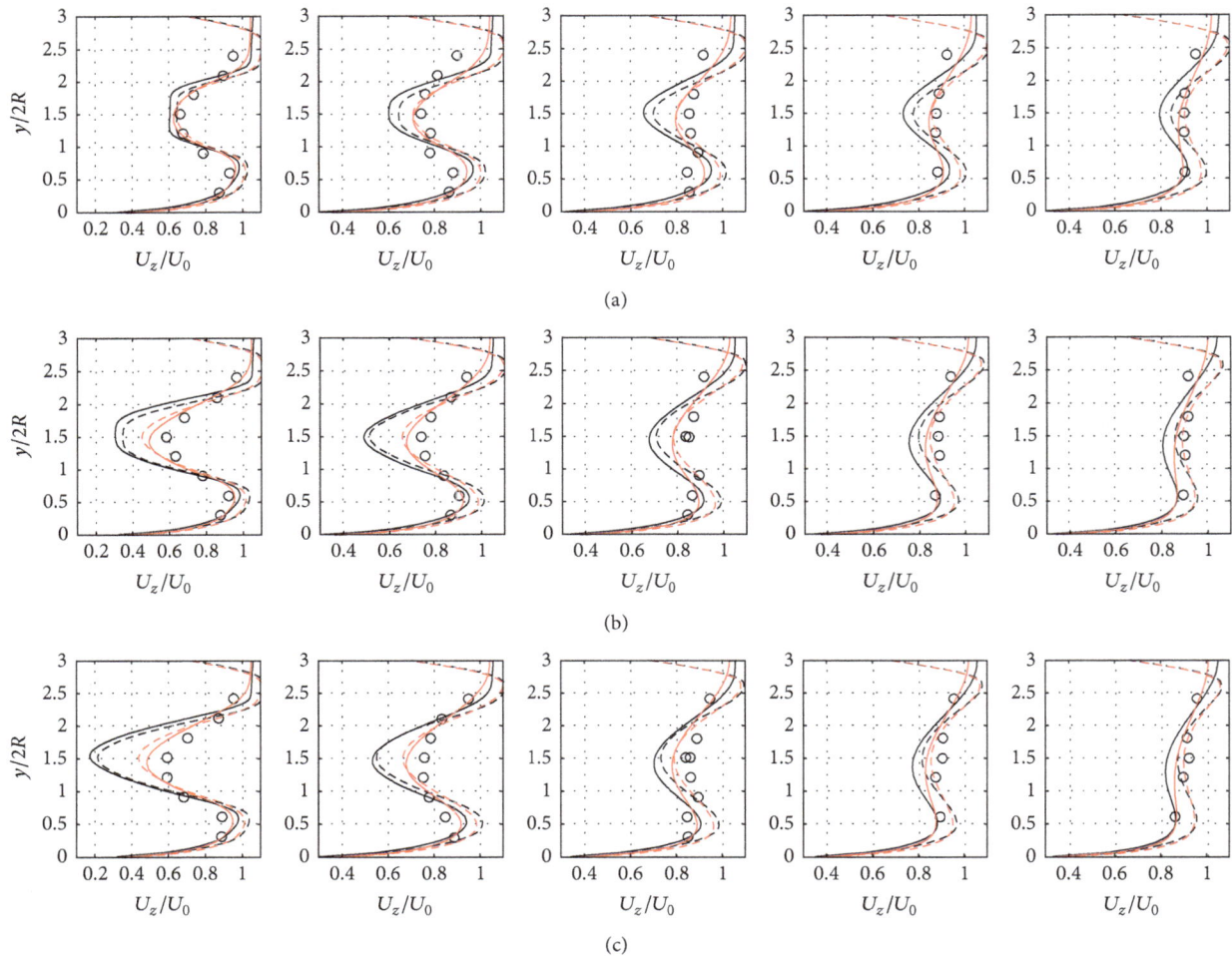

FIGURE 7: Normalized velocity of the channel (solid line), duct (dashed line), and experimental data (○) [11] at different C_T values of (a) 0.61, 0.62, (b) 0.86, 0.91, and (c) 0.97, 0.99 for this study and the experiment study, respectively [11]. The inlet 1 simulations are shown in black and inlet 2 simulations are in red. Each figure represents the distance downstream of the disc corresponding to $8R$, $14R$, $22R$, $30R$, and $40R$, respectively.

available ANSYS-CFX [9] to benchmark an actuator disc model without rotation with the 1D momentum theory [10] and previous numerical [3, 4] and experimental studies of porous discs [11, 12]. This study has compared four different boundary condition sets, a channel and a duct, each with two different turbulent inlets containing different actuator discs. The discs were simulated featuring different resistance coefficients to represent different porous discs used within the experimental study [11]. These simulations were found to be in good agreement with the 1D momentum equation [10] in terms of the velocity and pressure profiles. When compared to experimental data [11], model predictions deteriorated with respect of velocity and turbulence intensity magnitude, just behind the disc, although the agreement improved further downstream. This discrepancy can be attributed to small scale turbulence present in the experiments and the momentum extraction method employed by the models.

This paper shows that the model method was sufficient to predict the far field velocity characteristics of a porous disc.

Our future studies will model three-dimensional anisotropic effects at the disc by using variable momentum sources and include an additional turbulent source terms to account for the discrepancies found. Equally, more sophisticated modelling techniques such as adding rotation, the actuator line, surface model, or using a more sophisticated solver such as large-eddy simulation (LES) may produce closer agreement with field data. Moreover, such techniques might go some way to explaining the inherent poor prediction of turbulent intensity.

The main achievement of this study was demonstrating the usefulness of the duct boundary conditions (for computational ease) for representing an actuator disc in open channel flow when simulating far field effects, given the particular velocity profile, which is (1), applied at the inlet. It was found that the channel and duct simulations predicted very similar results with the duct predicting a slightly higher velocity magnitude for the majority of the domain. The main discrepancies observed when compared to the experimental study [11] can

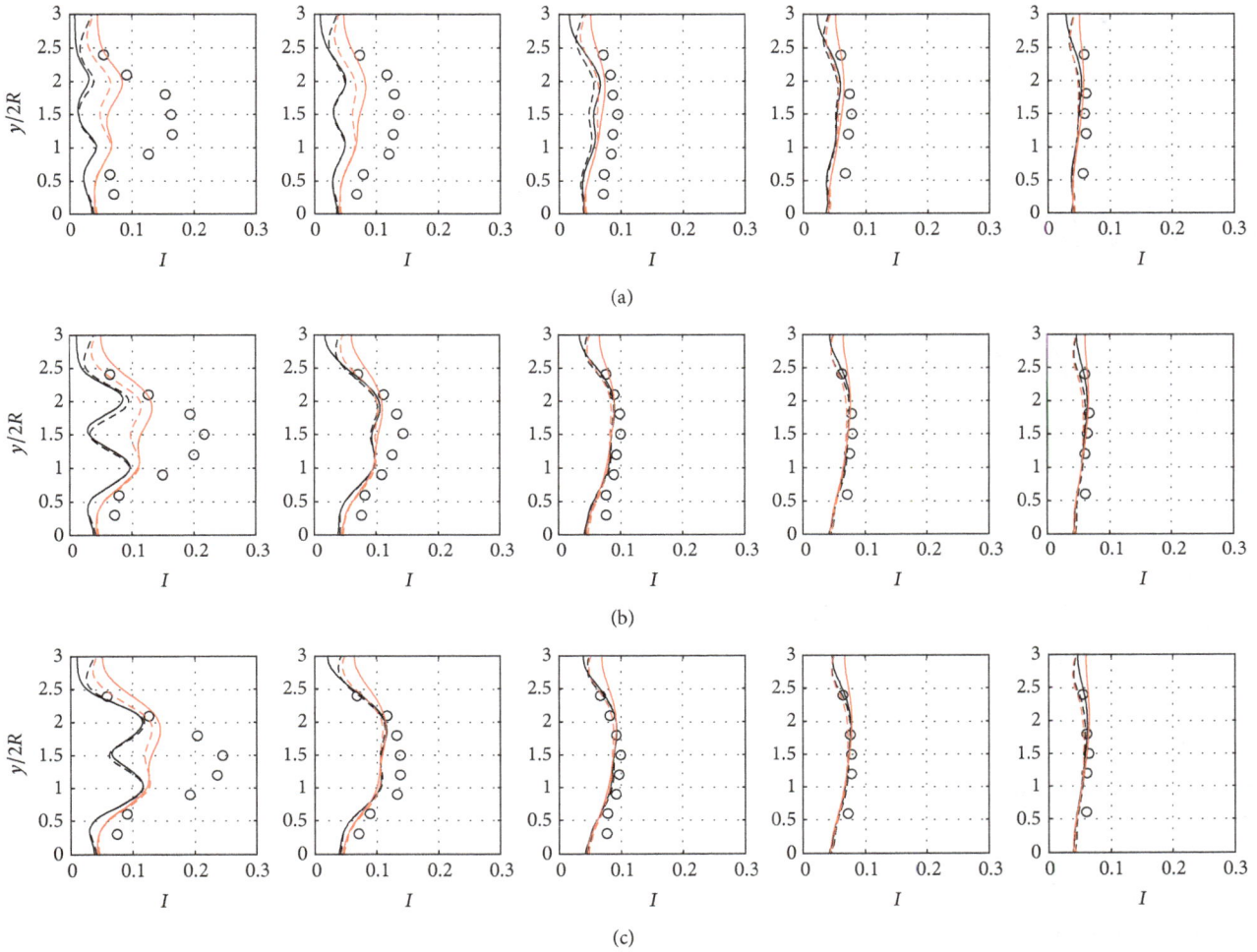

FIGURE 8: Turbulence intensity at $8R$, $14R$, $22R$, $30R$, and $40R$ downstream of the disc, with the channel (solid line), duct (dashed line), and experimental data (○) [11] at different C_T values of (a) 0.61, 0.62, (b) 0.86, 0.91, and (c) 0.97, 0.99 for this study and the experiment study, respectively [11]. The inlet 1 simulations are shown in black and inlet 2 simulations are in red.

be attributed to the definition of the momentum source, which explicitly extracts momentum from the flow rather than converting it into small scale turbulence. Moreover, only a unidirectional momentum source was used, which did not account for the three-dimensional effects of the real discs used in the experimental study [11].

Nomenclature

A: Boundary layer model constant
A_r: Area of the rotor
C_T: Thrust coefficient
D: Rotor diameter
I: Turbulence intensity
K: Resistance coefficient
k: Turbulence kinetic energy
p: Pressure
T: Thrust
U: Velocity
U_0: Free stream velocity

U_r: Velocity at the rotor
U_i: Velocity in ith ($=x$, y, or z) direction
U^*: Friction velocity
$\overline{U}$: Mean velocity
U_{in}: Inlet velocity
u': Root-mean-square of the turbulent velocity fluctuations
x: Distance in the widthwise direction
y: Distance in the depth-wise direction
z: Distance in the stream-wise direction
y_w: Water depth
ν: Kinematic viscosity
ρ: Density
d: Disc thickness
Δp: Change in pressure over the disc.

Abbreviations

CFD: Computational fluid dynamics
RANS: Reynolds averaged Navier-Stokes.

Conflict of Interests

The authors declare that there is no conflict of interests regarding the publication of this paper.

References

[1] A. F. Antoniadis, D. Drikakis, B. Zhong et al., "Assessment of CFD methods against experimental flow measurements for helicopter flows," *Aerospace Science and Technology*, vol. 19, no. 1, pp. 86–100, 2012.

[2] B. Sanderse, S. P. Van Der Pijl, and B. Koren, "Review of computational fluid dynamics for wind turbine wake aerodynamics," *Wind Energy*, vol. 14, no. 7, pp. 799–819, 2011.

[3] M. E. Harrison, W. M. J. Batten, L. E. Myers, and A. S. Bahaj, "A comparison between CFD simulations and experiments for predicting the far wake of horizontal axis tidal turbines," in *Proceedings of the 8th European Wave and Tidal Energy Conferences*, p. 10, Uppsala, Sweden, 2009.

[4] M. E. Harrison, W. M. J. Batten, L. E. Myers, and A. S. Bahaj, "Comparison between CFD simulations and experiments for predicting the far wake of horizontal axis tidal turbines," *IET Renewable Power Generation*, vol. 4, no. 6, pp. 613–627, 2010.

[5] B. Sanderse, "Aerodynamics of wind turbine wakes literature review," Tech. Rep. ECN E-09-016, ECN Wind Energy, 2009.

[6] S. Aubrun, S. Loyer, P. E. Hancock, and P. Hayden, "Wind turbine wake properties: comparison between a non-rotating simplified wind turbine model and a rotating model," *Journal of Wind Engineering and Industrial Aerodynamics*, vol. 120, pp. 1–8, 2013.

[7] F. Castellani and A. Vignaroli, "An application of the actuator disc model for wind turbine wakes calculations," *Applied Energy*, vol. 101, pp. 432–440, 2013.

[8] S. Ivanell, *Numerical computations of wind turbine wakes [Ph.D. thesis]*, KTH Royal Institute of Technology, 2009.

[9] ANSYS Inc, "ANSYS CFX Solver Theory Guide," 2000.

[10] M. O. L. Hansen, *Aerodynamics of Wind Turbines*, Earthscan, 2nd edition, 2008.

[11] L. E. Myers and A. S. Bahaj, "Experimental analysis of the flow field around horizontal axis tidal turbines by use of scale mesh disk rotor simulators," *Ocean Engineering*, vol. 37, no. 2-3, pp. 218–227, 2010.

[12] A. S. Bahaj, L. E. Myers, and G. Thompson, "Characterising the wake of horizontal axis marine current turbines," in *Proceedings of the 7th European Wave and Tidal Energy Conference*, p. 9, Porto, Portugal, 2007.

[13] F. M. White, *Fluid Mechanics*, McGraw-Hill, New York, NY, USA, 6th edition, 2009.

[14] F. R. Menter, "Zonal two-equation k-ω turbulence models for aerodynamic flows," AIAA Paper 93-2906, 1993.

[15] D. C. Wilcox, *Turbulence Modeling in CFD*, DCW, 2006.

5

Improved Cat Swarm Optimization for Simultaneous Allocation of DSTATCOM and DGs in Distribution Systems

Neeraj Kanwar, Nikhil Gupta, K. R. Niazi, and Anil Swarnkar

Department of Electrical Engineering, Malaviya National Institute of Technology Jaipur, Jaipur 302017, India

Correspondence should be addressed to Nikhil Gupta; nikhil2007_mnit@yahoo.com

Academic Editor: Joydeep Mitra

This paper addresses a new methodology for the simultaneous optimal allocation of DSTATCOM and DG in radial distribution systems to maximize power loss reduction while maintaining better node voltage profiles under multilevel load profile. Cat Swarm Optimization (CSO) is one of the recently developed powerful swarm intelligence-based optimization techniques that mimics the natural behavior of cats but usually suffers from poor convergence and accuracy while subjected to large dimension problem. Therefore, an Improved CSO (ICSO) technique is proposed to efficiently solve the problem where the seeking mode of CSO is modified to enhance its exploitation potential. In addition, the problem search space is virtually squeezed by suggesting an intelligent search approach which smartly scans the problem search space. Further, the effect of network reconfiguration has also been investigated after optimally placing DSTATCOMs and DGs in the distribution network. The suggested measures enhance the convergence and accuracy of the algorithm without loss of diversity. The proposed method is investigated on 69-bus test distribution system and the application results are very promising for the operation of smart distribution systems.

1. Introduction

The electric power industries have witnessed many reforms in recent years. The existing distribution systems are moving towards smart distribution systems to achieve larger socioeconomic and other nontangible benefits. The rise of smart grid is a boon not only to society as a whole but also to all who are involved in the electric power industry, its customers, and its stakeholders [1]. Building of such distribution systems requires local generation of reactive and active power using distributed energy resources (DERs) such as Distribution Static Compensators (DSTATCOMs) and Distributed Generations (DGs). DSTATCOM is a power electronic-based synchronous voltage generator capable of providing rapid and uninterrupted capacitive and inductive reactive power supply [2]. Various renewable and nonrenewable DG technologies are available on the market today, such as microturbines, fuel cells, combustion gas turbines, photovoltaic, wind turbines, and combined heat and power [3]. The integration of renewable DG technologies such as photovoltaic and wind turbines is becoming more popular in distribution systems on account of smart grid initiatives and strict environmental laws. These components allow increased efficiency, more reliability, and better quality of electric service. Moreover, they also facilitate effective utilization and life extension of existing distribution system infrastructure [1]. However, optimal placement and sizing of these components are the important issues to extract maximum possible benefits.

The optimal allocation of DSTATCOM and DGs in distribution systems is a highly nonlinear complex combinatorial problem which has to satisfy various equality and inequality constraints. In the recent past, several population-based metaheuristic techniques such as Genetic Algorithm (GA), Ant Colony Optimization (ACO), Immune Algorithm (IA), Differential Evolution Algorithm (DEA), Firefly Algorithm (FA), Particle Swarm Optimization (PSO), Teaching-Learning Based Optimization (TLBO), Artificial Bee Colony (ABC), Harmony Search Algorithm (HSA), and Cuckoo Search Algorithm (CSA) have shown their potential to solve optimal DSTATCOM placement problem [4–7] or optimal DG placement problem [8–12]. However, the simultaneous placement strategy can independently set and control the real and reactive power flow in distribution networks [13].

A lot of research work has been carried out to successfully optimize the siting and sizing problems of active and reactive components when allocated separately. However, only a few researchers have attempted simultaneous placement strategy. References [14–17] have shown mutual impact of these components on the performance of distribution networks. Abu-Mouti and El-Hawary [14] employed an ABC algorithm to determine the optimal size of DGs, power factor, and location to minimize power losses. A heuristic approach is suggested by Naik et al. [15] where a node sensitivity analysis is used to identify the candidate DER sites, and their optimal capacities are determined by suggesting heuristic curve fitting technique. Moradi et al. [17] proposed a combined imperialist competitive algorithm- (ICA-) GA method to solve this multiobjective optimization problem. In this method, first ICA is used to find the sites and sizing of DERs and then the operators of GA are employed to further refine these solutions.

The smart grid requires integrated solutions for available distributed resources that reflect their coexistence to achieve higher efficiency through loss minimization and good quality power supply. Distribution networks are reconfigured frequently with changing operating conditions, and it is one of the effective means to improve their performance. The network reconfiguration is a process that alters feeder topological structure by managing the open/close status of sectionalizing and tie-switches under contingencies or normal operating conditions [18]. Changing network topology by reallocating loads from one feeder to another may balance loads among the feeders and decrease the real power losses [19]. Therefore, this is another resource that can be utilized in conjunction with simultaneous placement of DSTATCOM and DG. This approach has possibly not been attempted till date.

Cat Swarm Optimization (CSO) is one of the recently established high performance computational techniques introduced by Chu and Tsai [20]. CSO is inspired by the natural behavior of cats where two major behaviors of the cats are modeled into two submodels: seeking mode and the tracing mode. In the seeking mode, the cat looks around and seeks the next position to move to, whereas, in the tracing mode, the cat tracks some targets [21]. The important property of CSO is that it provides local as well as global search capability simultaneously [22]. It converges better and shows a better performance in finding the global best solution [21]. It has been successfully applied to solve diverse engineering optimization problems such as linear antenna array synthesis [23], deployment of wireless sensors [21], IIR system identification [24], clustering [25], and linear phase FIR filter design [26]. However, the exploration potential of CSO needs to be enhanced while subjected to large dimension problems by reviewing its seeking mode. Further, convergence and accuracy of the algorithm can be improved by suitably placing tentative solutions in the problem search space during the iterative process.

Several researchers [10, 12, 15, 27], and many others, have squeezed the problem search space by restricting the number of candidate locations for placing these devices. They generated a node priority list using certain node sensitivity-based approach and then selecting top few nodes from it as the candidate sites to allocate these devices. This approach drastically reduces the problem search space and also the CPU time incurred. However, the sensitivities are normally calculated for the base case conditions, where no such devices are installed [28]. Furthermore, when selecting only top nodes as the sensitive components, it did not give the true picture of the entire distribution network [29]. Therefore, such approaches are unreliable and thus lead the algorithm to suboptimal solution.

In light of the above discussion, a new Improved CSO- (ICSO-) based method is proposed for the simultaneous allocation of DSTATCOM and DGs in radial distribution networks. The objective is to maximize power loss reduction while maintaining a better node voltage profile. The distribution network is reconfigured after the optimal placement of these devices to extract maximum possible benefits. The seeking mode of CSO is modified to enhance exploitation potential of the algorithm. In addition, an intelligent search is proposed to enhance the overall performance of the optimizing tool.

The remainder of the paper is organized as follows: The problem is formulated in Section 2. The description of the standard and proposed CSO algorithm is presented in Sections 3 and 4, respectively. Section 5 deals with simulation results and the analysis of results is discussed in Section 6. Finally, the conclusions drawn from this work are presented in Section 7.

2. Problem Formulation

The node voltage profile of distribution systems can be improved by installation of DERs as DGs and DSTATCOMs, network reconfiguration, tap changing transformers, and so forth. The proposed algorithm is installing these components, and then the distribution network is reconfigured. Therefore, a soft voltage constraint is treated as the part of objective function while installing DSTATCOM and DGs, and the solutions are accepted by imposing penalty so long as the voltage constraint violates within prespecified limits. However, a hard voltage constraint is necessary for the system operation. Thus it is employed while reconfiguring the distribution network after optimally placing these components. The amount of voltage profile improvement and the capacities of these components employed are not linearly related. Therefore, the advantage of soft voltage constraint employed in placing DSTATCOMs and DGs is that it results in lesser capacity allocation in the optimal solution. In other words, the whole burden of voltage profile improvement should not be imposed over these distributed components, as the network reconfiguration can successfully improve voltage profiles. Therefore, the objective function F_i is formulated to maximize power loss reduction while maintaining a better node voltage profile by proposing a voltage penalty factor approach as defined below:

$$\text{Max} \, F_i = \text{PF}_i \times \left(P_{\text{loss},bi} - P_{\text{loss},ai} \right); \quad \forall i \in L, \quad (1)$$

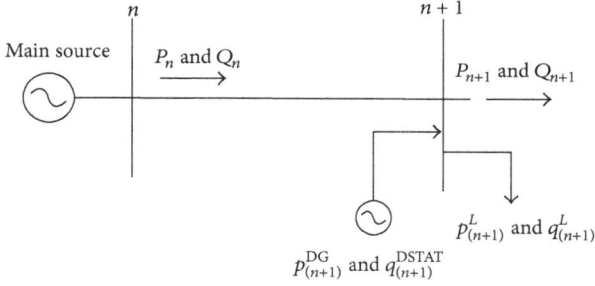

FIGURE 1: Single-line diagram of a two-bus system.

where PF_i is the node voltage deviation penalty factor which is given by

$$PF_i = \frac{1}{(1 + \text{Max}(\Delta V_{ni}))}, \quad (2)$$

where

$$\Delta V_{ni}$$
$$= \begin{cases} 1 - |V_{ni}|; & V_{\min S} < V_{ni} < V_{\min} \\ 0; & V_{\min} \le V_{ni} \le V_{\max} \\ \text{a very large number}; & \text{else} \end{cases}; \quad (3)$$

$$\forall n \in N, \ \forall i \in L,$$

subject to the following operational constraints.

(a) Power Flow Equations. The sum of the power purchased from utility grid and the total power generated by the different sources in the distribution system must be balanced by the local load demand and the power loss in the lines. For a radial network, a set of recursive equations are used to model the power flow in the network as shown by (4). A sample two-bus system including DG and DSTATCOM units is shown in Figure 1. Consider

$$P_{(n+1),i} = P_{n,i} - R_n \frac{P_{n,i}^2 + Q_{n,i}^2}{V_{n,i}^2} - P_{(n+1),i};$$

$$\forall n \in N, \ \forall i \in L,$$

$$Q_{(n+1),i} = Q_{n,i} - X_n \frac{P_{n,i}^2 + Q_{n,i}^2}{V_{n,i}^2} - q_{(n+1),i};$$

$$\forall n \in N, \ \forall i \in L,$$

$$V_{(n+1),i}^2 = V_{n,i}^2 - 2(R_n P_{n,i} + X_n Q_{n,i})$$
$$+ (R_n^2 + X_n^2) \frac{P_{n,i}^2 + Q_{n,i}^2}{V_{n,i}^2};$$

$$\forall n \in N, \ \forall i \in L,$$

$$P_{(n+1),i} = P_{(n+1),i}^L - P_{(n+1),i}^{DG}; \quad \forall n \in N, \ \forall i \in L,$$

$$q_{(n+1),i} = q_{(n+1),i}^L - q_{(n+1),i}^{DSTAT}; \quad \forall n \in N, \ \forall i \in L.$$

$$(4)$$

(b) Branch Current Limit. The current flow in each branch must satisfy the rated ampacity of each branch:

$$I_{ni} \le I_n^{\max}; \quad \forall n \in N, \ \forall i \in L. \quad (5)$$

(c) Active and Reactive Compensation Limit at a Node. The active and reactive power injected by DG and DSTATCOM at each node must be within their permissible ranges:

$$p_{n,\min}^{DG} \le p_n^{DG} \le p_{n,\max}^{DG}; \quad \forall n \in N,$$
$$q_{n,\min}^{DSTAT} \le q_n^{DSTAT} \le q_{n,\max}^{DSTAT}; \quad \forall n \in N. \quad (6)$$

(d) System Compensation Limit. The sum of active and reactive power injected by DGs and DSTATCOMs at all candidate nodes should be less than system nominal active and reactive power demand, respectively:

$$\sum q_n^{DSTAT} \le Q_D; \quad \forall n \in N,$$

$$\sum p_n^{DG} \le P_D; \quad \forall n \in N. \quad (7)$$

Equations (8) and (9) ensure nonrepetition of candidate nodes for DSTATCOM and DG allocation, respectively:

$$N_{DSTAT,a} \ne N_{DSTAT,b}; \quad a, b \in N, \quad (8)$$

$$N_{DG,a} \ne N_{DG,b}; \quad a, b \in N. \quad (9)$$

The distribution network is reconfigured after optimally placing these devices. The reconfiguration problem is solved to minimize real power loss while satisfying various network operational constraints. The mathematical formulation of reconfiguration problem for loss minimization is formulated below:

$$\text{Minimize, } P_{\text{loss},i} = \sum_{n=1}^{E} R_n \frac{P_{ni}^2 + Q_{ni}^2}{|V_{ni}|^2}; \quad (10)$$

$$\forall n \in N, \ \forall i \in L,$$

subject to constraints defined by (4)-(5) along with radial topology and node voltage constraints as given below.

(a) Radiality Constraint. The reconfigured network topology must be radial, that is, with no closed path:

$$\Phi_i(r) = 0; \quad \forall i \in L. \quad (11)$$

(b) Voltage Limit Constraint. A hard voltage constraint is employed during the network reconfiguration as it is one of

the important network operation strategies. All node voltages of the system must be maintained within acceptable operating limits during the optimization process:

$$V_{\min} \leq V_{ni} \leq V_{\max}; \quad \forall n \in N, \; \forall i \in L. \qquad (12)$$

In the present work the codification proposed in [18] is used to solve the network reconfiguration problem. This is a rule-based codification to check and correct infeasible radial topologies while attempting this problem.

3. A Brief Overview of CSO

Cats initiate their move very slowly and cautiously after sensing the presence of a prey and finally chase it very quickly. By observing these features, two modes of operation for CSO are simulated, that is, seeking and tracing modes. These modes have been mathematically modeled for solving optimization problems and are combined together by defining a mixture ratio (MR). The position of cats represents the set of tentative solutions in the problem search space. Every cat has its own position composed of D dimensions, velocity for each dimension, a fitness value, which represents the accommodation of the cat to the fitness function, and a flag to identify whether the cat is in seeking mode or tracing mode [20]. These two modes of operation can be described briefly as given below.

3.1. Seeking Mode. In this mode, random mutations within narrow range are employed at predefined dimensions on the cats selected according to MR. Some terms related to this mode can be defined as follows: Seeking Memory Pool (SMP) is the number of copies generated for each cat; Seeking Range of Selected Dimension (SRD) is the predefined range of each dimension being selected for mutation; CDC is the Counts of Dimension to Change, that is, the number of dimensions to be mutated. According to MR, a definite number of cats are selected for this mode. Making SMP copies of each cat: randomly select CDC dimensions for each copy. Each of these selected dimensions is varied in the range [−SRD, +SRD]. In this way each cat is mutated to facilitate local search. However, higher SRD (>2) causes unnecessarily larger search space; higher CDC (>0.6) causes much higher diversity of the individuals by mutation, thus delaying the convergence; lower MR (<0.2) causes the cats to spend most of their time resting and observing (seeking mode), thus avoiding local entrapment to suboptimal solutions [26]. The higher the value of SMP (>4) is, the better the exploration will be, but at the cost of more CPU time. Thus in seeking mode, the best values of SRD, CDC, MR, and SMP should be decided by the usual trade-off between accuracy and CPU time of the algorithm.

3.2. Tracing Mode. This mode corresponds to a local search technique for the optimization problem. Cat traces the target while spending high energy in this mode. The rapid chase of the cat is mathematically modeled as a large change in its position. Define position and velocity of kth cat in the D-dimensional search space as $X_k = (X_{k,1}, X_{k,2}, \ldots, X_{k,d})$ and

$V_k = (V_{k,1}, V_{k,2}, \ldots, V_{k,d})$. The global best position of the cat swarm is represented as $X_{\text{best}} = (X_{\text{best},1}, X_{\text{best},2}, \ldots, X_{\text{best},d})$ [22]. In this mode, the cat moves according to its own velocity among all its dimensions so that it can trace the prey, that is, the best fit cat. The velocity $V_{k,d}$ and position $X_{k,d}$ updates of the kth cat are governed by the following relations:

$$
\begin{aligned}
V_{k+1,d} &= V_{k,d} + C \times r(\cdot) \times (X_{\text{best},d} - X_{k,d}), \\
X_{k+1,d} &= X_{k,d} + V_{k+1,d},
\end{aligned}
\qquad (13)
$$

where C is a constant, usually taken as 2, and $r(\cdot)$ is a random number in the range [0, 1]. Whether a cat is in seeking or tracing mode, it survives, if its fitness is improved. At the end of each iteration, cats from both seeking and tracing modes merge together before initiating the next iteration.

4. Proposed ICSO

In CSO, the seeking mode provides local search whereas the tracing mode searches globally. However, it will be better if the current best cat is allowed to search locally. If it happens, the current best cat may upgrade its fitness, and later on this will positively influence the movement of all the cats going through the tracing mode. Moreover, it can also avoid possible local trappings. Therefore, seeking mode of the standard CSO is modified in the perception of the above facts as described below.

4.1. Modified Seeking Mode. Like other swarm optimization techniques, the philosophy of CSO is "to follow the leader." If the fitness of the current best cat is improved by some means, the convergence of CSO would be improved. It is therefore suggested that the current best cat is mandatorily selected for the seeking mode. Further, the way of generating cats around the current best cat is also different in the proposed seeking mode, as described below.

The operators of CSO inherently generate continuous decision variables. However, the decision variables are strictly integers in this problem. Therefore, while employing local search around the current best cat, two cats are generated from this cat by employing ceiling and flooring of its decision variables. All possible combinations of cats are generated from these two cats. An illustration of modified seeking mode is shown in Figure 2 as a flowchart. For simplicity, it is shown only for two dimensions. The fitness of all these cats is evaluated and the best fit cat is updated, if a better cat is found. However, other cats of the seeking mode are updated as in the standard CSO. The local search around the best cat not only avoids local trappings but also facilitates the exploration of new search points in the problem search space during the tracing mode.

4.2. Intelligent Search. While initializing or otherwise, it will be always better if all the tentative solutions spread in such a way that most of them lie near the promising region. But this is a difficult task. Nevertheless, a sufficient diversity is essential to explore new solution points in the problem search space. Several researchers, as mentioned earlier, have applied

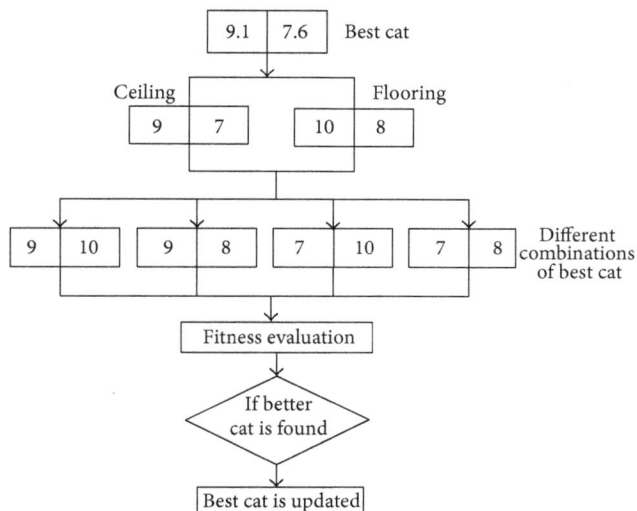

FIGURE 2: Modified seeking mode.

TABLE 1: Initial data of 69-bus system.

Particulars	Value
Sectionalizing switches	1–68
Base configuration (open lines)	69–73
Line voltage (kV)	12.66
Nominal active demand (kW)	3802.19
Nominal reactive demand (kVAr)	2694.6
P_{loss} at light/nominal/peak load (kW)	51.61/225/652.23
V_{min} at light/nominal/peak load (p.u.)	0.9567/0.9092/0.8445

TABLE 2: Selected design parameters.

Parameter	Value
loc	1
q_{max}^{DSTAT} (kVAr)/p_{max}^{DG} (kW)	2000/2000
Q_D (kVAr)/P_D (kW)	2694.6/3802.19
N_{DSTAT}/N_{DG}	1–69
I_n^{max} (n)	400 (1–9), 300 (46–50, 53–65), 200 (10–45, 51, 52, 66–73)
$V_{min}/V_{max}/V_{min}$S (p.u.)	0.95/1.05/0.90
MR/SMP/SRD/CDC	0.2/3/0.2/0.6
Population size (P)	10
Maximum iteration count	100

TABLE 3: Optimal solution for DSTATCOM and DG allocation.

Load levels	DSTATCOM in kVAr (node)/DG in kW (node)	Optimal configuration
Light	649 (61)/912 (61)	12, 20, 53, 69, 72
Nominal	1301 (61)/1828 (61)	12, 21, 69, 70, 72
Peak	1998 (61)/1999 (61)	12, 55, 69, 70, 73

perturbation-based node sensitivity approach to get the node priority list for optimal allocation of distributed resources and then selecting top few nodes from it to redefine the problem search space. However, none of the sensitivity-based approaches is foolproof so it is possible that the optimal node may not lie in the redefined problem search space. Therefore, such approaches are unreliable and may cause erroneous results. In the present work, candidate nodes are selected by proposing an intelligent search approach. The proposed approach is different than the conventional one in the sense that the node sensitivity is observed in terms of the change in objective function to be optimized, instead of the change in power loss. Moreover, the system nodes are arranged in the decreasing order of change in the objective function and then the candidate nodes are selected from this list using roulette wheel selection. In this way, the nodes are selected according to their probability of priority. Thus more chances are available to those nodes which are good for the allocation of these devices. However, this approach reserves the right of each system node to remain in the problem search space during the computational process. Therefore, using intelligent search the problem search space is virtually squeezed without loss of diversity. An illustration of intelligent search is shown in Figure 3.

4.3. Individual's Encoding. The structure of the individuals for the proposed methods is shown in Figure 4 which is composed of candidate nodes and sizing for the respective candidate DSTATCOM and DGs. The candidate nodes are allocated using intelligent search approach, whereas the sizing of distributed resources is selected randomly within their respective predefined bounds as described by (6)-(7).

4.4. Termination Criterion. Elitism is not required due to the intrinsic nature of CSO. Therefore, the termination criterion is taken as follows: "when either the maximum iteration count is exhausted or all cats acquire the same fitness,

the evolutionary process stops." This criterion is selected to reduce the CPU time of the algorithm. The flowchart of the proposed ICSO method is presented in Figure 5.

5. Simulation Results

The proposed CSO method is investigated on 69-bus test distribution system taken from [30]. The annual load profile is assumed to be piecewise segmented in three different load levels, that is, light, nominal, and peak, which are 50%, 100%, and 160% of the nominal system load, respectively, and the corresponding load durations are taken 2000, 5260, and 1500 hours as in [27]. It has been assumed that DSTATCOM exchanges only reactive power with the network and DG is dispatchable and operated at unity power factor. The initial data of this system are presented in Table 1 and various design parameters selected are given in Table 2. The Newton-Raphson power flow method is used for load flow of the distribution system.

The optimal solution obtained, for each load level, after 100 trials of proposed ICSO is presented in Table 3. The table shows the capacity of DSTATCOM and DG in kVAr and kW,

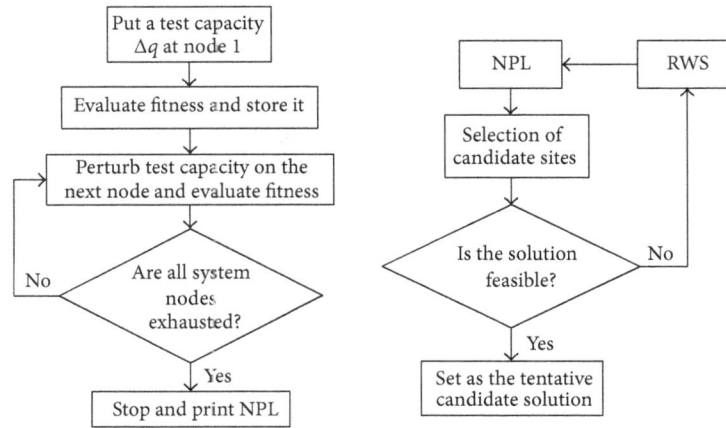

FIGURE 3: Intelligent search.

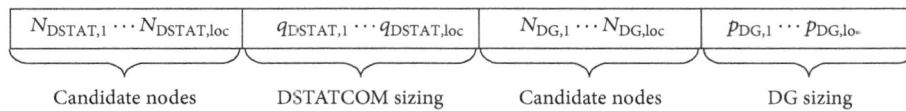

FIGURE 4: Cat's encoding for the proposed ICSO.

respectively. The table also shows that the optimal network configuration is affected by the presence of these devices. The performance of the distribution network obtained using this optimal solution is presented in Table 4.

The table depicts power loss reduction after optimal allocation of these devices for each load level. It can be observed from the table that annual energy losses for this system are reduced significantly by about 88%. This shows that simultaneous placement strategy is very useful for distribution systems. The feeder power losses are further reduced when distribution network is optimally reconfigured after optimally placing DSTATCOM and DG in the distribution system. This causes a net annual energy loss reduction of about 94% from the base case network which is substantial. Thus, the reconfiguration of distribution network can be fruitfully utilized in order to enhance the energy efficiency of distribution systems. It is interesting to note that the minimum node voltage obtained by optimally placing DSTATCOM and DG is found to be 0.9493 p.u. at peak load level. However, this voltage is enhanced to 0.9751 p.u. by network reconfiguration. This shows the utility of the proposed penalty factor approach where a soft voltage constraint is imposed to permit the solutions having minimum voltage below 0.95 p.u. during evolutionary process.

The voltage profile obtained using the proposed method during all load conditions is presented and compared with base case in Figure 6. It can be observed from the figures that the voltage profile is improved by simultaneous placement of DSTATCOM and DG which is then further enhanced by network reconfiguration. This is true for all load levels considered. All voltages are found to be within permissible limits while integrated approach is employed. Thus the integrated solution provided by the proposed ICSO successfully achieved the desired objectives.

Not much literature is available for the validation of the proposed method. A comparison result with PSO [2] is presented in Table 5. The only scenario is considered where the number of sites for locating each of these components is being restricted to one. The table shows that the power loss reduction and minimum voltage provided by the solution using the proposed method are much better than those obtained with PSO [2]. It happens because the penetration limits of DSTATCOMs and DGs considered in [2] are equal to the nominal load at the nodes where they have to be installed. This penetration limit is 0.888 MVAr and 1.244 MW, respectively, for node 61. However, PSO [2] explores only 0.1223 MW capacities for optimal DGs. Thus, the proposed ICSO is capable of generating a better solution than the existing PSO technique.

6. Discussion

The proposed ICSO has shown its potential to efficiently solve one of the large dimension problems of power system. It happens because of the proposed suggestions in the standard CSO. The comparison of the convergence characteristics of CSO and ICSO is presented in Figure 7. The figure also compares the convergence of PSO when applied for the problem with the same population size and maximum iterations. The figure shows a marked improvement in the convergence of ICSO, whereas the CSO and PSO seem to be getting trapped in local optima. It happens due to the poor exploitation of the search space. In ICSO, the intelligent search dispersed all individuals in the near vicinity of the promising region, without scarifying diversity. This causes better chances for good individuals to surf the problem search space amicably. Further, the modified seeking mode enhances

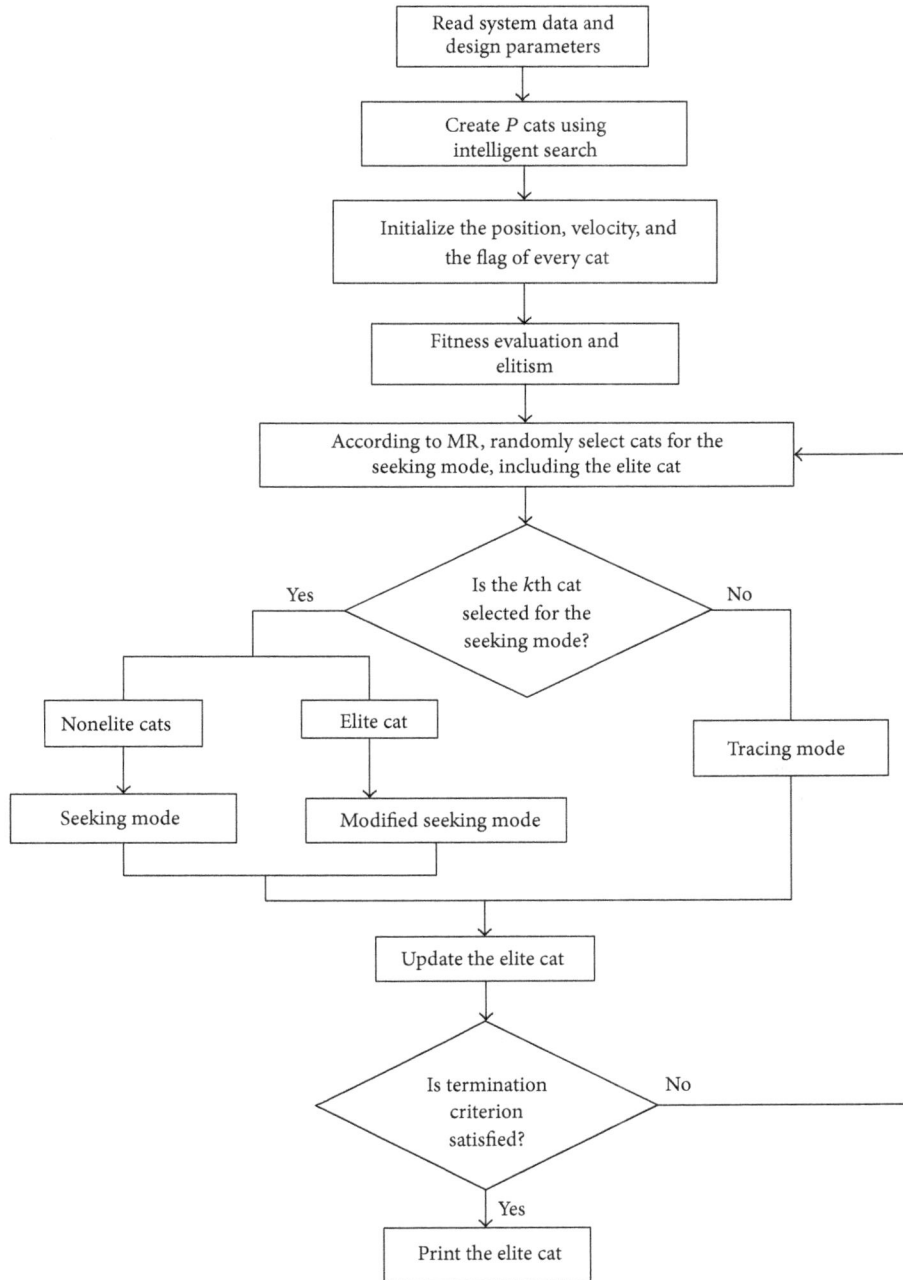

FIGURE 5: Flowchart of the proposed ICSO.

TABLE 4: Distribution network performance using the proposed method.

Particulars	After optimal allocation			Reconfiguration after optimal allocation		
	Light	Nominal	Peak	Light	Nominal	Peak
P_{loss} (kW)	5.69	23.17	92.23	2.95	12.06	47.80
Power loss reduction (%)	88.97	89.70	85.86	94.28	94.64	92.68
V_{min} (p.u.)	0.9864	0.9725	0.9493	0.9952	0.9887	0.9751
Annual energy loss (kWh)		271600			140908	
Annual energy loss reduction (%)		88.01			93.78	

(a)

(b)

(c)

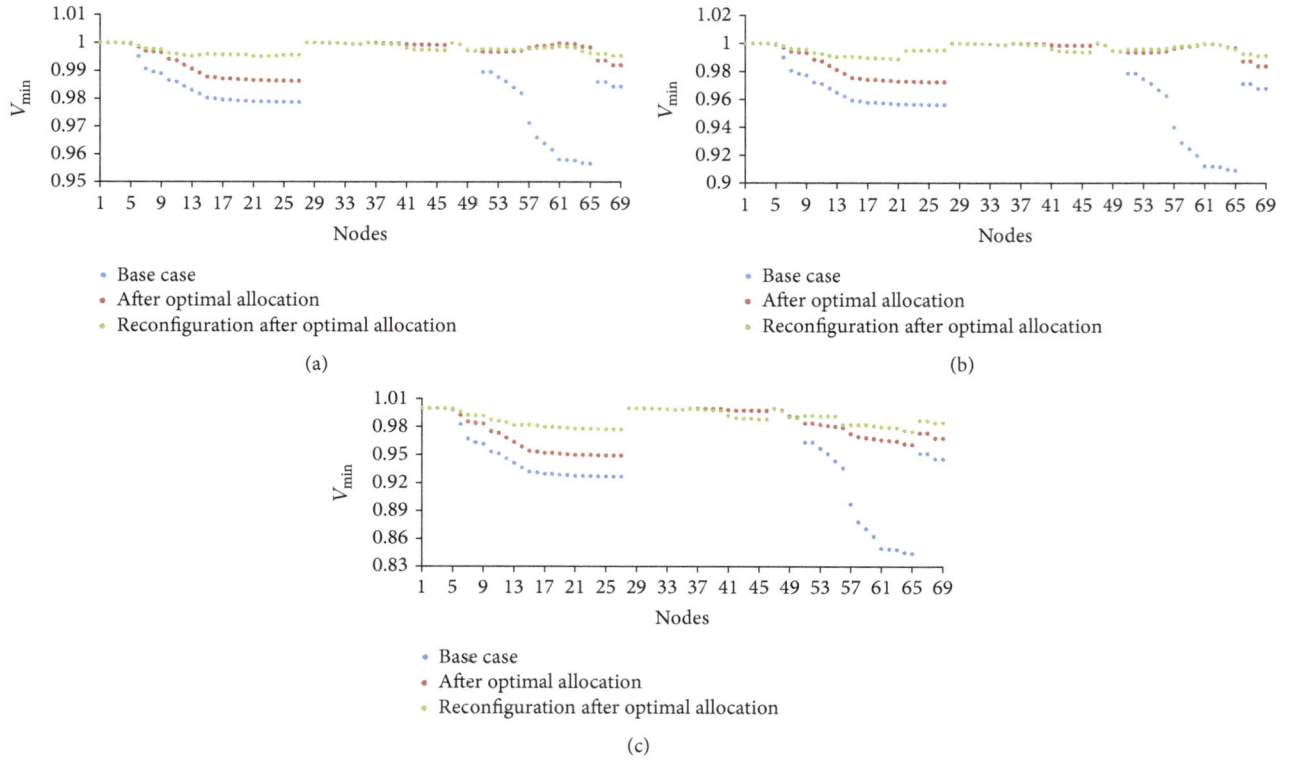

FIGURE 6: Comparison of voltage profiles for base case, after optimal allocation of devices and reconfiguration after optimal allocation of devices at (a) light load, (b) nominal load, and (c) peak load.

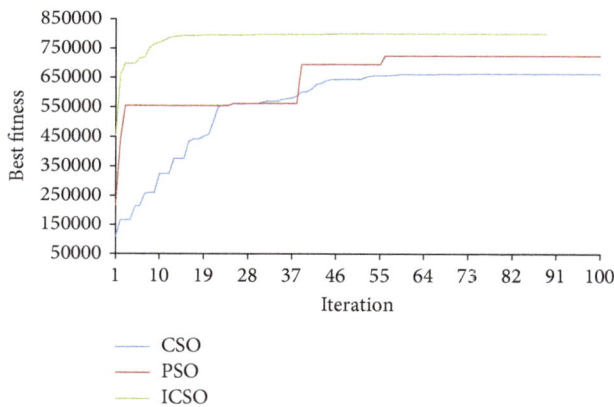

FIGURE 7: Comparison of convergence characteristics of CSO, PSO, and ICSO.

TABLE 5: Comparison results.

Parameters	PSO [2]	ICSO
Size of DSTATCOM in MVAr (node)	0.9045 (61)	1.301 (61)
Size of DG in MW (node)	0.1223 (61)	1.828 (61)
P_{loss} (kW)	141.14	23.17
Power loss reduction (%)	37.26	89.70
V_{min} (p.u.)	0.9291	0.9725

both exploration and exploitation potential of the swarm. This fact can be observed from the typical convergence of ICSO which is quite different than CSO or PSO. As a consequence ICSO facilitates obtaining global or near-global optima in less fitness evaluations.

Finally, a comparison for the solution qualities obtained by CSO, ICSO, and PSO for a sample of 100 independent solutions is presented in Table 6. The table shows best, mean, and worst fitness of the sampled solutions. It can be observed from the table that the solution quality obtained using ICSO

is better than either CSO or PSO on the basis of the obtained best, mean, and worst fitness of the sampled solutions. The table also shows statistical indices for the solution quality, that is, standard deviation (SD), coefficient of variation (COV), and error from the best (EFB) of sampled solutions. The smaller the value of these indices, the better the solution quality. It can be said that CSO is performing better than PSO on account of average and best fitness obtained, although it is more computationally demanding. However, ICSO is found to be substantially improved compared to CSO and is less computationally demanding. Therefore, the proposed ICSO not only is capable of generating better solution in less time but also provides good quality solutions. ICSO is taking more time than PSO because the seeking mode of CSO is inherently computationally demanding.

TABLE 6: Comparison of solution qualities.

Method	Best	Average	Worst	SD	COV	EFB	CPU time (s)
CSO	783613.33	749302.64	679973.57	30727.05	4.10	5.88	51.87
PSO	769719.63	737032.84	705699.83	21817.88	2.96	5.11	36.94
Proposed ICSO	799832.30	797367.69	778524.72	5472.46	0.69	0.75	47.31

7. Conclusions

An integrated approach for the optimal placement of distributed resources and network reconfiguration using improved variant of CSO technique is presented. The objectives of maximum power loss reduction and voltage profile enhancements have been successfully achieved using the proposed modeling. It has been observed that the proposed integrated approach of simultaneous placing of DSTATCOMs and DGs in the distribution network and then optimally reconfiguring it can enhance the performance of the systems by a good margin. The suggested seeking mode in ICSO provides local random walk to the best fit cat and the proposed intelligent search scans the problem search space efficiently without loss of diversity. This causes dispersion of all tentative solutions in the close vicinity of the promising region which helps to obtain the global or near-global optima. The comparison result validates that the proposed method is better than existing PSO and CSO methods. It has been observed that the proposed ICSO has better convergence and is also capable of generating fairly good quality solutions.

Nomenclature

C: Acceleration coefficient

D: Number of decision variables

d: Each dimension of selected cat $(1 \leq d \leq D)$

E: Total number of branches in the system

I_{ni}: Feeder current at the ith load level (p.u.)

$I_n^{\max}$: Rated feeder current (p.u.)

L: Set of load levels

loc: Total number of candidate locations for DSTATCOM/DG placement

N_{DSTAT}: Candidate nodes for DSTATCOM placement

N_{DG}: Candidate nodes for DG placement

N: Set of system nodes

$p_{n,\min}^{\text{DG}}/p_{n,\max}^{\text{DG}}$: Minimum/maximum limits of DG penetration at the nth node (kW)

$p_{n,i}^{\text{DG}}$: DG power generation at the nth node for the ith load level (kW)

$p_{n,i}^L/q_{n,i}^L$: Nominal active/reactive power demand at the nth node for the ith load level (kVAr/kW)

P_D: Nominal active power demand of the system (kW)

P_{loss}: Power loss (kW)

$P_{\text{loss},bi}$: Power loss for uncompensated system at the ith load level (kW)

$P_{\text{loss},ai}$: Power loss for compensated system at the ith load level (kW)

P_{ni}: Real power for sending end of the nth branch at the ith load level (kW)

$q_{n,i}^{\text{DSTAT}}$: Reactive power injection by DSTATCOM at the nth node for the ith load level (kVAr)

$q_{n,\min}^{\text{DSTAT}}/q_{n,\max}^{\text{DSTAT}}$: Minimum/maximum reactive compensation provided by DSTATCOMs at the nth node (kVAr)

Q_D: Nominal reactive power demand of the system (kVAr)

Q_{ni}: Reactive power for sending end of the nth branch at the ith load level (kVAr)

R_n: Line resistance of the nth branch (Ω)

$r(\cdot)$: Random number in the range of $[0, 1]$

$V_{\max}/V_{\min}$: Maximum/minimum permissible node voltage (p.u.)

$V_{\min S}$: Minimum specified node voltage (p.u.)

V_{ni}: Voltage of the nth node at the ith load level (p.u.)

V_k/V_{k+1}: Velocity of the kth/$(k+1)$th cat

ΔV_{ni}: Maximum node voltage deviation of the nth node at the ith load level (p.u.)

X_{best}: Position of the cat that has the best fitness value

X_k/X_{k+1}: Position of the kth/$(k+1)$th cat

X_n: Line reactance of the nth branch (Ω)

PF: Node voltage deviation penalty factor

Φ_i: Closed loop at the ith load level.

Conflict of Interests

The authors declare that there is no conflict of interests regarding the publication of this paper.

References

[1] M. E. El-Hawary, "The smart grid—state-of-the-art and future trends," *Electric Power Components and Systems*, vol. 42, no. 3-4, pp. 239–250, 2014.

[2] S. Devi and M. Geethanjali, "Optimal location and sizing determination of distributed generation and DSTATCOM using particle swarm optimization algorithm," *International Journal of Electrical Power & Energy Systems*, vol. 62, pp. 562–570, 2014.

[3] M. A. Darfoun and M. E. El-Hawary, "Multi-objective optimization approach for optimal distributed generation sizing and placement," *Electric Power Components and Systems*, vol. 43, no. 7, pp. 828–836, 2015.

[4] S. Jazebi, S. H. Hosseinian, and B. Vahidi, "DSTATCOM allocation in distribution networks considering reconfiguration

using differential evolution algorithm," *Energy Conversion and Management*, vol. 52, no. 7, pp. 2777–2783, 2011.

[5] S. A. Taher and S. A. Afsari, "Optimal location and sizing of DSTATCOM in distribution systems by immune algorithm," *International Journal of Electrical Power & Energy Systems*, vol. 60, pp. 34–44, 2014.

[6] M. Farhoodnea, A. Mohamed, H. Shareef, and H. Zayandehroodi, "Optimum D-STATCOM placement using firefly algorithm for power quality enhancement," in *Proceedings of the IEEE 7th International Power Engineering and Optimization Conference (PEOCO '13)*, pp. 98–102, Langkawi, Malaysia, June 2013.

[7] A. Bagherinasab, M. Zadehbagheri, S. Abdul Khalid, M. Gandomkar, and N. A. Azli, "Optimal placement of D-STATCOM using hybrid genetic and ant colony algorithm to losses reduction," *International Journal of Applied Power Engineering*, vol. 2, no. 2, pp. 53–60, 2013.

[8] R. Kollu, S. R. Rayapudi, and V. L. N. Sadhu, "A novel method for optimal placement of distributed generation in distribution systems using HSDO," *International Transactions on Electrical Energy Systems*, vol. 24, no. 4, pp. 547–561, 2014.

[9] J. A. Martín García and A. J. Gil Mena, "Optimal distributed generation location and size using a modified teaching—learning based optimization algorithm," *International Journal of Electrical Power and Energy Systems*, vol. 50, no. 1, pp. 65–75, 2013.

[10] R. S. Rao, K. Ravindra, K. Satish, and S. V. L. Narasimham, "Power loss minimization in distribution system using network reconfiguration in the presence of distributed generation," *IEEE Transactions on Power Systems*, vol. 28, no. 1, pp. 317–325, 2013.

[11] S. Kansal, V. Kumar, and B. Tyagi, "Optimal placement of different type of DG sources in distribution networks," *International Journal of Electrical Power and Energy Systems*, vol. 53, no. 1, pp. 752–760, 2013.

[12] D. K. Khatod, V. Pant, and J. Sharma, "Evolutionary programming based optimal placement of renewable distributed generators," *IEEE Transactions on Power Systems*, vol. 28, no. 2, pp. 683–695, 2013.

[13] H. Hedayati, S. A. Nabaviniaki, and A. Akbarimajd, "A method for placement of DG units in distribution networks," *IEEE Transactions on Power Delivery*, vol. 23, no. 3, pp. 1620–1628, 2008.

[14] F. S. Abu-Mouti and M. E. El-Hawary, "Optimal distributed generation allocation and sizing in distribution systems via artificial bee colony algorithm," *IEEE Transactions on Power Delivery*, vol. 26, no. 4, pp. 2090–2101, 2011.

[15] S. G. Naik, D. K. Khatod, and M. P. Sharma, "Optimal allocation of combined DG and capacitor for real power loss minimization in distribution networks," *International Journal of Electrical Power and Energy Systems*, vol. 53, pp. 967–973, 2013.

[16] S. M. Sajjadi, M.-R. Haghifam, and J. Salehi, "Simultaneous placement of distributed generation and capacitors in distribution networks considering voltage stability index," *International Journal of Electrical Power and Energy Systems*, vol. 46, no. 1, pp. 366–375, 2013.

[17] M. H. Moradi, A. Zeinalzadeh, Y. Mohammadi, and M. Abedini, "An efficient hybrid method for solving the optimal sitting and sizing problem of DG and shunt capacitor banks simultaneously based on imperialist competitive algorithm and genetic algorithm," *International Journal of Electrical Power and Energy Systems*, vol. 54, pp. 101–111, 2014.

[18] A. Swarnkar, N. Gupta, and K. R. Niazi, "A novel codification for meta-heuristic techniques used in distribution network reconfiguration," *Electric Power Systems Research*, vol. 81, no. 7, pp. 1619–1626, 2011.

[19] M. A. N. Guimarães, C. A. Castro, and R. Romero, "Distribution systems operation optimisation through reconfiguration and capacitor allocation by a dedicated genetic algorithm," *IET Generation, Transmission & Distribution*, vol. 4, no. 11, pp. 1213–1222, 2010.

[20] S.-C. Chu and P.-W. Tsai, "Computational intelligence based on the behavior of cats," *International Journal of Innovative Computing, Information and Control*, vol. 3, no. 1, pp. 163–173, 2007.

[21] S. Temel, N. Unaldi, and O. Kaynak, "On deployment of wireless sensors on 3-D terrains to maximize sensing coverage by utilizing cat swarm optimization with wavelet transform," *IEEE Transactions on Systems, Man, and Cybernetics: Systems*, vol. 44, no. 1, pp. 111–120, 2014.

[22] P. M. Pradhan and G. Panda, "Solving multiobjective problems using cat swarm optimization," *Expert Systems with Applications*, vol. 39, no. 3, pp. 2956–2964, 2012.

[23] L. Pappula and D. Ghosh, "Linear antenna array synthesis using cat swarm optimization," *AEU—International Journal of Electronics and Communications*, vol. 68, no. 6, pp. 540–549, 2014.

[24] G. Panda, P. M. Pradhan, and B. Majhi, "IIR system identification using cat swarm optimization," *Expert Systems with Applications*, vol. 38, no. 10, pp. 12671–12683, 2011.

[25] Y. Liu, X. Wu, and Y. Shen, "Cat swarm optimization clustering (KSACSOC): a cat swarm optimization clustering algorithm," *Scientific Research and Essays*, vol. 7, no. 49, pp. 4176–4185, 2012.

[26] S. K. Saha, S. P. Ghoshal, R. Kar, and D. Mandal, "Cat swarm optimization algorithm for optimal linear phase FIR filter design," *ISA Transactions*, vol. 52, no. 6, pp. 781–794, 2013.

[27] D. Das, "Optimal placement of capacitors in radial distribution system using a Fuzzy-GA method," *International Journal of Electrical Power & Energy Systems*, vol. 30, no. 6-7, pp. 361–367, 2008.

[28] R. A. Gallego, A. J. Monticelli, and R. Romero, "Optimal capacitor placement in radial distribution networks," *IEEE Transactions on Power Systems*, vol. 16, no. 4, pp. 630–637, 2001.

[29] V. Haldar and N. Chakraborty, "Power loss minimization by optimal capacitor placement in radial distribution system using modified cultural algorithm," *International Transactions on Electrical Energy Systems*, vol. 25, no. 1, pp. 54–71, 2015.

[30] M. E. Baran and F. F. Wu, "Optimal capacitor placement on radial distribution systems," *IEEE Transactions on Power Delivery*, vol. 4, no. 1, pp. 725–734, 1989.

6

The Study of Kinetic Properties and Analytical Pyrolysis of Coconut Shells

Mahir Said,[1] Geoffrey John,[1] Cuthbert Mhilu,[1] and Samwel Manyele[2]

[1]Department of Mechanical and Industrial Engineering, University of Dar es Salaam, Dar es Salaam, Tanzania
[2]Department of Chemical and Mining Engineering, University of Dar es Salaam, Dar es Salaam, Tanzania

Correspondence should be addressed to Mahir Said; mahir@udsm.ac.tz

Academic Editor: Hasan Ferdi Gercel

The kinetic properties of coconut shells during pyrolysis were studied to determine its reactivity in ground form. The kinetic parameters were determined by using thermogravimetric analyser. The activation energy was 122.780 kJ/mol. The pyrolysis products were analyzed using pyrolysis gas chromatography/mass spectrometry (Py-GC/MS). The effects of pyrolysis temperature on the distribution of the pyrolytic products were assessed in a temperature range between 673 K and 1073 K. The set time for pyrolysis was 2 s. Several compounds were observed; they were grouped into alkanes, acids, ethers and alcohols, esters, aldehydes and ketones, furans and pyrans, aromatic compounds, and nitrogen containing compounds. The product compositions varied with temperature in that range. The highest gas proportion was observed at high temperature while the acid proportion was observed to be highest in coconut shells, thus lowering the quality of bio-oil. It has been concluded that higher pyrolysis temperature increases the amount of pyrolysis products to a maximum value. It has been recommended to use coconut shell for production of gas, instead of production of bio-oil due to its high proportion of acetic acid.

1. Introduction

The consumption of biomass fuel is increasing worldwide due to concerns over energy shortage of fossil fuels and due to increasing of carbon dioxide emission. Sustainable production of biomass fuel from forest and agriculture products can displace fossil fuels. Although burning of biomass fuels releases carbon dioxide, the regrowth of the sustainable managed trees offset that release, a property not possible with fossil fuels. This forest fuel can supply energy virtually without net contribution to greenhouse gas levels. Many African countries depend on agricultural activities and have abundant forest resources and agrowaste that can benefit them in energy production.

Tanzania is one among the African countries, which is located in the equatorial region. It has large source of biomass material. The agricultural sector has about 10 million hectares [1]. Biomass makes 88% of the primary energy consumption in Tanzania [2].

The biomass always contains high moisture content, irregular shapes and size, and low bulk density, when collected from the field. This makes the biomass have low energy density and become expensive and difficult to manage. It is dried and used for combustion directly, but in developed countries after drying the biomass is densified for making pellets [3].

The biomass material constitutes mainly cellulose, hemicellulose, and lignin. The variation proportions of these constituents in different biomass species have been studied in the past, the hemicellulose varies from 25 to 40%wt, cellulose varies from 30 to 40%wt, and lignin is between 5 and 15%wt [4]. The main products during decomposition of lignin are aromatic compounds, while the decomposition products of cellulose and hemicellulose are linear compounds which have the same functional groups.

Although biomass is used as a solid fuel, the demand of liquid and gaseous fuel has made several researchers convert biomass to liquid and gaseous fuel through relevant processes. Pyrolysis is one among the thermal processes that is used to convert solid biomass to liquid fuel. It is the process by which biomass is heated without oxidizing agent. Three main products are obtained as gas, liquid, and solid phases.

TABLE 1: Proximate, ultimate analysis, and higher heating value of biomass materials.

	Proximate analysis				Ultimate analysis			
MC	Volatile matter[db]	Fixed carbon[db]	Ash[db]	C	H	O[db]	N	Higher heating value (MJ/kg)
(%wt)	(%wt)	(%wt)	(%wt)	(%wt)	(%wt)	(%wt)	(%wt)	
10.70	79.18	20.26	0.56	47.94	6.41	45.56	0.10	17.35

Note: db = dry basis; df = by difference.

The gas mainly contains carbon dioxide, carbon monoxide, methane, and hydrogen [5]. The solid part is char, which is made up of fixed carbon and ash. The third and main product is liquid, also known as bio-oil, which contains several compounds. The compounds formed depend on the type of biomass species, elemental composition of biomass, and pyrolysis conditions [4].

In this study, the formations of compounds which are constituted in bio-oil were studied by using PY-GC/MS at different pyrolysis temperature, the biomass material used being coconut shells.

2. Methodology

2.1. Biomass Preparation.
The coconut shells were obtained from local farmers of Tanzania. The coconut shells were obtained after removing husks and kernels of the coconut. The coconut shells were sun-dried until they became brittle, followed by grinding to less than 0.25 mm to increase surface area yielding bio-oil during pyrolysis [6]. The grinding machine used was Retsch GmbH 5657 HAAN, type SK1, Nor 72307.

2.2. Coconut Shell Characterization

2.2.1. Determination of Ultimate and Proximate Analysis.
Important characteristics of biomass materials were done by determining the proximate and ultimate analysis. The proximate analysis was done to observe the moisture, volatile, fixed carbon, and ash content of the biomass material. Ultimate analysis was done to observe the elemental analysis of biomass materials, such as carbon, hydrogen, oxygen, and nitrogen. The important components of fuels are carbon and hydrogen.

2.2.2. Determination of Gross Calorific Value.
The gross calorific value of the biomass material was obtained by using bomb calorimeter model CAB001.AB1.C. The standard test method for biomass analysis was ASTM E870-82 [7]. These analyses are also considered for determination of combustion properties of a fuel [8].

2.2.3. Thermogravimetric Analysis.
The biomass materials were dried for two hours in the oven at 378 K (105°C) to constant weight before subjecting the samples to thermogravimetric analysis. During thermogravimetric analysis, 20 mg (±0.03 mg) of biomass material was put in the crucible and then kept in the thermogravimetric analyzer type NETZSCH STA 409 PC Luxx. The TG was run under nonisothermal conditions and the temperature was raised from room temperature 303 K to 1100 K and the heating medium was nitrogen at a flow rate of 50 mL/min. The standard method used was ASTM E1131-08 [9]. The heating rate applied was 10 K/min.

2.3. Analytical Pyrolysis.
The pyrolysis of biomass samples and analysis of product compositions were done by using Py-GC/MS (Agilent Technology 5975C, model number G3174A, serial number US12504A05). A 1 mg (±0.01 mg) sample of biomass material was kept on the filament and heated at a constant temperature for 2 s. The reaction temperatures applied were 673 K (400°C), 773 K (500°C), 873 K (600°C), 973 K (700°C), and 1073 K (800°C); the pyrolysis medium was helium. The gas and volatiles produced during pyrolysis were analyzed by using GC/MS. The solid residues left on the filament were measured after each experiment. Each experiment was done three times to ensure it can be repeated.

3. Results and Discussion

3.1. Biomass Characterization.
Test results of the proximate and ultimate analysis of the coconut shells and higher heating value are shown in Table 1. It has been observed that coconut shells contain high carbon about 50%wt and hydrogen content about 6%wt. The presence of nitrogen and high amount of oxygen about 46%wt reduces the energy content of biomass materials, because these elements do not support combustion [10]. The higher heating value of coconut shell was about 17 MJ/kg; this calorific value is lower than that of conventional fossil fuels such as coal which is about 30 MJ/kg [11]. Therefore, it is important to convert coconut shells to a high energy fuel through thermochemical process such as pyrolysis. The proximate, ultimate analysis and higher heating value obtained in this study are in agreement with other biomass characterization studies done [12].

Table 2 shows the cellulose, hemicellulose, and lignin of coconut shells as reported by different researchers [13, 14]. When coconut shells are compared to the other wood biomass such as pine, it has been reported that coconut shells contain higher composition of hemicellulose and small amount of cellulose [14].

3.2. Thermogravimetric (TG) Analysis of Biomass.
The TG and DTG profiles of biomass sample were drawn (shown in Figures 1(a) and 1(b)). The profile is divided into three regions. The first region is formed between room temperature and 500 K; this is due to moisture removal; there is no chemical reaction taking place in this region. The second region is devolatilization process. In this region, a high mass loss is observed; it started above 500 K depending on type of

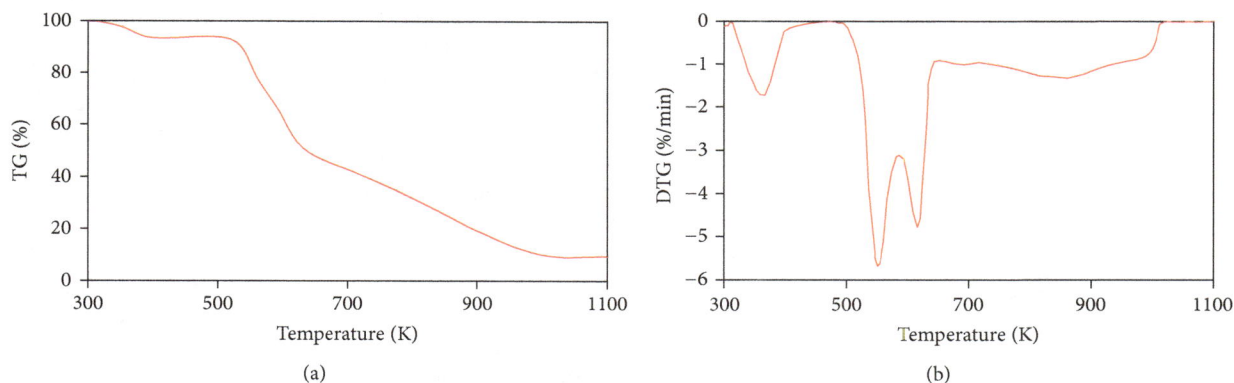

FIGURE 1: (a) Thermogravimetric (TG) analysis of coconut shells. (b) Differential thermogravimetric (DTG) analysis of coconut shells.

TABLE 2: Biomass composition.

Biomass	Cellulose	Hemicellulose	Lignin	Reference
Coconut shells	19.8	50.1	30.1	[13]

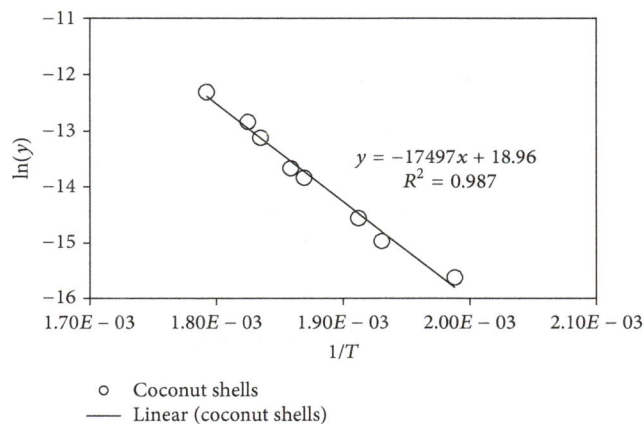

FIGURE 2: The graphs for determining kinetic parameters.

biomass material [15]. The devolatilization of coconut shell is between 500 K and 1000 K; the curve has a kink at about 600 K which differentiates between light and heavy volatiles [16]. The light volatiles are released at low temperature, while heavy volatiles are released at high temperature. Furthermore, the DTG (Figure 1(b)) curves show the degradation of biomass components (hemicellulose, cellulose, and lignin). The first peak shows the degradation of hemicellulose, followed by cellulose, while lignin has very short peak but wide peak which spread to overlap hemicelluloses and cellulose. However, the peaks of cellulose and hemicellulose overlap in the DTG of coconut shells, but they can be differentiated, the peak of cellulose appeared at 640 K, and the hemicellulose peak appeared at 550 K. The peak height of hemicellulose in coconut shells is larger than that of cellulose, since coconut shells contain higher amount of hemicellulose as shown in Table 2. The lignin degradation starts above 500 K, but the final temperature depends on the type of the biomass material. The final temperature for lignin degradation for coconut shells is about 1000 K. This means that the pyrolysis temperature for any biomass material should be above its final temperature of cellulose, because at that temperature all volatiles from biomass will be already removed.

The kinetic parameters were determined in order to understand the biomass degradation behavior. The method used was Coats and Redfern method ((1) and (2)). Equation (1) is applied when the thermodegradation of biomass is first order ($n = 1$) and (2) when the reaction is not first order ($n \neq 1$) [17]. The graph in Figure 2 was obtained after solving (1), since biomass degradation was observed to fit on first order.

The activation energy for coconut shells is 122.780 kJ/mol, preexponential factor is 2.177×10^9/s, and the reaction is first order. Consider

$$\ln(y) = \ln\left[\frac{-\ln(1-\alpha)}{T^2}\right]$$
$$= \ln\left[\frac{AR}{\beta E_a}\left(1 - \frac{2RT}{E_a}\right)\right] - \frac{E_a}{RT} \quad n = 1, \tag{1}$$

$$\ln(y) = \ln\left[\frac{(1-\alpha)^{1-n} - 1}{(n-1)T^2}\right]$$
$$= \ln\left[\frac{AR}{\beta E_a}\left(1 - \frac{2RT}{E_a}\right)\right] - \frac{E_a}{RT} \quad n \neq 1. \tag{2}$$

3.3. Biomass Pyrolysis by Using PY-GC/MS. PY-GC/MS uses isothermal process whereby the biomass sample is heated at a constant temperature. The PY-GC/MS cannot measure the yield of products produced, but it provides the peak area (intensity) of the products. The intensity corresponds to the amount of products produced during pyrolysis [18]. There are several chemical compounds formed during biomass

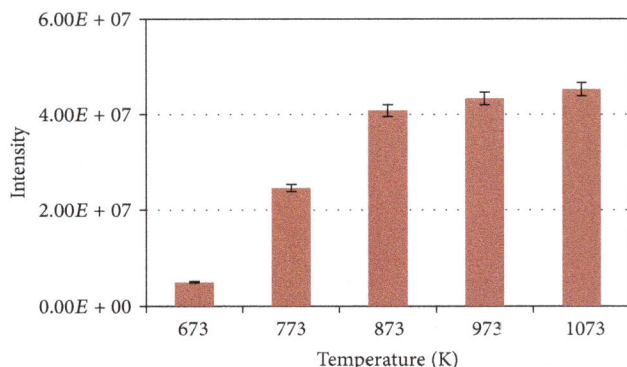

FIGURE 3: The gas production from pyrolysis of biomass.

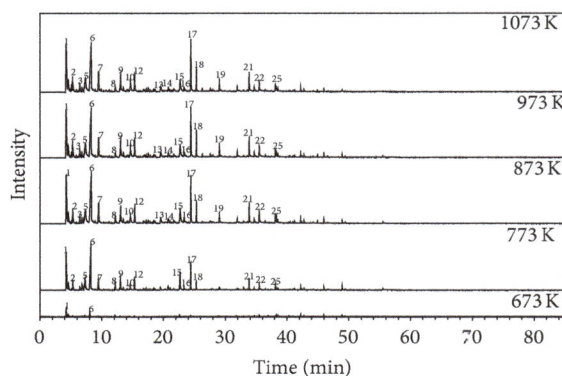

FIGURE 4: Chromatograms of pyrolysis of coconut shells at different temperatures.

pyrolysis. But the compounds that will be discussed in this study are those that vary with pyrolysis temperature according to their chromatograms. The chromatograms of biomass depend on both types of biomass and pyrolysis temperature.

The analysis of biomass pyrolysis products shows that there are several chemical compounds produced during pyrolysis. The details of chemical compounds observed during pyrolysis of biomass material are shown in the chromatograms (Figure 4). The origins of the compounds are derived from three components of biomass; these are cellulose, hemicellulose, and lignin. The products that are derived from cellulose and hemicellulose are grouped into gas, alkane, alcohol, furan, carboxylic acid, ketone, aldehyde, and pyran [19, 20]. Phenols and benzene related compounds are derived from lignin [21, 22].

3.3.1. The Gas. It has been observed that the gas yield increases as pyrolysis temperature increases as shown in Figure 3, with this also being reported by other researchers such as Bridgwater [23]. The main components of the gas are carbon dioxide (CO_2), carbon monoxide (CO), methane (CH_4), and hydrogen (H_2). The increase of the yield of gaseous products is due to decarboxylation and decarbonylation [24, 25].

3.3.2. Alkane. The alkane produced during biomass pyrolysis is pentane, the amount which increases with increasing temperature. The alkanes are important compounds in bio-oil production, since it is very combustible. The alkane is produced through a cracking of high hydrocarbons in biomass during the pyrolysis process and the yield is between 2 and 6%, while in catalytic pyrolysis the yield can reach up to 24.66% [26].

3.3.3. Acids. The acid observed is only acetic acid. The acids reduce the pH of the bio-oil produced during pyrolysis [27]. The acids are produced when hemicellulose and cellulose are decomposing [5, 28]. The acids are formed by removing acetyl groups in xylose [29], but Güllü and Demirbaş [30] reported that acids are also formed due to decomposition of all three wood components (cellulose, hemicellulose, and lignin). Furthermore, high proportion of acid content (about 20%) was observed in coconut shells attributable to its high amount of hemicellulose, while the other biomass produces acetic acid in the range of 5 to 15% [31] as observed in Table 3 and in the DTG curve (Figure 1(b)). The detailed analysis of acids is shown in Table 3.

3.3.4. Ethers, Esters, and Alcohols. The ethers, esters, and alcohols were produced during biomass pyrolysis as shown in Table 3. The ethers produced are in the form of oxirane, methyl-,(S)- and 1,3-dioxolane,2-3 ethenyl-4-methyl-. The alcohol is in the form of propyl alcohol. The esters formed during pyrolysis of biomass are acetic acid methyl ester and propanoic acid,2-oxo-,methyl ester. The formation of ethers, esters, and alcohols is due to decomposition of cellulose, hemicellulose, and lignin [29].

3.3.5. Furans and Pyrans. Furans and pyrans are formed from all biomass during thermal decomposition. They are derived from cellulose, while pyrans are produced from the destruction of hemicelluloses [32, 33]. The furfural and 2H-pyran-2,6(3H)-dione were produced at high temperature only. The proportions of compounds were not affected by increasing temperature as shown in Table 3.

3.3.6. Aldehydes and Ketones. The aldehydes and ketones are derived from the degradation of cellulose, as proposed by Piskorz et al. [34]. Aldehydes are formed through dehydration of cellulose, while Gao et al. [29] reported that ketones are formed by breaking the molecular bonds between C2 and C3 of glucose monomers and hemiacetal group loops. Generally, the temperature has least effect on the production of aldehydes and ketones. The aldehyde compound formed during pyrolysis of coconut shells is acetaldehyde hydroxyl, which is produced at all temperature ranges. While the compounds for ketones are 2-propanone, 1-hydroxy-, 1,2-cyclopentanedione, 2-pentanone, and 3-pentanone. They are produced at high temperature. The temperature has less effect on production of ketone compounds in all biomass materials.

3.3.7. Aromatic Compounds. The main compounds formed during decomposition of lignin are aromatic hydrocarbons

TABLE 3: Compounds produced during biomass pyrolysis.

PN	Compounds	RT, min	673 K	773 K	873 K	973 K	1073 K
	Pentane						
2	Pentane (wt, %)	5.321	2.10	2.92	3.18	3.97	3.18
	Acids, alcohol, esters, and ethers						
6	Acetic acid (wt, %)	8.077	20.41	20.66	19.56	18.31	17.52
8	Propargyl alcohol (wt, %)	12.118	2.74	3.1	1.88	1.13	1.18
9	Acetic acid, methyl ester (wt, %)	13.021	3.92	4.23	4.5	4.18	4.75
10	Propanoic acid, 2-oxo-methyl ester (wt, %)	14.611	—	—	1.69	1.96	1.82
11	Oxirane, methyl-,(S)-(wt, %)	14.732	—	—	1.17	1.41	1.5
24	1,3-Dioxolane, 2-ethenyl-4-methyl-(wt, %)	36.955	3.29	—	—	—	—
	Furans and pyrans						
12	Furfural (wt, %)	15.337	—	3.41	4.06	4.53	3.8
16	2H-Pyran-2,6(3H)-dione (wt, %)	23.401	—	—	1.50	1.28	1.20
	Aldehydes and ketones						
3	2-Pentanone (wt, %)	6.467	—	—	1.79	1.7	1.89
4	3-Pentanone (wt, %)	6.838	—	1.92	0.99	1.01	0.99
5	Acetaldehyde hydroxy-(wt, %)	7.290	4.15	6.85	6.1	7.07	7.72
7	2-Propanone, 1-hydroxy-(wt, %)	9.478	—	3.65	4.8	4.19	4.41
13	1,2-Cyclopentanedione (wt, %)	19.598	—	—	2.33	2.27	2.23
	Aromatic compounds						
17	Phenol (wt, %)	24.380	6.7	6.17	7.98	7.58	8.32
18	Phenol,2-methoxy-(wt, %)	25.305	—	2.12	4.07	4.69	3.7
19	Creosol (wt, %)	29.058	—	—	2.18	1.95	2
21	2-Methoxy-4-vinylphenol (wt, %)	33.863	—	2.78	3.31	3.6	3.21
23	Phenol,2,6-dimethoxy-(wt, %)	35.438	2.57	2.18	1.89	2.25	2.44
25	trans-Isoeugenol (wt, %)	38.126	—	1.48	1.03	1.13	1.24
26	3-Hydroxy-4-methoxybenzoic acid (wt, %)	38.298	3.27	—	—	—	—
27	Vanillin (wt, %)	38.567	3.18	—	—	—	—
28	Phenol,2,6-dimethoxy-4-(2-propenyl)-(wt, %)	42.768	1.87	—	—	—	—
29	2-Propionic acid, 3-(4-hydroxy-3-methoxyphenyl (wt, %)	45.987	3.33	—	—	—	—
30	4-((1E)-3-Hydroxy-1-propenyl)-2-methoxyphenol (wt, %)	48.928	3.32	—	—	—	—
	Nitrogen containing compounds						
14	2-Imidazolidinone (wt, %)	20.814	—	1.45	1.20	1.18	1.08
15	2-Methyliminoperhydro-1,3-oxazine (wt, %)	22.597	4.68	7.61	5.56	4.31	4.14

such as phenolic, guaiacyl, and syringyl compounds. The product distribution and the yield of products are strongly dependent on the type of biomass [35]. These show that there is no uniform trend for methoxy phenols; this was also observed by Niemz et al. [36]. The relation of phenol formation is not the same to all biomass; each type of biomass behaves differently. Brebu and Vasile [35] reported that guaiacol and syringols are intermediate degradation products, they are decreasing with increasing pyrolysis temperature, and they form vinyl phenols by cleavage of the O–C (alkyl) and O–C (aryl) bonds and other small molecular compounds. The proportions of aromatic compounds are shown in Table 3. Coconut shell has a high proportion of aromatic compounds at lower temperature.

3.3.8. Nitrogen Compounds. Furthermore, it has been observed that there is a formation of organic compounds containing nitrogen as shown in Table 3. This is the evidence

that nitrogen in the biomass is in the form of compounds such as protein [37, 38]. Literature has reported the formation of nitrogen containing compounds during pyrolysis [37, 38]. The analysis of different studies [37, 38] revealed that the formation of nitrogen containing compounds during pyrolysis is due to amino groups present in the biomass. Ammonia and nitrogen oxides can be released during pyrolysis; this means the compounds containing nitrogen were decomposed into ammonia and other low molecular weight compounds. In this study, it has been observed that the nitrogen containing compounds decrease by increasing reaction temperature. The nitrogen containing compounds formed were 2-imidazolidinone and 2-methyliminoperhydro-1,3-oxirane.

3.4. The Formation of Aromatic Compound during Pyrolysis. Table 3 revealed that the aromatic compounds that are formed at low temperature (<773 K) are different to

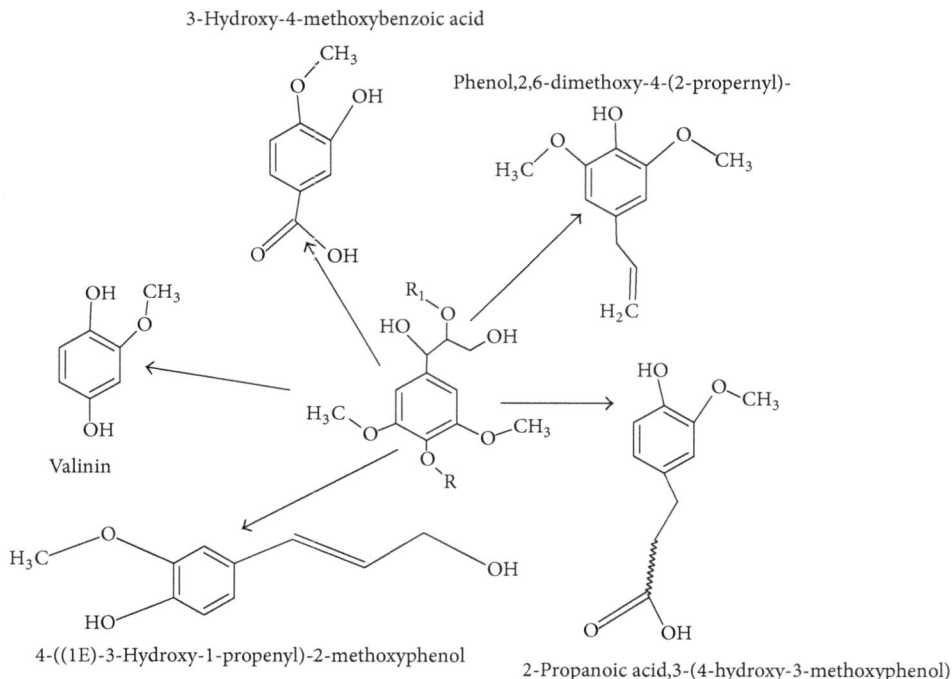

FIGURE 5: The schematic diagram of the lignin decomposition at low temperature (<773 K).

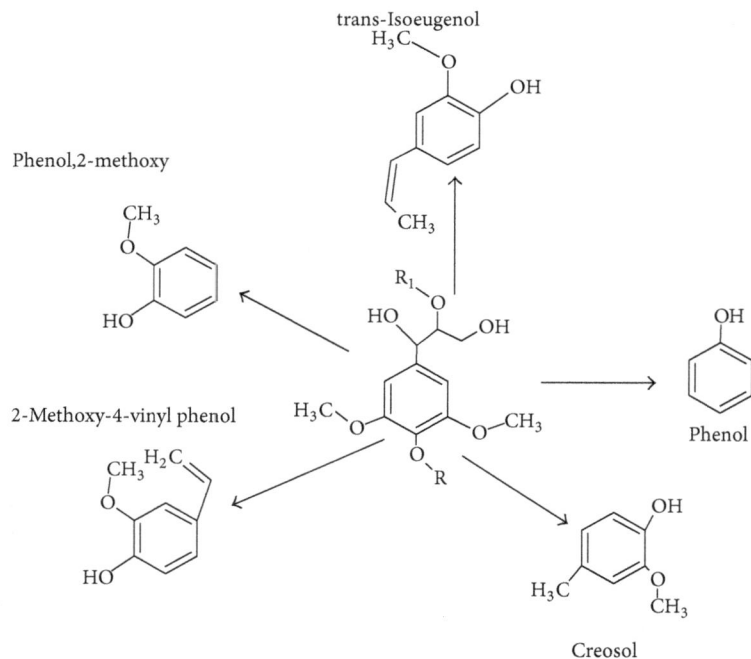

FIGURE 6: The schematic diagram of the lignin decomposition at high temperature (>773 K).

those formed at higher temperature (>773 K). The aromatic compounds are produced due to the decomposition of lignin.

When lignin is heated at low temperature, the compounds formed are 4-((1E)-3-hydroxy-1-propenyl)-2-methoxyphenol, 2-propernoic acid,3-(4-hydroxy)-3-methoxyphenyl, Phenol-2-6-dimethoxy-4-(2-propenyl), 3-hydroxy-4-methoxybenzoic acid and valinin. Generally, these aromatic compounds have long braches as shown in Figure 5. This can be explained that the temperature was not high enough to break the bond of linear compound from benzene ring.

The aromatic compounds that are formed at high temperature were phenol, trans-isoeugenol, phenol,2-methoxy, and 2-methoxy-4-vinylphenol as shown in Figure 6. These aromatic compounds have shorter branches of linear compounds than the former ones formed at low temperature. This can be seen that the linear compounds that attached to the benzene rings are removed when the high temperature is applied to the lignin. The linear compounds that have been removed from lignin increase the amount of other liner compounds such as, shown in Figure 4, acetic acid methyl ester, propanoic acid,2-oxo-,methyl ester, oxirane, methyl-,(S)-, 2-pentanone, acetaldehyde hydroxy-, 1,2-cyclopentanedione, and some gaseous compounds as shown in Table 3. Also, it has been observed that the amount of phenol increases as temperature increases for the same reasons.

4. Conclusion

The characterization of coconut shell shows that the activation energy of coconut shell was 122.780 kJ/mol; this resembles other biomass materials which are in the range of 60 to 200 kJ/mol [39].

The incremental reaction temperature during pyrolysis increases the intensity of the pyrolysis products, which reveal the increment of the amount of the products. Some compounds only appear at a certain temperature range. There are several compounds that were formed at low temperature (673 K) such as vanillin, phenol,2,6-dimethoxy, and 4-((1E)-3-hydroxy-1-propenyl)-2-methoxyphenol, and creosol. Some compounds were also produced at high temperature (above 773 K) such as 1,2-cyclopentanedione and 2H-pyran-2,6(3H)-dione.

It has been observed that amounts of some linear compounds and phenol increase at high temperature due to the removal of linear compounds of aromatic compounds.

The coconut shell cannot be a good source of liquid (bio-oil) fuel, since it has a high proportion of acetic acid and nitrogen containing compounds, lower pentane and aldehyde, and lower furans, but it contains higher proportion of phenolic compounds. Instead, it can be a good source of gaseous fuel (such as syngas).

Nomenclature

A: Preexponential factor (s^{-1})
E_a: Activation energy (J/mol)
R: Gas constant (J/molK)
T: Absolute temperature (K)
β: Heating rate (K/s)
α: Mass ratio.

Conflict of Interests

The authors declare that there is no conflict of interests regarding the publication of this paper.

Acknowledgments

The authors wish to thank Swedish International Development Cooperation Agency (Sida) and Swedish Energy Agency for their generous financial and material support rendered to allow this research work to be conducted.

References

[1] L. Wilson, W. Yang, W. Blasiak, G. R. John, and C. F. Mhilu, "Thermal characterization of tropical biomass feedstocks," *Energy Conversion and Management*, vol. 52, no. 1, pp. 191–198, 2011.

[2] M. A. Kusekwa, *Biomass Conversion to Energy in Tanzania: A Critique, New Developments in Renewable Energy*, P. H. Arman, Ed., InTech, 2003.

[3] R. G. Fernández, C. P. García, A. G. Lavín, J. L. B. de las Heras, and J. J. Pis, "Influence of physical properties of solid biomass fuels on the design and cost of storage installations," *Waste Management*, vol. 33, no. 5, pp. 1151–1157, 2013.

[4] D. Mohan, C. U. Pittman Jr., and P. H. Steele, "Pyrolysis of wood/biomass for bio-oil: a critical review," *Energy & Fuels*, vol. 20, no. 3, pp. 848–889, 2006.

[5] S. Yaman, "Pyrolysis of biomass to produce fuels and chemical feedstocks," *Energy Conversion and Management*, vol. 45, no. 5, pp. 651–671, 2004.

[6] J. Shen, X.-S. Wang, M. Garcia-Perez, D. Mourant, M. J. Rhodes, and C.-Z. Li, "Effects of particle size on the fast pyrolysis of oil mallee woody biomass," *Fuel*, vol. 88, no. 10, pp. 1810–1817, 2009.

[7] ASTM, *E870-82. Standard Test Methods for Analysis of Wood Fuels*, ASTM International, West Conshohocken, Pa, USA, 2006.

[8] A. Demirbas, "Combustion characteristics of different biomass fuels," *Progress in Energy and Combustion Science*, vol. 30, no. 2, pp. 219–230, 2004.

[9] ASTM International, *E1131-08, Standard Method for Compositional Analysis by Thermogravimetry*, ASTM International, West Conshohocken, Pa, USA, 2014.

[10] S. Czernik and A. V. Bridgwater, "Overview of applications of biomass fast pyrolysis oil," *Energy & Fuels*, vol. 18, no. 2, pp. 590–598, 2004.

[11] S. A. Channiwala and P. P. Parikh, "A unified correlation for estimating HHV of solid, liquid and gaseous fuels," *Fuel*, vol. 81, no. 8, pp. 1051–1063, 2002.

[12] A. J. Tsamba, W. Yang, and W. Blasiak, "Pyrolysis characteristics and global kinetics of coconut and cashew nut shells," *Fuel Processing Technology*, vol. 87, no. 6, pp. 523–530, 2006.

[13] W. M. A. W. Daud and W. S. W. Ali, "Comparison on pore development of activated carbon produced from palm shell and coconut shell," *Bioresource Technology*, vol. 93, no. 1, pp. 63–69, 2004.

[14] M. J. Ramsden and F. S. R. Blake, "A kinetic study of the acetylation of cellulose, hemicellulose and lignin components in wood," *Wood Science and Technology*, vol. 31, no. 1, pp. 45–50, 1997.

[15] M. Carrier, A. Loppinet-Serani, D. Denux et al., "Thermogravimetric analysis as a new method to determine the lignocellulosic composition of biomass," *Biomass and Bioenergy*, vol. 35, no. 1, pp. 298–307, 2011.

[16] K. M. Isa, S. Daud, N. Hamidin, K. Ismail, S. A. Saad, and F. H. Kasim, "Thermogravimetric analysis and the optimisation of bio-oil yield from fixed-bed pyrolysis of rice husk using response surface methodology (RSM)," *Industrial Crops and Products*, vol. 33, no. 2, pp. 481–487, 2011.

[17] M. K. Baloch, M. J. Z. Khurram, and G. F. Durrani, "Application of different methods for the thermogravimetric analysis of polyethylene samples," *Journal of Applied Polymer Science*, vol. 120, no. 6, pp. 3511–3518, 2011.

[18] Q. Lu, X.-C. Yang, C.-Q. Dong, Z.-F. Zhang, X.-M. Zhang, and X.-F. Zhu, "Influence of pyrolysis temperature and time on the cellulose fast pyrolysis products: analytical Py-GC/MS study," *Journal of Analytical and Applied Pyrolysis*, vol. 92, no. 2, pp. 430–438, 2011.

[19] I. Pastorova, R. E. Botto, P. W. Arisz, and J. J. Boon, "Cellulose char structure: a combined analytical Py-GC-MS, FTIR, and NMR study," *Carbohydrate Research*, vol. 262, no. 1, pp. 27–47, 1994.

[20] D. K. Shen, S. Gu, and A. V. Bridgwater, "Study on the pyrolytic behaviour of xylan-based hemicellulose using TG–FTIR and Py–GC–FTIR," *Journal of Analytical and Applied Pyrolysis*, vol. 87, no. 2, pp. 199–206, 2010.

[21] J. Ralph and R. D. Hatfield, "Pyrolysis-GC-MS characterization of forage materials," *Journal of Agricultural and Food Chemistry*, vol. 39, no. 8, pp. 1426–1437, 1991.

[22] C. A. Mullen and A. A. Boateng, "Catalytic pyrolysis-GC/MS of lignin from several sources," *Fuel Processing Technology*, vol. 91, no. 11, pp. 1446–1458, 2010.

[23] A. V. Bridgwater, "Review of fast pyrolysis of biomass and product upgrading," *Biomass and Bioenergy*, vol. 38, pp. 68–94, 2012.

[24] P. A. Horne and P. T. Williams, "Influence of temperature on the products from the flash pyrolysis of biomass," *Fuel*, vol. 75, no. 9, pp. 1051–1059, 1996.

[25] P. T. Williams and S. Besler, "The influence of temperature and heating rate on the slow pyrolysis of biomass," *Renewable Energy*, vol. 7, no. 3, pp. 233–250, 1996.

[26] E. Pütün, F. Ateş, and A. E. Pütün, "Catalytic pyrolysis of biomass in inert and steam atmospheres," *Fuel*, vol. 87, no. 6, pp. 815–824, 2008.

[27] K. Sipilä, E. Kuoppala, L. Fagernäs, and A. Oasmaa, "Characterization of biomass-based flash pyrolysis oils," *Biomass and Bioenergy*, vol. 14, no. 2, pp. 103–113, 1998.

[28] X.-S. Zhang, G.-X. Yang, H. Jiang, W.-J. Liu, and H.-S. Ding, "Mass production of chemicals from biomass-derived oil by directly atmospheric distillation coupled with co-pyrolysis," *Scientific Reports*, vol. 3, article 1120, 2013.

[29] N. Gao, A. Li, C. Quan, L. Du, and Y. Duan, "TG–FTIR and Py–GC/MS analysis on pyrolysis and combustion of pine sawdust," *Journal of Analytical and Applied Pyrolysis*, vol. 100, pp. 26–32, 2013.

[30] D. Güllü and A. Demirbaş, "Biomass to methanol via pyrolysis process," *Energy Conversion and Management*, vol. 42, no. 11, pp. 1349–1356, 2001.

[31] E. Kantarelis, W. Yang, and W. Blasiak, "Production of liquid feedstock from biomass via steam pyrolysis in a fluidized bed reactor," *Energy & Fuels*, vol. 27, no. 8, pp. 4748–4759, 2013.

[32] S. Wang, X. Guo, T. Liang, Y. Zhou, and Z. Luo, "Mechanism research on cellulose pyrolysis by Py-GC/MS and subsequent density functional theory studies," *Bioresource Technology*, vol. 104, pp. 722–728, 2012.

[33] R. J. Evans, C. C. Elam, M. Looker, and M. Nimlos, "Formation of aromatic hydrocarbons due to partial oxidation reactions in biomass gasification," in *Proceedings of the 218th National Meeting of the American Chemical Society*, Abstracts of Papers of the American Chemical Society, p. U810, New Orleans, La, USA, August 1999.

[34] J. Piskorz, D. Radlein, and D. S. Scott, "On the mechanism of the rapid pyrolysis of cellulose," *Journal of Analytical and Applied Pyrolysis*, vol. 9, no. 2, pp. 121–137, 1986.

[35] M. Brebu and C. Vasile, "Thermal degradation of lignin—a review," *Cellulose Chemistry & Technology*, vol. 44, no. 9, pp. 353–363, 2010.

[36] P. Niemz, T. Hofmann, and T. Rétfalvi, "Investigation of chemical changes in the structure of thermally modified wood," *Maderas: Ciencia y Tecnologia*, vol. 12, no. 2, pp. 69–78, 2010.

[37] C.-Z. Li and L. L. Tan, "Formation of NO_x and SO_x precursors during the pyrolysis of coal and biomass. Part III. Further discussion on the formation of HCN and NH_3 during pyrolysis," *Fuel*, vol. 79, no. 15, pp. 1899–1906, 2000.

[38] Q. Ren, C. Zhao, X. Chen, L. Duan, Y. Li, and C. Ma, "NO_x and N_2O precursors (NH_3 and HCN) from biomass pyrolysis: co-pyrolysis of amino acids and cellulose, hemicellulose and lignin," *Proceedings of the Combustion Institute*, vol. 33, no. 2, pp. 1715–1722, 2011.

[39] C. Di Blasi, "Modeling chemical and physical processes of wood and biomass pyrolysis," *Progress in Energy and Combustion Science*, vol. 34, no. 1, pp. 47–90, 2008.

7

Feasibility and Optimal Design of a Stand-Alone Photovoltaic Energy System for the Orphanage

Vincent Anayochukwu Ani

Department of Electronic Engineering, University of Nigeria (UNN), Nsukka 410001, Nigeria

Correspondence should be addressed to Vincent Anayochukwu Ani; vincent_ani@yahoo.com

Academic Editor: Nuri Azbar

Access to electricity can have a positive psychological impact through a lessening of the sense of exclusion, and vulnerability often felt by the orphanages. This paper presented the simulation and optimization study of a stand-alone photovoltaic power system that produced the desired power needs of an orphanage. Solar resources for the design of the system were obtained from the National Aeronautics and Space Administration (NASA) Surface Meteorology and Solar Energy website at a location of 6°51'N latitude and 7°35'E longitude, with annual average solar radiation of $4.92\,kWh/m^2/d$. This study is based on modeling, simulation, and optimization of energy system in the orphanage. The patterns of load consumption within the orphanage were studied and suitably modeled for optimization. Hybrid Optimization Model for Electric Renewables (HOMER) software was used to analyze and design the proposed stand-alone photovoltaic power system model. The model was designed to provide an optimal system configuration based on an hour-by-hour data for energy availability and demands. A detailed design, description, and expected performance of the system were presented in this paper.

1. Introduction

Isolated (remote) sites are locations far from the places where most people live and often lack grid power supply. The price of conventional energy sources in remote areas, such as candles, paraffin, gas, coal, and batteries, is often more expensive than in places where most people live because of the remoteness of retailers. Providing grid electricity in remote areas is often associated with higher costs to the grid supplier. Power may be supplied through stand-alone systems (serving just one or two users). These systems can provide power for domestic uses such as lighting, cooling, TV, radio, and communication. The power may be generated from various resources, using diesel, biomass, wind, PV, or small hydrogenerators, or hybrid combinations of these resources. Depending on the characteristics of a specific use (i.e., the load profile) and the local supply options, the least cost solution for an orphanage may consist of any of the above options. The attraction of these sources lies primarily in their abundance and ready access. Many of the isolated areas lying remotely from the grid have a high potential of renewable energy with solar energy being the most abundant.

Solar home system (SHS) typically includes a photovoltaic (PV) module, a battery, a charge controller, wiring setup, and a DC/AC inverter. A standard small SHS can operate several lights, a television (black-and-white or coloured), a radio or cassette player, and a small fan. SHS can eliminate or reduce the need for candles, kerosene, liquid propane gas, and/or battery charging and provide increased convenience and safety, improved indoor air quality, and a higher quality of light than kerosene lamps for reading [1]. The size of the system (typically 10 to 100 Watts peak (Wp)) determines the number of "light hours" or "TV-hours" available. For example, a 35 Wp SHS provides enough power for four hours of lighting from four 7 W lamps each evening, as well as several hours of television. There are more than 500,000 SHS now installed in rural areas of developing countries [2–7].

Orphanages are often located in an isolated area and access to electricity can bring tangible social and economic benefits to them. The possible benefits can include household

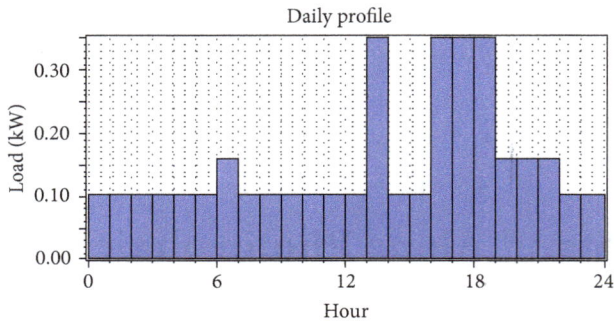

FIGURE 1: Load daily profile of typical orphanage electricity consumption.

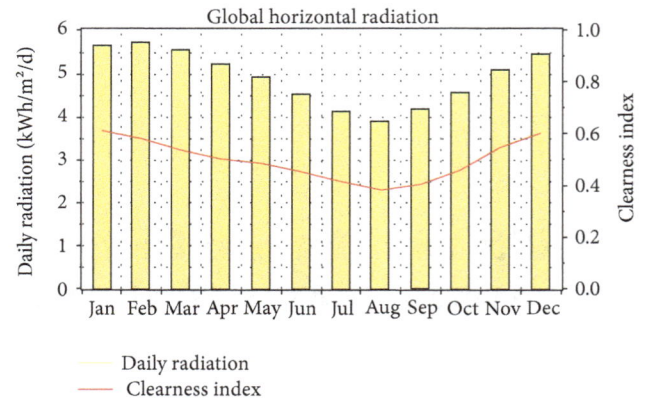

FIGURE 2: Graphics of monthly solar radiation profile for Nsukka.

(orphanage) lighting, the ability to refrigerate food and make washing clothes more convenient. The presence of electricity in an orphanage also can result in better reading culture. Finally, electricity can have a positive psychological impact through a lessening of the sense of exclusion and vulnerability often felt by orphanages, hence the need for the provision of an alternative sustainable electric power supply system. It is always convenient to perform a thorough simulation of the energy system to obtain an optimal output using the natural resources around it before its construction. Therefore, the purpose of this paper is to simulate and optimize a renewable (PV-battery) energy system that will produce the desired power needs of the orphanage, and the optimization parameter proposed here as a base is the offered service.

2. Methodology

In order to design a power system, one has to provide some information from the remote location of the orphanage such as the load profile that should be met by the system, solar radiation for PV generation, the initial cost of each component (PV panels, a charge controller, battery, and inverter), annual interest rate, and project lifetime.

2.1. The Reference Orphanage. From the acquired data, a profile of the orphanage was created. This profile consists of the orphanage load variations and electrical usage patterns within the orphanage. Figure 1 shows the daily profile electricity consumption in an orphanage in Nsukka (Enugu State, Nigeria). The orphanage in Nsukka is simple and does not require large quantities of electrical energy used for lighting and electrical appliances. Table 1 shows an estimation of each appliance's rated power, its quantity, and the hours of use by the orphanage in a single day.

2.2. The Pattern of Using Electricity Power within the Orphanage. The lights in the orphanage will always be on as from 6 am (06:00 h) to 7 am (07:00 h). By this time (6 am to 7 am) the orphans start preparing for school. They leave the orphanage to school by 8 am (08:00 h) and come back to the orphanage by 2 pm (14:00 h). By 7 am, the light will go off, since the rays of light come in through the windows during

day time [7 am–7 pm (07:00 h–19:00 h)]. The light comes on again by 7 pm (19:00 h) till 10 pm (22:00 h) to enable them to read their books. Once it is 10 pm, there will be light out and they will go to bed. The light out will be there till 6 am before the light comes in again. Meanwhile, between 1 pm (13:00 h) and 2 pm (14:00 h), when the radiation is at the apex, the washing machine will be used to wash orphans clothes. As from 4 pm (16:00 h), the orphans will be in the waiting room watching television (programmes from the satellite dish), while the television, the satellite decoder, and the fans will all be ON till 7 pm. Once it is 7 pm, they will go and read their books till 10 pm.

2.3. Study Area. This research focuses on the simulation of photovoltaic power generation system for an orphanage sited in Nsukka located in a valley with poor wind but good solar energies. It is geographically located at 6°51′N latitude and 7°35′E longitude with annual average solar radiation of 4.92 kWh/m²/d. The data for solar resource were obtained from the National Aeronautics and Space Administration (NASA) Surface Meteorology and Solar Energy website [8]. For this study, only solar PV technology was considered. Figure 2 shows the solar resource profile of this location. February is the sunniest month of the year. During this month, the solar energy resource is 5.7 kWh/m²/d while in August it is only 3.9 kWh/m²/d. In the months of September, October, November, December, January, and February, the solar radiation increases with differences from month to month as (0.28), (0.38), (0.54), (0.35), (0.22), and (0.06), respectively, whereas in the months of March, April, May, June, July, and August, the solar radiation decreases with differences from month to month as (0.17), (0.32), (0.31), (0.4), (0.4), and (0.23), respectively. These differences will be considered during system sizing.

3. Modeling of Energy System Components

The mathematical model of the proposed energy system components contains photovoltaic system with battery storage

TABLE 1: The electrical load data.

Description of item	Qty	Load (watts per unit)	Load (watts) total	Daily hours of actual utilization (hr. per day)
Television	1	80	80	3 hrs. (16:00 hr.–19:00 hr.)
Satellite decoder	1	20	20	3 hrs. (16:00 hr.–19:00 hr.)
Fan	2	75	150	3 hrs. (16:00 hr.–19:00 hr.)
Electric bulb (lighting)	4	15	60	4 hrs. (19:00 hr.–22:00 hr.); (06:00 hr–07:00 hr.)
Refrigerator	1	100	100	24 hrs. (0:00 hr.–23:00 hr.)
Washing machine	1	250	250	1 hr. (13:00 hr.–14:00 hr.)

system. The theoretical aspects are given below and based on [9–11].

Mathematical Model of Solar Photovoltaic. Using the solar radiation available, the hourly energy output of the PV generator (E_{PV}) can be calculated according to the following equation [9, 12, 13]:

$$E_{PV} = G(t) \times A \times P \times \eta_{PV},\qquad(1)$$

where $G(t)$ is the hourly irradiance in kWh/m^2, A is the surface area in m^2, P is the PV penetration level factor, and η_{PV} is the efficiency of PV generator.

Mathematical Model of Charge Controller. To prevent overcharging of a battery, a charge controller is used to sense when the batteries are fully charged and to stop or decrease the amount of energy flowing from the energy source to the batteries. The model of the charge controller is presented below [9]:

$$E_{CC\text{-}OUT}(t) = E_{CC\text{-}IN}(t) \times \eta_{CC}$$
$$E_{CC\text{-}IN}(t) = E_{SUR\text{-}DC}(t),\qquad(2)$$

where $E_{CC\text{-}OUT}(t)$ is the hourly energy output from charge controller, kWh, $E_{CC\text{-}IN}(t)$ is the hourly energy input to charge controller, kWh, η_{CC} is the efficiency of a charge controller, and $E_{SUR\text{-}DC}(t)$ is the amount of surplus energy from DC sources, kWh.

Mathematical Model of Battery Bank. The battery state of charge (SOC) is the cumulative sum of the daily charge/discharge transfers. The battery serves as an energy source entity when discharging and a load when charging. At any hour t the state of the battery is related to the previous state of charge and to the energy production and consumption situation of the system during the time from $t - 1$ to t.

During the charging process, when the total output from renewable sources exceeds the load demand, the available battery bank capacity at hour t can be described by [9, 12–14]

$$E_{BAT}(t) = E_{BAT}(t-1) - E_{CC\text{-}OUT}(t) \times \eta_{CHG},\qquad(3)$$

where $E_{BAT}(t)$ is the energy stored in the battery at hour t, kWh, $E_{BAT}(t-1)$ is the energy stored in the battery at hour $t - 1$, kWh, and η_{CHG} is the battery charging efficiency.

On the other hand, when the load demand is greater than the available energy generated, the battery bank is in discharging state. Therefore, the available battery bank capacity at hour t can be expressed as [9, 12–14]

$$E_{BAT}(t) = E_{BAT}(t-1) - E_{Needed}(t),\qquad(4)$$

where $E_{Needed}(t)$ is the hourly load demand or energy needed at a particular period of time.

Let d be the difference between minimum allowable SOC voltage limit and the maximum SOC voltage across the battery terminals when it is fully charged which is equal to $1 - DOD/100$.

So, the depth of discharge (DOD) is as follows:

$$DOD = (1 - d) \times 100.\qquad(5)$$

DOD is a measure of how much energy has been withdrawn from a storage device, expressed as a percentage of full capacity. The maximum value of SOC is 1, and the minimum SOC measured in percentage is determined by maximum depth of discharge (DOD):

$$SOC_{Min} = 1 - \frac{DOD}{100}.\qquad(6)$$

Mathematical Model of Inverter. In the proposed scheme, the PV panel and battery systems are connected with DC bus while the electric loads are connected with AC bus as shown in Figure 3.

The inverter models for photovoltaic and battery bank are given below [15]:

$$E_{PV\text{-}INV,BAT\text{-}INV}(t)$$
$$= \left(E_{PV}(t) + \frac{E_{BAT}(t-1) - E_{LOAD}(t)}{\eta_{INV} \times \eta_{DCHG}} \right) \times \eta_{REC},\qquad(7)$$

where $E_{PV\text{-}INV,BAT\text{-}INV}(t)$ is the hourly energy output from inverter kWh, $E_{BAT}(t-1)$ is the energy stored in the battery

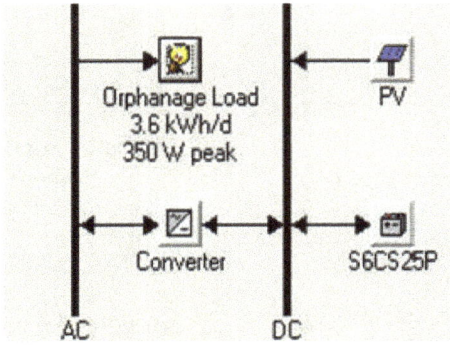

FIGURE 3: Schematic diagram of photovoltaic energy system.

TABLE 2: Simulation results of electricity production, consumption, losses, and excess (kWh/yr).

Component	Quantity of electricity (kWh/yr)
Production from PV array	1,916
Losses from the battery	122
Losses from the inverter	234
Other losses such as cables	52
Consumption from AC load	1,329
Excess electricity	179

at hour $t - 1$, kWh, $E_{\text{Load}}(t)$ is the hourly energy consumed by the load side, kWh, η_{INV} is the efficiency of inverter, and η_{DCHG} is the battery discharging efficiency.

3.1. Power Generation Model.
Total power generated at any time t is given by [9, 12–14]

$$P(t) = \sum_{\text{PV}=1}^{N_P} P_{\text{PV}}, \tag{8}$$

where N_P are number of PV cells. This generated power will feed the loads and when this generated power exceeds the load demand, then the surplus of energy will be stored in the battery bank. This energy (battery) will be used when a deficiency of power occurs to meet the load. The charged quantity of the battery bank has the constraint $\text{SOC}_{\text{min}} \leq \text{SOC}(t) \leq \text{SOC}_{\text{max}}$. The SOC_{min} is at 40%, while that of SOC_{max} is at 80%. The approach involves the minimization of a cost function subject to a set of equality and inequality constraints.

3.2. Cost Model (Economic and Environmental Costs) of Energy Systems.
The equation for estimating the level of optimization of photovoltaic energy solution being considered for the orphanage and a location is derived as economic and environmental cost (carbon credit of CO_2) model of running solar-photovoltaic + batteries and calculated as [15]

$$C_{\text{ann,tot},s+b} = \sum_{s=1}^{N_s} \left(C_{\text{acap},s} + C_{\text{arep},s} + C_{\text{aop},s} + C_{\text{emissions}} \right)$$

$$+ \sum_{b=1}^{N_b} \left(C_{\text{acap},b} + C_{\text{arep},b} + C_{\text{aop},b} + C_{\text{emissions}} \right), \tag{9}$$

where $C_{\text{acap},s}$ is annualized capital cost of solar power, $C_{\text{arep},s}$ is annualized replacement cost of solar power, $C_{\text{aop},s}$ is annualized operating cost of solar power, $C_{\text{emissions}}$ is cost of emissions, $C_{\text{acap},b}$ is annualized capital cost of batteries power, $C_{\text{arep},b}$ is annualized replacement cost of batteries power, and $C_{\text{aop},b}$ is annualized operating cost of batteries power.

The mathematical model derived in (9) estimates the life-cycle cost of the systems (solar-photovoltaic), which is the total cost of installing and operating the system over its lifetime. The output when run with HOMER software/tool will give the optimal configuration of the energy system that takes into account technical and economic performance of supply options.

Net Present Cost (NPC) for Energy Systems. The total net present cost (NPC) of a system is the present value of all the costs that it incurs over its lifetime, minus the present value of all the revenue that it earns over its lifetime. Revenues include salvage value and grid sales revenue. The net present cost (NPC) for each component is derived using [9, 12–14, 16, 17]

$$C_{\text{NPC}} = \frac{C_{\text{ann,tot}}}{\text{CRF}\left(i, R_{\text{proj}}\right)}, \tag{10}$$

where the capital recovery factor is [9, 12–14, 16, 17]

$$\text{CRF} = \frac{i \cdot (1 + i)^N}{(1 + i)^N - 1}. \tag{11}$$

The economic optimization identifies the most financially attractive solution. For this research paper, HOMER version 2.8 beta has been used as the sizing and optimization software tool. It contains a number of energy component models and evaluates suitable technology options based on cost and availability of resources [18].

3.3. Configuration and Optimization of Stand-Alone Photovoltaic Energy System.
Stand-alone photovoltaic system typically has an electricity generation device equipped with a wiring setup and supporting structure as well as the necessary BOS (balance of system) components (i.e., the battery bank, the charge controller, and the DC/AC inverter). The selection of components of energy system is done using Hybrid Optimization Model for Electric Renewables (HOMER) design software developed by the National Renewable Energy Laboratory, accurate enough to reliably predict system performance. HOMER is an optimization model, which performs many hundreds or thousands of approximate simulations in order to design the optimal system. The diagram of the completed stand-alone photovoltaic energy system can be seen in Figure 3.

TABLE 3: Simulation results of economic cost.

Component	Capital ($)	Replacement ($)	O and M ($)	*Salvage ($)	Total NPC ($)
PV	2,800	0	1,606	0	4,406
Surrette 6CS25P	54,960	23,855	28	−4,989	73,853
Converter	200	0	0	0	200
System	57,960	23,855	1,633	−4,989	78,459

*Salvage value is the value remaining in a component of the power system at the end of the project lifetime; that is, the salvage value of a component is directly proportional to its remaining life.

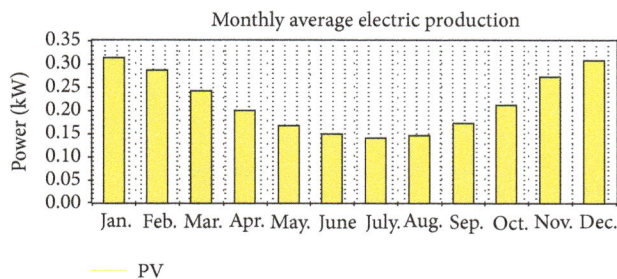

FIGURE 4: Electrical production of PV energy system.

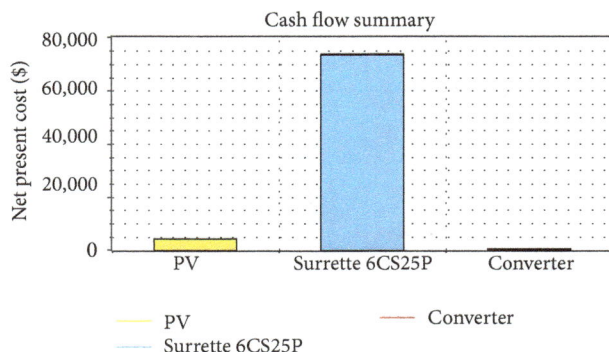

FIGURE 6: Net present cost of component of PV energy system.

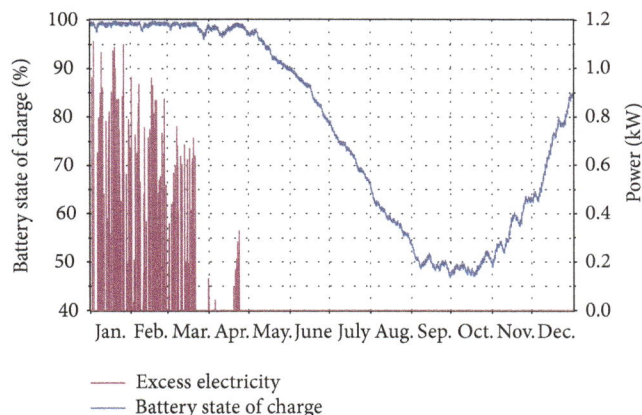

FIGURE 5: Battery state of charge versus excess electricity.

4. Results and Discussion

4.1. Results. The optimization result shows that sixteen solutions were simulated; one was feasible, which is PV-battery option with 1.4 kW PV, 48 Surrette 6CS25P battery, and 1 kW inverter; fifteen were infeasible due to the capacity shortage constraint. Twenty-four were omitted (twenty-two due to infeasibility, one for lacking a converter, and the remaining one for having an unnecessary converter). The obtained results provide information concerning the electricity production, consumption, losses, excess, and economic costs of the feasible system and are given in Tables 2 and 3 and shown in Figures 4, 5, and 6.

4.2. Discussion

Electricity Production. The PV array in this orphanage generates 1,916 kWh of electricity per year which effectively powers the load demand of 1,329 kWh per year with little excess electricity of 179 kW per year as shown in Table 2, and the electrical production of PV energy system is shown in Figure 4.

Losses from the System. A battery is used to store excess energy for later use. The conversion efficiency of batteries is not perfect and energy is usually lost as heat during chemical reaction, that is, during charging or recharging. Also, the amount of energy that will be delivered from the battery is managed by the inverter. The inverter connects directly to the battery bank and converts the direct current (DC) electrical energy from the battery bank to alternative current (AC) electrical energy, which is the energy that orphanages and most residential homes use. During the conversion, energy is also lost. Other losses, such as cables, were calculated and the amount of energy that is lost from the system was tabulated. From Table 2, it was shown that losses from the battery have a total of 122 kWh/yr, losses from the inverter have a total of 234 kWh/yr, and other losses have 52 kWh/yr, making a grand total of 408 kWh/yr energy losses from the system as shown in Table 2.

Excess Electricity. Excess electricity always occurs when the battery state of charge (SOC) is at 98% upwards and this is between Januaries and Aprils. As of May when the solar radiation is low, the battery is at 96% downward and discharges much and there will be no excess electricity from

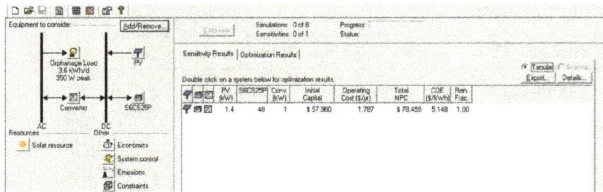

FIGURE 7: HOMER simulator diagram of photovoltaic energy system and the optimization results.

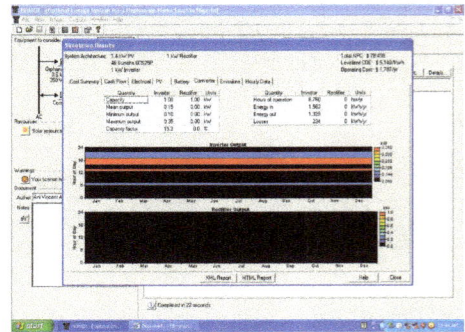

FIGURE 8: HOMER showing the simulation results of economic cost of component of PV energy system.

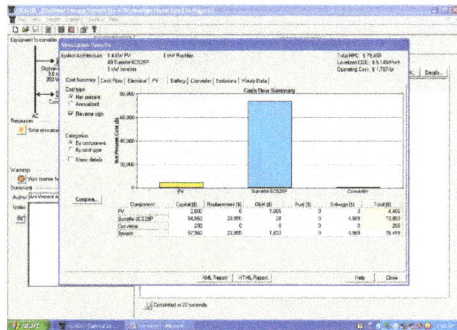

FIGURE 9: HOMER showing the electricity production of PV energy system.

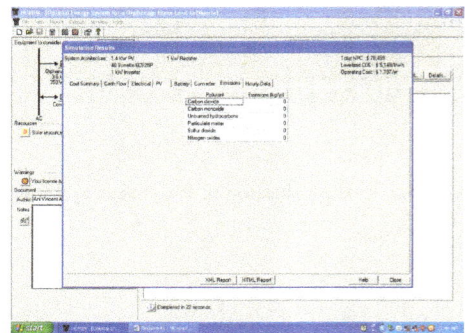

FIGURE 10: HOMER showing the battery state of charge and losses.

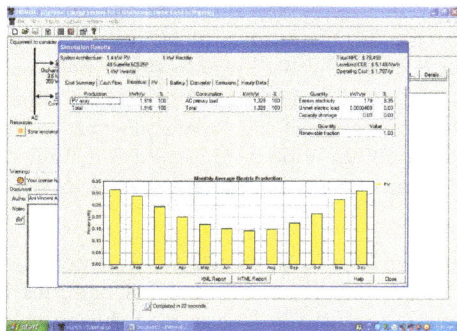

FIGURE 11: HOMER showing the inverter losses.

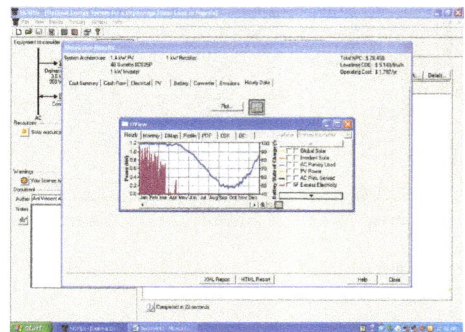

FIGURE 12: HOMER showing the result of the emissions.

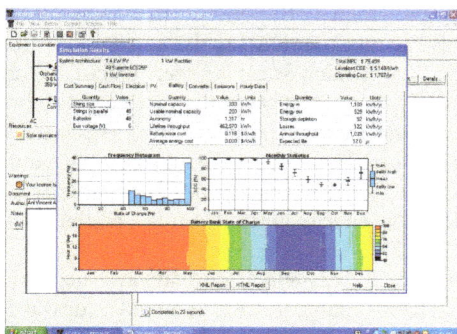

FIGURE 13: HOMER showing the battery state of charge versus excess electricity.

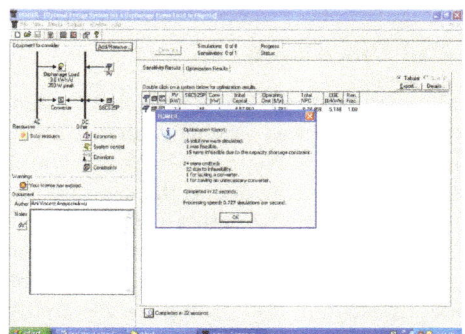

FIGURE 14: HOMER showing the optimization report.

this point downward. The battery state of charge versus excess electricity is shown in Figure 5.

Economic Costs. Batteries are considered as a major cost factor in small-scale stand-alone power systems [15]. The optimization of the system is carried out by modifying the size of the batteries until a configuration that ensures sufficient autonomy was achieved with the least net present cost (NPC). The salvage value was used to calculate the annualized replacement cost. Battery is the only component that has replacement cost (23,855$) and therefore has salvage value (−4,989$) because it did not last till project lifetime and the replacement extended the estimated project lifetime which was deducted from the system cost (73,853$) as shown in Table 3 and the net present cost of component of PV energy system is shown in Figure 6.

The software solutions showing the running program with the results are shown in Figures 7, 8, 9, 10, 11, 12, 13, and 14.

5. Conclusion

The optimal design of PV/battery energy system was carried out minimizing the net present cost (NPC) by varying the size of the batteries until a configuration that produces the desired power needs of the orphanage is achieved. This optimization study indicates that energy requirements to provide electricity for an orphanage in Nigeria can be accomplished by 1.4 kW PV, 48 Surrette 6CS25P battery, and 1 kW inverter. The PV system is in significant mode during the day time, particularly in the dry season, but, at night and other cloudy days, the battery compensates. Due to the abundance of solar resource in Nigeria and having no environmental impact in terms of CO_2, solar energy can be a choice for green power solutions in powering the orphanages located in remote areas.

Abbreviations

NASA: National Aeronautics and Space Administration
HOMER: Hybrid Optimization Model for Electric Renewables
SHS: Solar home system
PV: Photovoltaic
DC: Direct current
AC: Alternate current
SOC: State of charge
DOD: Depth of discharge
NPC: Net present cost
BOS: Balance of system
Min: Minimum
Max: Maximum.

Symbols

Wp: Watts peak
A: Surface area
kWh: Kilowatts hour

m^2:   Meter square
d: Day
t: Time.

Greek Symbols

η: Efficiency.

Conflict of Interests

The author declares that there is no conflict of interests regarding the publication of this paper.

Acknowledgment

The author would like to thank Professor Chinedu Ositadinma Nebo of Ministry of Power, Nigeria, for his useful discussion on the subject.

References

[1] H. Von, "Mini-grid system for rural electrification in the great Mekong sub-regional countries," in *Renewable Energies and Energy Efficiency*, vol. 6, University of Kassel, Kassel, Germany, 2007.

[2] F. Gerald, "Photovoltaic applications in rural areas of the developing world," Tech. Rep. no. 304, World Bank, Washington, DC, USA, 1995.

[3] A. Cabraal, M. Cosgrove Davies, and L. Schaeffer, "Best practices for photovoltaic household electrification programs: lessons from experiences in selected countries," Tech. Rep. no. 324, World Bank, Washington, DC, USA, 1996.

[4] A. Cabraal, M. Cosgrove Davies, and L. Schaeffer, "Accelerating sustainable photovoltaic market development," *Progress in Photovoltaics: Research and Applications*, vol. 6, no. 5, pp. 297–306, 1998.

[5] D. Kammen, "Promoting appropriate energy technologies in the developing world," *Environment*, vol. 41, no. 5, pp. 11–15, 34–41, 1999.

[6] K. Kapadia, "Off-grid in Asia: the solar electricity business," *Renewable Energy World*, vol. 2, no. 6, pp. 22–33, 1999.

[7] G. Loois and B. van Hemert, *Stand-Alone Photovoltaic Applications: Lessons Learned*, James & James, London, UK, 1999.

[8] NASA, 2013, https://eosweb.larc.nasa.gov/.

[9] V. A. Ani, "Optimal energy system for single household in Nigeria," *International Journal of Energy Optimization and Engineering*, vol. 2, no. 3, 26 pages, 2013.

[10] S. Ashok, "Optimised model for community-based hybrid energy system," *Renewable Energy*, vol. 32, no. 7, pp. 1155–1164, 2007.

[11] A. Gupta, R. P. Saini, and M. P. Sharma, "Steady-state modelling of hybrid energy system for off grid electrification of cluster of villages," *Renewable Energy*, vol. 35, no. 2, pp. 520–535, 2010.

[12] D. K. Lal, B. B. Dash, and A. K. Akella, "Optimization of PV/Wind/Micro-Hydro/diesel hybrid power system in homer for the study area," *International Journal on Electrical Engineering and Informatics*, vol. 3, no. 3, pp. 307–325, 2011.

[13] K. Sopian, A. Zaharim, Y. Ali, Z. M. Nopiah, J. A. Razak, and N. S. Muhammad, "Optimal operational strategy for hybrid

renewable energy system using genetic algorithms," *WSEAS Transactions on Mathematics*, vol. 4, no. 7, pp. 130–140, 2008.

[14] H. Abdolrahimi and H. K. Karegar, "Optimization and sensitivity analysis of a hybrid system for a reliable load supply in Kish Iran," *International Journal of Advanced Renewable Energy Research*, vol. 1, no. 4, pp. 33–41, 2012.

[15] V. A. Ani, *Energy optimization at telecommunication base station sites [Ph.D. dissertation]*, University of Nigeria, Nsukka, Nigeria, 2013.

[16] V. A. Ani and A. N. Nzeako, "Energy optimization at GSM base station sites located in rural areas," *International Journal of Energy Optimization and Engineering*, vol. 1, no. 3, 31 pages, 2012.

[17] T. Lambert, "HOMER: The HybridOptimization Model for Electrical Renewables," 2009, http://www.nrel.gov/international/tools/HOMER/homer.html.

[18] HOMER, 2013, http://www.nrel.gov/international/tools/HOMER/homer.html.

8

The Technical and Economic Study of Solar-Wind Hybrid Energy System in Coastal Area of Chittagong, Bangladesh

Shuvankar Podder, Raihan Sayeed Khan, and Shah Md Ashraful Alam Mohon

Department of Electrical and Electronic Engineering, Bangladesh University of Engineering and Technology (BUET), Dhaka 1000, Bangladesh

Correspondence should be addressed to Shuvankar Podder; podder.shuvankar@gmail.com

Academic Editor: Jing Shi

The size optimization and economic evaluation of the solar-wind hybrid renewable energy system (RES) to meet the electricity demand of 276 kWh/day with 40 kW peak load have been determined in this study. The load data has been collected from the motels situated in the coastal areas of Patenga, Chittagong. RES in standalone as well as grid connected mode have been considered. The optimal system configurations have been determined based on systems net present cost (NPC) and cost of per unit energy (COE). A standalone solar-wind-battery hybrid system is feasible and economically comparable to the present cost of diesel based power plant if 8% annual capacity shortage is allowed. Grid tied solar-wind hybrid system, where more than 70% electricity contribution is from RES, is economically comparable to present grid electricity price. Moreover, grid tied RES results in more than 60% reduction in greenhouse gases emission compared to the conventional grid. Sensitivity analysis has been performed in this study to determine the effect of capital cost variation or renewable resources variation on the system economy. Simulation result of sensitivity analysis has showed that 20% reduction of installation cost results in nearly 9%–12% reductions in cost of per unit energy.

1. Introduction

The Government of Bangladesh has issued its Vision and Policy Statement in February 2008, to bring the entire country under electricity service by the year 2020 [1]. Presently, more than 40% of the total population do not have access to electricity [2]. The installed electricity generation capacity of the country is about 10,445 MW in 2014 [3]. At present, electricity demand growth in Bangladesh is about 10% annually which is expected to increase in the coming years. Due to gas shortage and inadequate addition of new generation plants in the past few years, demand of electricity has outpaced generation capacity resulting in persistent load-shedding [4].

Present electricity generation scenario in Bangladesh is shown in Figure 1. As the figure shows, 65% of total generated electricity comes from gas.

Each year millions of tons of greenhouse gases (GHGs) are being emitted from fossil fuel based power plants. Although the per capita CO_2 emission of Bangladesh is very low compared to developed countries [5], a report showed that the CO_2 emission has increased from 0.1374 tons/capita in 1990 to 0.2667 tons/capita in 2007 [6]. Annual energy shortage and greenhouse gas emission, the two pressing problems, can be solved by installing renewable energy system (RES) such as solar, wind, biogas, and hydro.

The present share of renewable energy in Bangladesh is only 1% [7]. This is due to the high initial cost compared to fossil fuel based system. But renewable energy based options become economically viable when environmental cost, health hazard, and lower operating cost are taken into consideration [1].

The Government of Bangladesh has also declared its target of meeting five and ten percent of the total power demand using renewable energy by 2015 and 2020, respectively [1]. With this target at hand, government and some nongovernment organizations are conducting surveys. A joint survey by Local Government Engineering Department (LGED) and Chittagong University of Engineering and Technology (CUET) during early 2000 revealed that the average wind speed in the coastal areas of the country remains

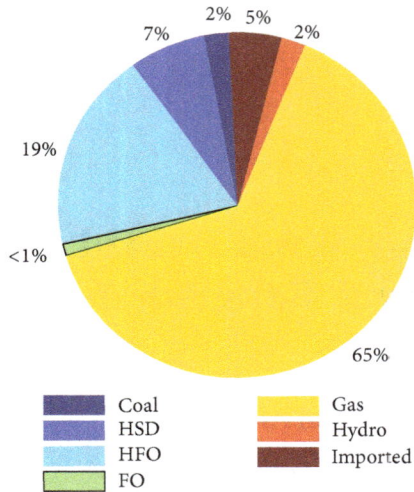

FIGURE 1: Present electricity generation scenario in Bangladesh.

TABLE 1: Average wind speed and average solar radiation at six coastal stations.

Location	Average wind speed (m/s)	Average solar radiation (kWh/m^2/day)
Kuakata	4.52	4.55
Patenga	3.8	4.35
Charfassion (Bhola)	4.07	4.52
Cox's Bazar	3.34	4.69
Teknaf	2.94	4.56
Noakhali	2.96	4.56

between 3 and 4.5 m/s in the months of March to September and 1.7–2.3 m/s for remaining period of the year [8]. This survey showed that the coastal areas and islands possess good potential for generating electricity from wind energy during March to September. The average sunshine data of these areas show that the period of bright (i.e., more than 200 W/m^2 intensity) sunshine hours in the coastal region of Bangladesh varies from 3 to 11 hours daily and the global radiation varies from 3.8 kWh/m^2/day to 6.4 kWh/m^2/day [9]. This amount of radiation is sufficient to produce electricity using solar cells. Average wind speed at 25 m height studied by LGED under Solar and Wind Energy Resource Assessment (SWERA) project [8] and average solar radiation measured by NASA Surface Meteorology and Solar Energy (SSE) [10] for six coastal stations are shown in Table 1.

Generating power from RES costs more than conventional fossil fuel based plants but the cost can be minimized by combining two or more renewable energy sources. A number of studies [11–14] showed that renewable energy based standalone hybrid systems are compatible with grid electricity for remote areas where grid extension is not feasible. The optimum size of RES to supply desired load is determined in studies [15, 16]. Alam Hossain Mondal and Sadrul Islam studied [17] the potential and viability of grid connected solar PV system in Bangladesh and found that cost of generating electricity from grid connected PV is comparable to grid connected fossil fuel based system. The optimization of hybrid energy systems for minimizing excess electricity and cost of energy is studied by Razak et al. [18].

A solar-wind hybrid power system uses solar insolation and wind energy to produce electricity. As both solar radiation and wind speed vary throughout the year, neither solar nor wind based system can provide reliable electricity individually. Wind speed remains fairly high during June to August when solar insolation is low due to cloud cover. On the other hand, wind speed remains quite low during December to February when solar radiation on earth surface

is fairly high to generate electricity. Thus hybridizing solar-wind system can be an alternative and reliable source of energy round the year. This hybrid system backed by storage elements/medium can supply electricity consistently and reliably. Wind-PV charge controllers regulate the charging of the energy before it is stored in the battery banks. An inverter converts the DC output of storage into AC of desired voltage and frequency.

In this study mainly two issues are examined regarding solar-wind grid connected as well as standalone hybrid power system.

(1) Is solar-wind hybrid power system feasible with available renewable energy resources at the target location?

(2) Which configuration (grid only, grid and RES hybrid, and RES only) provides the most economically viable solution, using net present cost (NPC) as the basis of comparison?

2. Tools

Hybrid Optimization of Multiple Energy Resources (HOMER) software [19] is chosen as the primary simulation tool for this study. This software is designed to simulate long-term operation of a combination of micropower system configurations that could include components like solar PV, wind, hydro, diesel generators and storage devices like battery banks. It can also model grid connected systems. The reason behind choosing this software is that it can perform the three major tasks. They are simulation, optimization, and sensitivity analysis. After simulating a number of combinations based on the data supplied by the user, HOMER suggests the optimal configuration based on the lowest net present cost (NPC). Sensitivity analysis enables the designer to determine the best combination of system components under different conditions.

3. Location of the Project

This study is done at Patenga (22.26°N, 91.8°E) which is a coastal area under Chittagong District. After assessing the potential of electricity generation from wind and solar resources in Chittagong coastal region, BPDB has planned to implement 50–200 MW wind power project at Parky beach area, Anwara, which is 32.2 km away from Patenga [6]. This study is done considering community like office and hospitals where continuous electricity supply is essential and grid electricity is not reliable. Also, standalone RES can provide electricity to remote areas like motels along coastal areas where grid electricity is not available.

4. Renewable Resources

4.1. Solar Resources. Solar radiation data are essential elements in PV output calculation. The accuracy of the radiation data helps to provide accurate results. The target place is Patenga, Chittagong. But direct solar radiation data for Patenga is not available from Bangladesh Meteorological Department (BMD) or any other institution. A number of studies estimated solar radiation from bright sunshine hours [23–26], while other studies estimated solar irradiance from the measurement of cloud cover in the sky [27, 28]. As the data of bright sunshine hours of Patenga is available at BMD, in this study, solar radiation is calculated from hours of bright sunshine during a day.

The popularly known Angstrom-Prescott regression equation [29] relating ratio of monthly average daily solar radiation to average daily extraterrestrial radiation and ratio of bright sunshine hours to total day length in hours is given by

$$\frac{\overline{H}}{\overline{H}_0} = a + b\frac{\overline{n}}{N}, \tag{1}$$

where $\overline{H}$ is monthly average daily global radiation (Wh/m^2/day), $\overline{H}_0$ is monthly average daily extraterrestrial radiation (Wh/m^2/day), n is actual sunshine duration in a day (hours), N is monthly average maximum possible bright sunshine duration in a day (hours), and a, b are empirical coefficients.

The values of monthly average daily global radiation $\overline{H}$ are obtained from daily measurement covering period 1983–2005 by NASA Surface Meteorology. The monthly average of daily hours of bright sunshine n is obtained from Bangladesh Meteorological Department (BMD) and monthly average of the maximum possible daily hours of sunshine or day length N is obtained from a weather based website [30].

The method of least squares was used to obtain the constants a and b. The estimated global solar radiation according to (1) and monthly solar radiation measured by NASA over Chittagong are shown in Figure 2. From the figure it is evident that measured radiation reported by NASA from the analysis of satellite captured image and estimated radiation are close enough. In this study estimated solar radiation is used in optimizing solar panel size.

4.2. Wind Resources. Just like solar radiation data, wind speed data are also essential for hybrid system analysis. Wind speed data are obtained from BMD [31]. HOMER requires four parameters to generate hourly wind speed from provided monthly wind speed data [32]. They are listed below.

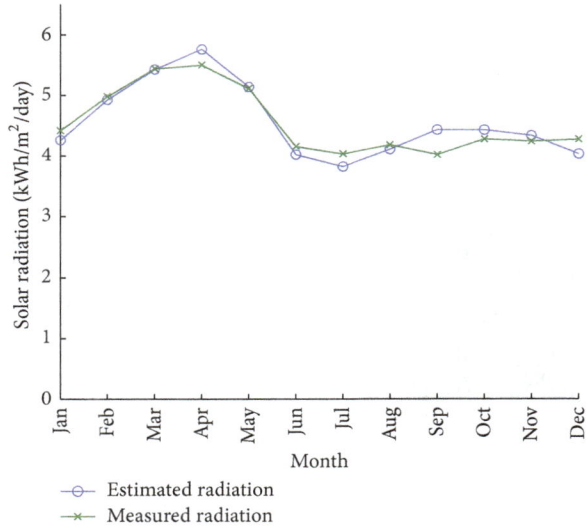

FIGURE 2: Estimated and measured solar radiation at Patenga.

(i) *Weibull Value.* The Weibull value k is a measure of distribution of wind speed over the year. In this study, the value of k has taken 2.

(ii) *Autocorrelation Factor.* The autocorrelation factor measures the randomness of the wind. Higher value indicates that the wind speed during an hour tends to depend strongly on the wind speed during the previous hour. Lower value means that the wind speed tends to fluctuate in a more random fashion from hour to hour. The autocorrelation factor value has taken 0.78.

(iii) *Diurnal Pattern Strength.* The diurnal pattern strength indicates how strongly wind speed tends to depend on the time of a day. To measure this, average wind speed at each of the 24 hours over the year is calculated. HOMER then fits a cosine function to this diurnal profile. The diurnal pattern strength is equal to the ratio of the amplitude of the cosine wave to the average wind speed. The range of diurnal pattern strength is normally 0 to 0.4. In this study, considering wind speed moderately depends on the time of a day, the value of this parameter is chosen as 0.3.

(iv) *Hour of Peak Wind Speed.* The time of day tends to be windiest on average throughout the year. In this study, 3 pm is used as the hour of peak wind speed. Figure 3 shows the monthly average wind speed, measured at the height of 10 m above ground level.

5. Electrical Load

Seasonal load profile of proposed area is shown in Figure 4. Energy consumed by the hypothetical community is 276 kWh/day with a peak demand of 40 kW. Load profile is prepared by conducting survey on electricity consumption of tourist motels at Patenga. It is considered that day to

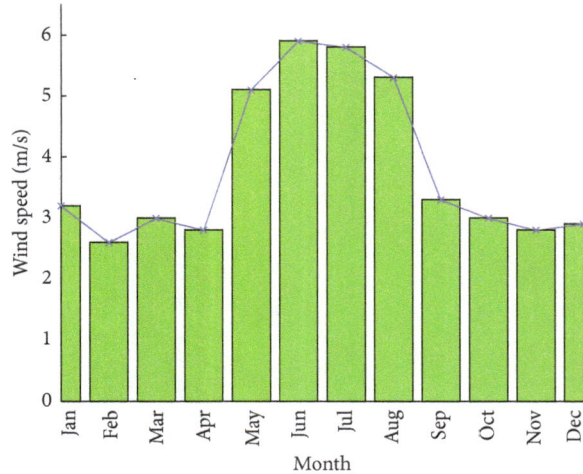

FIGURE 3: Monthly average wind speed.

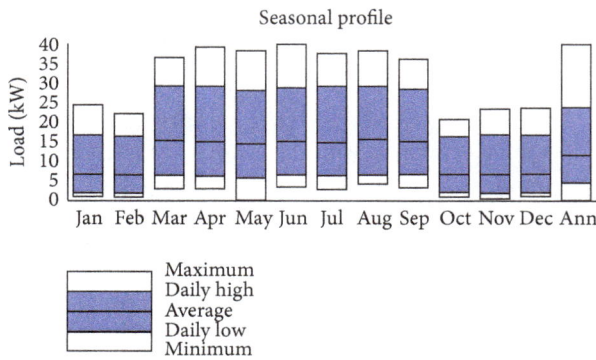

FIGURE 4: Seasonal load profile of proposed area.

TABLE 2: Electricity consumption charge for different sectors by BPDB.

Consumer	Time (local time)	Rate type	Rate (BDT/kWh)
Nonresidential Commercial Office	11.00 am–5.00 pm	Flat	9
	5.00 pm–12.00 am	Peak	11.85
	12.00 am–11.00 am	Off-peak	7.22

TABLE 3: Technical parameters of PV array.

Output current	DC
Lifetime	20 years
Azimuth angle	0 degrees
Ground reflection	20%
Slope	22.05 degrees
Derating factor	85.50%

BPDB during different demand hours are shown in Table 2 [33].

6.2. Photovoltaic Array. The optimal PV array size in kW for concerned load is determined in this study. PV array size is dependent on technical and economical parameters of PV panel. It is assumed that PV panel output is linearly proportional to incident radiation. Technical parameters of PV array are shown in Table 3. Economical parameters are shown in Table 6. Charge controller is required to control charging and discharging of battery by PV array current and as HOMER does not provide this option the price of charge controller is merged with PV panel price.

6.3. Wind Turbine. This study uses 10 kW wind turbine manufactured by Yangzhou Shenzhou Wind-Driven Generator Co., Ltd., China [34]. Technical parameters of the wind turbine are shown in Table 4. Economical parameters are shown in Table 6. Power curve of this turbine is shown in Figure 5. Lifetime of considered wind turbine has taken 9 years while the lifetime of wind turbines made by Bergey Excel, Vestas, spans from 15 to 20 years. Though replacement cost for this low lifetime turbine is high, it is found from HOMER simulation that, due to low capital cost, considered turbine provides low net present cost (NPC) compared to high quality wind turbines over project lifetime. Surface area requirements for wind turbines are assessed for this study using the National Renewable Energy Laboratory (NREL) wind turbine area measurement [35]. According to NREL, average land capacity density (capacity per unit area) is 3.0 ± 1.7 MW/km^2. Therefore area required for 1 kW wind turbine is roughly 0.00025 km^2 or 250 m^2.

6.4. Battery and Inverter. As renewable resources like solar radiation and wind speed are seasonal, energy storage medium is required to get continuous supply of energy from RES. In case of grid connected RES electricity can be supplied reliably by using backup of

day random variability is 15.2% and time step to time step variability is 20.4%. Demand is high during summer (March–September) and low during winter (October–February). Distribution system would be required for electric service regardless of whether the service is provided by grid or standalone RES. At present, utility is providing 11 kv line for new connection. So a 11 kv/220 v step-down transformer is required for grid connected system. Also an Automatic Transfer Switch (ATS) is required so that, in case of inadequate supply from RES, load is being connected automatically to standby grid electricity.

6. Hybrid System Component

6.1. Grid. In this study, grid connected renewable energy system is compared with grid only and RES only system. In the grid connected RES, grid is used as standby supply. When energy supply is inadequate from RES due to bad weather, electricity from grid is consumed. Distribution companies impose different charges for consuming electricity from grid during peak-hours, off-peak hours, and flat rate hours. For commercial and office buildings, charges determined by

TABLE 4: Technical parameters of the wind turbine.

Rated power	10 kW
Hub height	25
Lifetime	9 years
Rotor diameter	8 m
Number of blades	3
Efficiency	0.85%

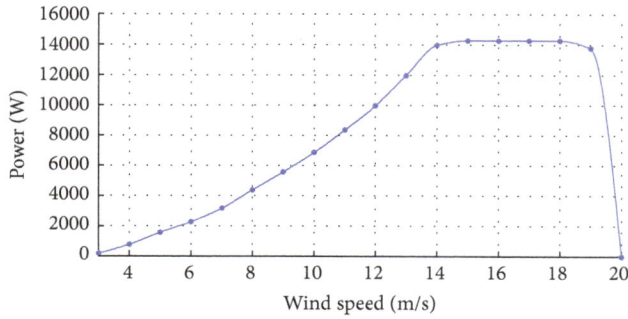

FIGURE 5: Power curve of the wind turbine.

TABLE 5: Technical parameters of electrosolar lead acid battery.

Nominal capacity	130 Ah
Nominal voltage	12 V
Round trip efficiency	80%
Minimum state of charge	40%
Float life	6 yrs
Maximum charge rate	0.4 A/Ah
Maximum charge current	52 A
Lifetime throughput	2230 kWh
Suggested value	2230 kWh

(1) grid supply,

(2) deep cycle lead acid batteries.

Electrosolar deep cycle, lead acid battery manufactured by Electro Solar Power Ltd., Bangladesh [36], is used in this study. Two batteries are connected in series to give 24 V DC bus. Technical parameters of this battery are given in Table 5. Economic parameters of this battery are shown in Table 6.

Converter sizing is roughly in proportion to the size of the load it serves. Sizing for the case-study converters therefore ranged from 20 to 45 kW. Economic parameters of converter are shown in Table 6.

7. System Economics, Constraints

The project lifetime is considered to be 25 years and the annual real interest rate 7.25% according to Bangladesh Bank [37]. The annual capacity shortage penalty and system fixed operation and maintenance costs are not considered. There are two types of dispatch strategies in HOMER, load following and cycle charging strategy. In the load following strategy, generators or grid supplies the power just to meet

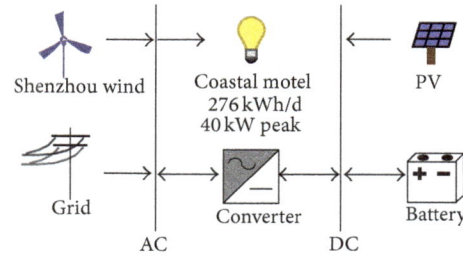

FIGURE 6: Grid-RES hybrid system.

the demand that is not fulfilled by RES. On the other hand, in the cycle charging strategy, generators run at full capacity and any excess electricity is stored in batteries [38]. In the present work, load following strategy is considered. System fixed capital cost is assumed to be $1000 considering land cost, LT (low tension) grid side connection cost, and so forth. It is required to consider operating reserve in designing power system because electric load can fluctuate to a level above rated demand suddenly. A system that includes wind and solar power sources requires additional operating reserve to guard against such random increase in demand or shortage in renewable supply. Operating reserve is considered 10% of hourly load as recommended by Cotrell and Pratt [39].

8. Results

In this study, several system configurations are considered. Off-grid-RES and on-grid-RES are compared on the basis of economy with grid only system. The basic system configuration with all system elements considered in this study is shown in Figure 6. For off-grid system, two scenarios regarding annual electricity shortage are considered. In the first scenario, no annual shortage is considered and, in the second scenario, 8% annual electricity shortage is considered. In the subsequent sections, optimization result and sensitivity result are shown for different configurations.

8.1. Grid Only System. In this case, the NPC is $129,209 over the project lifetime of 25 years. The cost of per unit energy (COE) is $0.113 at the 2014 electricity purchase rate. NPC increases in direct proportion to increase in electricity price. Greenhouse gas (GHG) emissions are 63,668 kg per year. The total electricity purchase from distribution companies for 1 year is 100,740 kWh. Figure 7 shows the monthly purchase of electricity from grid.

8.2. RES Only System. HOMER simulation shows that a RES only configuration is technically feasible with available renewable resources. Search space (sizes and quantities to consider that is given as input) is shown in Table 7. This search space is created after many trials and errors, eliminating inefficient entries to reduce simulation time.

HOMER predicts that a configuration of 54 kW PV, 17 wind turbines (10 kW each), 40 kW converter, and 290 nos twelve (12) V batteries is the only economically feasible

TABLE 6: Economic parameters of system components.

Characteristics	PV module	Wind turbine	Battery	Converter
Model	Typical	EW10000	Electrosolar	Typical
Power	250 W	10 kW	Nominal voltage 12 V Nominal capacity 130 Ah	2 kW
Lifetime	25 years	9 years	Lifetime throughput 2230 kWh	15 years
Capital	$355 [20]	$10000 [21]	$260	$730 [22]
Replacement	$300	$9,000	$260	$730
Maintenance	$7	$13	$12.50	$7

TABLE 7: Search space.

PV array (kW)	Wind turbine (EW10k) (quantity)	Battery (ElecSol12) (strings)	Converter (kW)
0	0	0	22
5000	9	40	25
15000	10	45	28
20000	11	50	30
25000	12	55	35
30000	13	60	40
35000	14	65	
40000	16	70	
50000	17	80	
60000	18	90	

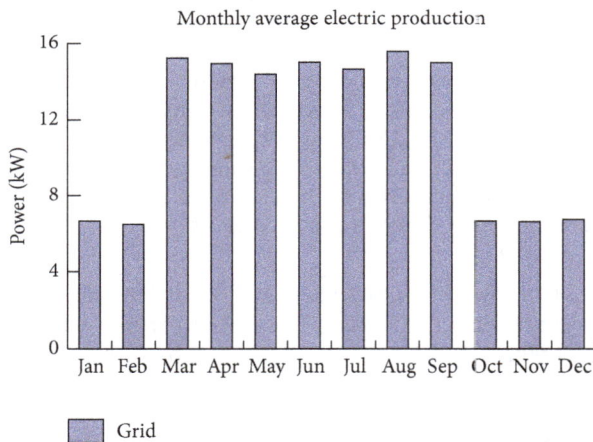

FIGURE 8: Monthly average electricity production in PV-WT configuration at 8% shortage.

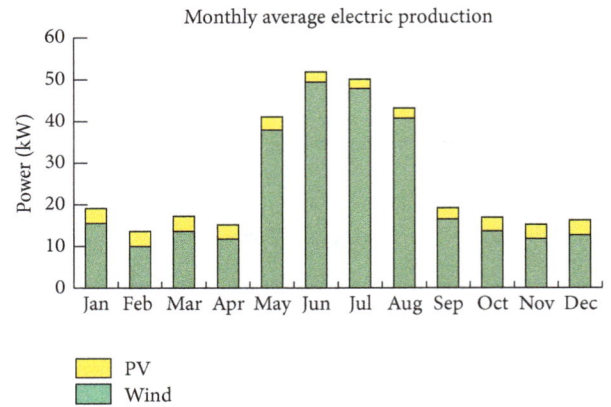

FIGURE 7: Monthly purchase of grid electricity.

Considering 125 Wp solar panel size 64.25 × 38.82 square inches [32] or 1.60915 square meters, 1 kW PV array covers 12.8732-square-meter area. So footprint required for PV-wind-battery hybrid system would be roughly 28050 square meters including spacing for battery banks and converters. Footprints required for wind-battery-converter hybrid system would be roughly 35250 square meters. The monthly average electricity production from PV-wind-battery hybrid system at 8% annual electricity shortage is shown in Figure 8.

8.3. Grid-RES Hybrid. From HOMER simulations it is found that grid tied RES is technically feasible with present renewable resource. It is also economically lucrative. Optimization result of simulation for this system configuration is shown in Table 9. From Table 9 it is found that 6 grid tied wind turbines give an NPC of $147,899 which is 13.85% higher than that of the grid only configuration (Table 9, option 2). Renewable fraction (RF) for the configuration is 75%. COE is $0.129. Battery and inverter are not required in this configuration, as the grid network acts as the backup reserve. Greenhouse gas (GHG) emissions are 23,523 kg per year. Excess electricity is 23.4%. The addition of 5 kW of PV with 6 wind turbines increases the RF to 77% (option 4 in comparison to option 1). GHG emission in this option is 22,446 kg/year which is 64.75% less than grid only system. A comparison of GHGs emission for grid only, grid-wind, and grid-PV-wind system

solution at zero annual electricity shortage. NPC of this configuration is $642,262 and COE is $0.56. On the other hand, if 8% annual electricity shortage is considered, a configuration of 11 wind turbines (10 kW each), 20 kW PV, a 25 kW converter, and 120 nos twelve (12) V batteries provides the lowest NPC of $317,591 (Table 8, option 1). Another configuration of 14 WT with 25 kW converter and 130 nos batteries gives NPC of $338,676 (Table 8, option 2). If 125 Wp solar panel is considered, then 8 panels are needed to yield 1 kW PV array.

TABLE 8: RES only configuration optimization result.

🖤	⊀	🖳	🔋	PV (kW)	EW10k	ElecSol12	Conv. (kW)	Initial capital	Operating cost ($/yr)	Total NPC	COE ($/kWh)	Ren. frac.	Capacity shortage
🖤	⊀	🖳	🔋	20	11	120	25	$179,725	12,098	$317,591	0.292	1.00	0.08
	⊀	🖳	🔋		14	130	25	$183,925	13,580	$338,676	0.311	1.00	0.08

TABLE 9: Grid-RES configuration optimization result.

🏭	🖤	⊀	🖳	🔋	PV (kW)	EW10k	ElecSol12	Conv. (kW)	Grid (kW)	Initial capital	Operating cost ($/yr)	Total NPC	COE ($/kWh)	Ren. frac.
🏭									10000	$1000	11,312	$129,909	0.113	0.00
🏭	🖤		🔋		5			22	10000	$16,255	11,241	$144,351	0.126	0.07
🏭		⊀				6			10000	$61,125	7,615	$147,899	0.129	0.75
🏭	🖤	⊀	🔋		5	6		22	10000	$76,255	7,882	$166,079	0.145	0.77

TABLE 10: Comparison of greenhouse gas emissions of 3 systems.

Pollutant	Emissions (kg/yr)		
	Grid only system	Grid-wind system	Grid-PV-wind system
Carbon dioxide	63668	23523	22446
Carbon monoxide	0	0	0
Unburned hydrocarbons	0	0	0
Particulate matter	0	0	0
Sulfur dioxide	276	102	97.3
Nitrogen oxides	135	49.9	47.6

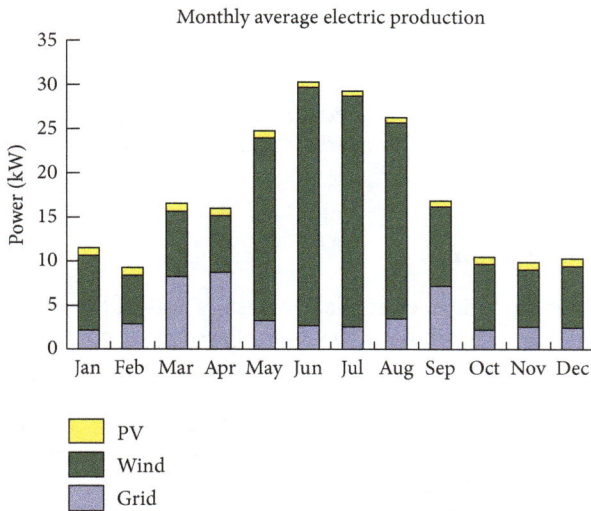

FIGURE 9: Monthly average electricity production in grid-PV-WT configuration.

is shown in Table 10. The monthly average electricity production from grid-PV-wind hybrid system (option 4, Table 9) is shown in Figure 9.

A comparison between system configurations on the basis of economy is provided in Table 11.

9. Sensitivity Analysis

In sensitivity analysis, system's control parameters can be varied to examine the effect of control parameters on system's performance. It helps the system designer to choose most economical system configuration for a given range of control parameters. In this study, two types of sensitivity variables are chosen to conduct sensitivity analysis.

Firstly, solar and wind resources are varied from annual average value to determine the effect of resource variation on system economy. Secondly, the capital costs of PV panel and wind turbine are varied.

It is known that the nature of renewable resource is intermittent. Average solar radiation on Patenga varies from year to year, though not significantly. The same is for wind speed, which fluctuates over the year. Random variability factor is also provided in resource input. Resources are varied to make suggestion about system economy if project location is changed.

On the other hand, installation cost of PV array and wind turbine is decreasing over the decades. National Renewable Energy Laboratory (NREL) researchers showed that installation cost of renewable power system is undergoing a downward trend [40, 41]. For these reasons, PV and wind turbine capital costs are varied to determine the effect of capital cost variation on system economy.

9.1. Renewable Resource Variation. In Figure 10, solar and wind resources are varied from annual average value to determine the effect of resource variation on system economy. In the x-axis we have plotted solar radiation and varied it from $3.8 \, kWh/m^2/day$ to $6 \, kWh/m^2/day$. The reason behind choosing this range is that, according to Figure 2, the annual solar radiation varies within this range annually. The same goes for wind speed. According to Figure 3, the wind speed at target location varies from 2.5 m/s to 6 m/s annually. So this range for wind speed variation is used.

Now the sensitivity analysis in the HOMER software finds out the most cost effective system for a given set of resources. In Figure 10, when solar radiation is up to $4 \, kWh/m^2/day$, the wind-PV/battery system is more cost effective than

TABLE 11: Economic comparison between different configurations.

System	PV (kW)	WT (10 kW) (quantity)	Battery (number)	Converter (kW)	NPC (BDT)	COE (BDT)	Renewable fraction
Grid only					129,909	0.113	0
PV-wind hybrid	20	11	120	25	317,591	0.292	1 (8% shortage)
Grid-wind hybrid		6			147,899	0.128	0.75
Grid-PV-wind hybrid	5	6		22	166,079	0.145	0.77

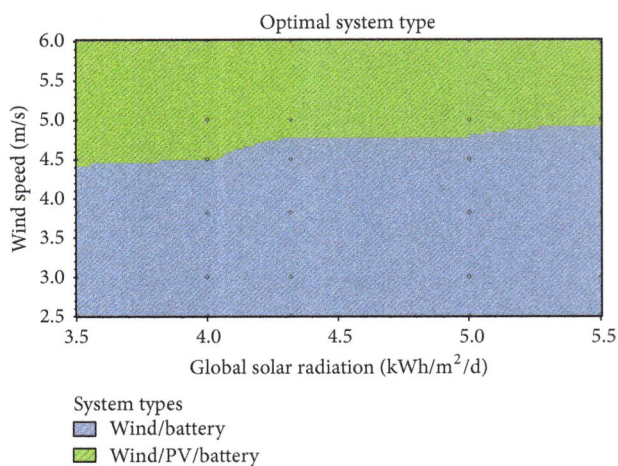

FIGURE 10: Sensitivity result.

TABLE 12: COE of different system configurations with 3 different multipliers.

System	Capital multiplier	COE ($)
Grid/wind	1	0.129
	0.8	0.112
	0.6	0.096
Grid/PV/wind/inverter	1	0.145
	0.8	0.127
	0.6	0.109
Grid/PV/battery/inverter (5 kW PV, 22 kW inverter)	1	0.173
	0.8	0.171
	0.6	0.17
Grid/PV/wind/battery/inverter	1	0.192
	0.8	0.174
	0.6	0.156
PV/wind/battery/inverter	1	0.292
	0.8	0.254
	0.6	0.215

wind/battery system for wind speed up to 4.4 m/s. If the wind speed is higher than that, the wind/battery system becomes more cost effective. Similarly, at solar radiation around 4.3 kWh/m^2/day, the wind-PV/battery system is more cost effective than wind/battery system if the wind speed is less than or equal to 4.8 m/s. All the analyses are done with the consideration that maximum annual capacity shortage is 8%.

9.2. Capital Cost Variation. HOMER provides option of cost variation of system components. In this study, capital and replacement costs of PV and wind turbine are multiplied by scaling factors. When cost multipliers are provided, these factors are multiplied by given costs in the cost table. HOMER then simulates system with these varying prices and calculates NPC.

The study [41] showed that per watt installation price of PV panel decreased by around 15% from 2010 to 2011 and around 30% from 2010 to 2012. On the other hand, another study [40] showed that cost of energy production in MWh from wind turbine reduces by around 62% from 1990 to 2000 and around 20% from 2000 to 2005. To include this trend in system size optimization, capital cost of PV panel and wind turbine is multiplied by 0.8 and 0.6, for all system configurations shown in Table 12. COE are reported when both PV panels and wind turbines capital cost are multiplied by the same factor.

From the table it is evident that 20% reduction of installation cost results in nearly 9%–12% reductions in cost of per unit energy.

10. Conclusion

From this study it is clear that, in case of off-grid system, the optimized PV-wind-battery hybrid system is more cost effective compared to *wind-alone system, PV alone system,* and wind-PV hybrid system for the load with 8% annual capacity of shortage for this hypothetical system at the proposed site. From the sensitivity analysis, it is also clear that the major portion of the energy comes from wind. The sensitivity analysis also predicts that the reduction of installation cost of PV or wind energy system results in per unit electricity cost that is comparable to the grid electricity price. Furthermore, RES system can reduce GHG emission by a significant amount, thus being friendly to the environment. The analysis can be further improved and system economy can be determined more precisely if more related data like minute-wise load curve, land price, variable interest rate, and environmental hazard effects are taken into consideration. Nevertheless, the initial analysis suggests that grid-RES hybrid system is promising enough to justify further effort to collect additional data to perform a deeper analysis.

Conflict of Interests

The authors declare that there is no conflict of interests regarding the publication of this paper.

Acknowledgment

The authors are grateful to Dr. Enamul Basher, Professor at the Department of Electrical and Electronic Engineering, BUET, for directing and thoroughly supervising the research work.

References

[1] Power Division and Ministry of Power, *Energy and Mineral Resources*, Renewable Energy Policy of Bangladesh, Government of the People's Republic of Bangladesh, 2008.

[2] Power Division, Ministry of Power, Energy and Mineral Resources, and Government of the People's Republic of Bangladesh, "Investment potentials," http://www.pd.gov.bd/user/brec/104/110.

[3] Bangladesh Power Development Board (BPDB), "Power generation units (fuel type wise)," 2014, http://www.bpdb.gov.bd/bpdb/index.php?option=com_content&view=article&id=150&Itemid=16.

[4] Bangladesh Power Development Board, *Annual Report*, Bangladesh Power Development Board, 2012-2013.

[5] M. Ahiduzzaman and A. K. M. S. Islam, "Greenhouse gas emission and renewable energy sources for sustainable development in Bangladesh," *Renewable and Sustainable Energy Reviews*, vol. 15, no. 9, pp. 4659–4666, 2011.

[6] Wikipedia, "List of countries by carbon dioxide emissions per capita," 2014, http://en.wikipedia.org/wiki/Listofcountriesbycarbondioxideemissionspercapita.

[7] Bangladesh Power Development Board (BPDB), *Development of Renewable Energy Technologies by BPDB*, Bangladesh Power Development Board (BPDB), 2014, http://www.bpdb.gov.bd/bpdb/index.php.

[8] S. M. Formanul Islam, S. Aziz, and S. A. Chowdhury, "Renewable energy initiatives by infrastructure development company limited in Bangladesh," in *Proceedings of the IEEE International Conference on the Developments in Renewable Energy Technology*, pp. 202–206, December 2009.

[9] "Wind energy resource mapping (WERM) in Bangladesh," Project Report, Local Government Engineering Department (LGED), Government of the People's Republic of Bangladesh, Dhaka, Bangladesh.

[10] Surface meteorology and Solar Energy, a renewable energy resource website, sponsored by NASAs Earth Science Enterprise Program, 2014, https://eosweb.larc.nasa.gov/sse/.

[11] A. H. Mondal and M. Denich, "Hybrid systems for decentralized power generation in Bangladesh," *Energy for Sustainable Development*, vol. 14, no. 1, pp. 48–55, 2010.

[12] M. A. H. Mondal, L. M. Kamp, and N. I. Pachova, "Drivers, barriers, and strategies for implementation of renewable energy technologies in rural areas in Bangladesh—an innovation system analysis," *Energy Policy*, vol. 38, no. 8, pp. 4626–4634, 2010.

[13] J. L. Bernal-Agustín and R. Dufo-López, "Simulation and optimization of stand-alone hybrid renewable energy systems," *Renewable and Sustainable Energy Reviews*, vol. 13, no. 8, pp. 2111–2118, 2009.

[14] S. K. Nandi and H. R. Ghosh, "A wind-PV-battery hybrid power system at Sitakunda in Bangladesh," *Energy Policy*, vol. 37, no. 9, pp. 3659–3664, 2009.

[15] L. Xu, X. Ruan, C. Mao, B. Zhang, and Y. Luo, "An improved optimal sizing method for wind-solar-battery hybrid power system," *IEEE Transactions on Sustainable Energy*, vol. 4, no. 3, pp. 774–785, 2013.

[16] D. B. Nelson, M. H. Nehrir, and C. Wang, "Unit sizing and cost analysis of stand-alone hybrid wind/PV/fuel cell power generation systems," *Renewable Energy*, vol. 31, no. 10, pp. 1641–1656, 2006.

[17] M. Alam Hossain Mondal and A. K. M. Sadrul Islam, "Potential and viability of grid-connected solar pv system in Bangladesh," *Renewable Energy*, vol. 36, no. 6, pp. 1869–1874, 2011.

[18] J. A. Razak, K. Sopian, and Y. Ali, "Optimization of renewable energy hybrid system by minimizing excess capacity," *International Journal of Energy*, vol. 1, no. 3, 2007.

[19] T. Lambert and L. P. Homer, "The micro-power optimization model," software produced by NREL.

[20] LG solar panel model LG-295N1C, 2014, http://www.wholesalesolar.com/products.folder/module-folder/LG/LG250S1C-G2.html.

[21] 10KW on-grid and off-grid wind turbine generator system, 2014, http://shenzhougenerator.en.alibaba.com/product/555911860-221402430/10KWongridandoffgridwindturbinegeneratorsystem.html.

[22] 2kw grid tie power inverter, 2014, http://www.alibaba.com/product-detail/2kw-grid-tie-power-inverter-off_664680451.html.

[23] S. Rangarajan, M. S. Swaminathan, and A. Mani, "Computation of solar radiation from observations of cloud cover," *Solar Energy*, vol. 32, no. 4, pp. 553–556, 1984.

[24] A. Nyberg, "Determination of global radiation with the aid of observations of cloudiness," *Acta Agriculturae Scandinavica*, vol. 27, no. 4, pp. 297–300, 2009.

[25] M. R. Rietveld, "A new method for estimating the regression coefficients in the formula relating solar radiation to sunshine," *Agricultural Meteorology*, vol. 19, no. 2-3, pp. 243–252, 1978.

[26] K. K. Gopinathan, "A general formula for computing the coefficients of the correlation connecting global solar radiation to sunshine duration," *Solar Energy*, vol. 41, no. 6, pp. 499–502, 1988.

[27] R. B. Benson, M. V. Paris, J. E. Sherry, and C. G. Justus, "Estimation of daily and monthly direct, diffuse and global solar radiation from sunshine duration measurements," *Solar Energy*, vol. 32, no. 4, pp. 523–535, 1984.

[28] H. Ögelman, A. Ecevit, and E. Tasdemiroğlu, "A new method for estimating solar radiation from bright sunshine data," *Solar Energy*, vol. 33, no. 6, pp. 619–625, 1984.

[29] J. A. Prescott, "Evaporation from water surface in relation to solar radiation," *Transactions of The Royal Society of South Australia*, vol. 40, pp. 114–118, 1940.

[30] Weatherspark, a weather based website https//weatherspark.com/.

[31] M. G. Uddin, Bangladesh Meteorological Department (BMD), 2014.

[32] A. Demiroren and U. Yilmaz, "Analysis of change in electric energy cost with using renewable energy sources in Gökceada,

Turkey: an island example," *Renewable and Sustainable Energy Reviews*, vol. 14, no. 1, pp. 323–333, 2010.

[33] Bangladesh Power Development Board, "Retail tariff rate," 2014.

[34] L. Shenzhou Wind-driven Generator Co., http://f-n.cn.

[35] P. Denholm, M. Hand, M. Jackson, and S. Ong, "Land use measurement of modern wind power," Tech. Rep., National Renewable Energy Laboratory, 2009, http://www.nrel.gov/docs/fy09osti/45834.pdf.

[36] Electro Solar Power, http://www.electrosolarbd.com/.

[37] Bangladesh Bank, "Interest rates," December 2014, https://www.bb.org.bd/econdata/intrate.php.

[38] S. Lal and A. Raturi, "Techno-economic analysis of a hybrid mini-grid system for Fiji islands," *International Journal of Energy and Environmental Engineering*, vol. 3, article 10, 2012.

[39] J. Cotrell and W. Pratt, "Modeling the feasibility of using fuel cells and hydrogen internal combustion engines in remote renewable energy systems," Tech. Rep., National Renewable Energy Laboratory, 2003, http://www.nrel.gov/docs/fy03osti/34648.pdf.

[40] E. Lantz, R. Wiser, and M. Hand, "The past and future cost of wind energy," IEA Wind Task, 2012.

[41] D. Feldman, R. Barbose, R. Margolis, R. Wiser, N. Darghouth, and A. Goodrich, "Photovoltaic system pricing trends: historical, recent, and near-term projections," Tech. Rep., National Renewable Energy Laboratory, 2013.

9

Dynamic Stability Improvement of Grid Connected DFIG Using Enhanced Field Oriented Control Technique for High Voltage Ride Through

V. N. Ananth Duggirala[1] and V. Nagesh Kumar Gundavarapu[2]

[1]Department of EEE, Viswanadha Institute of Technology and Management, Visakhapatnam 531173, India
[2]Department of EEE, GITAM University, Visakhapatnam, Andhra Pradesh 530045, India

Correspondence should be addressed to V. Nagesh Kumar Gundavarapu; drgvnk14@gmail.com

Academic Editor: Shuhui Li

Doubly fed induction generator (DFIG) is a better alternative to increased power demand. Modern grid regulations force DFIG to operate without losing synchronism during overvoltages called high voltage ride through (HVRT) during grid faults. Enhanced field oriented control technique (EFOC) was proposed in Rotor Side Control of DFIG converter to improve power flow transfer and to improve dynamic and transient stability. Further electromagnetic oscillations are damped, improved voltage mitigation and limit surge currents for sustained operation of DFIG during voltage swells. The proposed strategy has advantages such as improved reactive power control, better damping of electromagnetic torque oscillations, and improved continuity of voltage and current from stator and rotor to grid during disturbance. In EFOC technique, rotor flux reference changes its value from synchronous speed to zero during fault for injecting current at the rotor slip frequency. In this process, DC-Offset component of stator flux is controlled so that decomposition during overvoltage faults can be minimized. The offset decomposition of flux will be oscillatory in a conventional FOC, whereas in EFOC it is aimed to be quick damping. The system performance with overvoltage of 1.3 times, 1.62 times, and 2 times the rated voltage occurring is analyzed by using simulation studies.

1. Introduction

The doubly fed induction generator (DFIG) is preferred due to its small size with higher MVA ratings available in the market, low power ratings of converters, variable generator speed and constant frequency operation, robust four-quadrant reactive power control, and improved performance during the high and low voltage ride through (HVRT and LVRT). However, DFIG is sensitive to external disturbances like voltage swell and sag. If grid voltage falls or rises suddenly due to any reason, large surge currents enter the rotor terminals and induce the voltage significantly. So, the rotor side converter (RSC) gets damaged due to exceeding voltage or the current rating. Apart from this, there will be huge electromagnetic torque pulsations and increase in rotor speed which may reduce burden on gears of the wind turbine-generator. Such phenomenon reduces lifetime of overall system.

A study of LVRT and HVRT issue for DFIG with rotor current dynamics is given in [1]. In this paper, rotor open and short circuit analysis and differential equations are derived and analyzed to improve the capability of DFIG during disturbances. Hybrid current controllers are used to improve the capability of DFIG to deliver desired reactive power and withstand capability during LVRT and HVRT in [2]. In [3], authors discuss the problems and possible measures to improve performance during HVRT by proposing active voltage control. A fast coordinated control scheme based on the characteristics of DFIG was proposed to improve performance during LVRT and HVRT [4]. Effects of asymmetrical HVRT are analyzed and control strategy for improving performance during overvoltages is done in [5]. A grid side converter (GSC) based control strategy to improve voltage changes in DC link capacitor during HVRT is shown in [6]. FACTS devices like STATCOM [7] and dynamic voltage restorer [8] are used to mitigate voltage swell without phase angle deviation and to investigate HVRT issues. Energy storage devices like SMES to meet the requirement of transmission system operators (TSOs) for wind generation system of HVRT through two grid codes have been discussed in [9]. The critical comparison between HVRT and LVRT is drawn in [10]. The use of active resistance to suppress rotor current and torque oscillations for enhanced operation HVRT is described in [11].

Some external passive elements and active sources are used in conjunction to improve stability and thereby leading to a better LVRT operation of DFIG during symmetrical and asymmetrical faults. Single phase crowbar, supercapacitor energy storage system [12], fault current limiter (FCL) [13], and superconducting FCL with magnetic energy storage devices [14] are connected in coordination with DFIG system to enhance system LVRT behavior during severe faults.

The STATCOM [15–22], familiar shunt device from FACTS family, is used to surmount the system to lose synchronism due to external disturbances like large symmetrical and asymmetrical faults, voltage rise, variation in wind speed, and subsynchronous resonance. Two sequence components with dual voltage control during grid faults for DFIG-STATCOM system are used in [22, 23]. The DVR with high temperature superconductor fault current limiter (SCFCL) application is used for controlling balanced and unbalanced grid faults for DFIG [24]. Eigenanalysis with frequency domain approach used for sea-shore wind farm to improve system stability for DFIG and SCIG is discussed in [25] by applying an UPFC controller. The process of energy storage with the help of superconducting coil by integrating it to the DC link back-to-back converters for DFIG system is discussed in [26].

The research on the behavior of DFIG system during HVRT is still in its nascence. There are a few research papers available. As per modern grid codes for the countries like Australia and Spain, DFIG system must withstand to 1.3 times the rated voltage without losing synchronism. It is generally observed that, with sudden increase in grid voltage, stator and rotor voltages increase and current decreases. The electromagnetic torque (EMT) increases when rotor speed decreases. The rotor voltage and current frequency decrease as slip frequency decreases (sf) due to decrease in slip value. The rotor current decrease for DFIG depends on rotor and stator winding parameters, increase in voltage magnitude at grid, and decrease in speed of rotor. If rotor and stator flux change is controlled, EMT, rotor speed, and machine currents can be controlled effectively. To achieve this, the first step is that a new reference synchronous speed has to be chosen based on change in speed during fault. The second step is that DC offset component of flux must be eliminated, oscillations must be damped, and change in magnitude of q-axis flux components must be controlled. This methodology is termed as enhanced field oriented control (EFOC). The efficacy of EFOC is analyzed for a standard DFIG system for improving voltage and current profile of stator and rotor with stable torque, speed, and flux control mechanism.

The system performance with overvoltage at grid terminals during 0.8 to 1.2 seconds with an electromagnetic torque of 200 Nm is analyzed. In the analysis, the grid fault voltage is considered in three cases: case 1, 1.3 times; case 2: 1.62 times; and case 3: 2 times the grid voltage with other parameters remaining constant. The voltage and current parameters at rotor, stator, grid voltage, and DFIG electromagnetic torque (EMT) and speed of the rotor are compared and analyzed for all the three cases. The proposed EFOC method can be applicable to both LVRT and HVRT issues, which help in improving current and voltage profile at stator and rotor terminals during disturbance. Better performance DFIG operation is expected when using EFOC technique in contrast to the conventional FOC method. In addition, there is no need to use robust PIR or any other sophisticated controllers. Generally, when a severe grid overvoltage occurs, oscillations arise with the increase of stator terminal voltage, DC link capacitor voltage increases, speed of rotor decreases, and electromagnetic torque increases.

In Section 2, design of converters for EFOC is explained. In Section 3, mathematical modeling of wind turbine and generator converters for the grid connected DFIG is explained during transient state. In this section, effect of system during symmetrical fault, EFOC control technique, and behavior of mechanical and electrical system with variation in rotor speed are explained in subsections. Simulation results are described in Section 4 of overvoltage of 1.3, 1.62, and 2 times the rated voltage in the MATLAB environment. The conclusion is presented in Section 5 followed by the Appendix and References.

2. Design of Rotor Side Converter Control for EFOC

RSC controller helps in improving reactive power demand at grid and in extracting maximum power from the machine by making the rotor run at optimal speed. The optimal speed of the rotor is decided from machine's real power and rotor speed characteristic curves from MPPT algorithm. The stator active and reactive power control is possible with the RSC controller strategy through i_{qr} and i_{dr} components controlling, respectively. The rotor voltage in a stationary reference frame [11] and further analysis from [27] is given by

$$V_r^s = V_{0r}^s + R_r i_r^s + \sigma L_r \frac{di_r^s}{dt} - j\omega i_r^s, \tag{1a}$$

where $\sigma = 1 - L_m^2/L_s L_r$, ω is the rotor speed, i_r^s is the rotor current in a stationary frame of reference, L_s, L_r, and L_m are stator, the rotor, and mutual inductance parameters in Henry or in pu, and

$$V_{0r}^s = \frac{L_m}{L_s}\left(\frac{d}{dt} - j\omega_s\right)\Phi_s^s \tag{1b}$$

is the voltage induced in the stator flux with

$$\Phi_s^s = L_s i_s^s + L_m i_s^s,$$
$$\Phi_r^s = L_r i_r^s + L_m i_r^s. \tag{2}$$

The d- and q-axis rotor voltage equations ((1a), (1b)) and (2) in the synchronous rotating reference frame are given by

$$V_{dr} = \frac{d\Phi_{dr}}{dt} - (\omega_s - \omega)\Phi_{qr} + R_r i_{dr},$$

$$V_{qr} = \frac{d\Phi_{qr}}{dt} + (\omega_s - \omega)\Phi_{dr} + R_r i_{qr}. \tag{3}$$

The stator and rotor two-axis fluxes are

$$\Phi_{dr} = (L_{lr} + L_m)i_{dr} + L_m i_{ds},$$

$$\Phi_{qr} = (L_{lr} + L_m)i_{qr} + L_m i_{qs},$$

$$\Phi_{ds} = (L_{ls} + L_m)i_{ds} + L_m i_{dr}, \tag{4}$$

$$\Phi_{qs} = (L_{ls} + L_m)i_{qs} + L_m i_{qr},$$

where $L_r = L_{lr} + L_m$, $L_s = L_{ls} + L_m$, and $\omega_r = \omega_s - \omega$.

By substituting (4) into (3) and by rearranging the terms, then

$$V_{dr} = \left(R_r + \frac{dL'_r}{dt}\right)i_{dr} - s\omega_s L'_r i_{qr} + \frac{L_m}{L_s}V_{ds},$$

$$V_{qr} = \left(R_r + \frac{dL'_r}{dt}\right)i_{qr} + s\omega_s L'_r i_{dr} \tag{5}$$

$$+ \frac{L_m}{L_s}\left(V_{qs} - \omega\Phi_{ds}\right),$$

where ω is rotor speed, $\omega_{\Phi s}$ is speed of stator flux, and ω_s is synchronous speed.

The MATLAB and SIMULINK based on the control circuit of RSC for enhancing performance for LVRT issues are presented in Figure 1. The right side corner with subsystem 2 is a subcircuit of the controller for EFOC technique and its design is shown later in Figure 5.

The above equations (5) can be rewritten in terms of decoupled parameters and are designed for RSC controller as in the following:

$$\sigma V_{dr} = \sigma L_r \frac{dI_{dr}}{dt} - \omega_s \Phi_{qr}$$

$$+ \frac{L_m}{L_s}\left(V_{ds} - R_s I_{ds} + \omega_1 \Phi_{qs}\right), \tag{6}$$

$$\sigma V_{qr} = \sigma L_r \frac{dI_{qr}}{dt} + \omega_s \Phi_{dr} - \frac{L_m}{L_s}\left(R_s I_{qs} + \omega_1 \Phi_{ds}\right). \tag{7}$$

In this paper, the rotor speed is represented with ω_r and the synchronous speed of stator is with ω_s. But this synchronous frequency has to be changed from ω_s to a new synchronous speed value as described in flowchart in Figure 4, ω'_s as it is represented commonly by ω_1. Under ideal conditions, reference stator d-axis flux Φ^*_d is zero and q-axis flux Φ^*_q is equal to the magnitude of stator flux Φ_s for given

back emf and rotor speed. The transient rotor dq-axis current is given by

$$\frac{di_{dr}}{dt} = \frac{-R_r}{\sigma L_r}i_{dr} + s\omega_s i_{qr} + \frac{1}{\sigma L_r}V_{dr}, \tag{8a}$$

$$\frac{di_{qr}}{dt} = \frac{-1}{\sigma}\left(\frac{R_r}{L_r} + \frac{R_s L_m^2}{L_s^2 L_r}\right)i_{qr} - s\omega_s i_{dr} + \frac{1}{\sigma L_r}V_{qr}. \tag{8b}$$

The reference rotor voltages in dq transformation can be rewritten from (6) and (7) and from the control circuit from Figure 1 are given below. This is the output voltage from rotor windings during normal and transient conditions:

$$V^*_{qr} = \left(i^*_{dr} + \frac{1}{\sigma}\left(\frac{R_r}{L_r} + \frac{R_s L_m^2}{L_s^2 L_r}\right)i_{qr} + s\omega_s i_{dr}\right)\sigma L_r, \tag{9a}$$

$$V^*_{qr} = \left(i^*_{dr} + \frac{1}{\sigma}\left(\frac{R_r}{L_r} + \frac{R_s L_m^2}{L_s^2 L_r}\right)i_{qr} + s\omega_s i_{dr}\right)\sigma L_r. \tag{9b}$$

The overall block diagram of RSC is presented in Figure 2(a). The rotor speed is multiplied with pole numbers and is subtracted from angular grid synchronous frequency. It is integrated and given a 90° phase shift to get rotor slip injection frequency angles (θ_s). At this slip frequency, RSC converter injects current into the rotor circuit to control the rotor speed for optimum value and to control grid reactive power. The stator voltage magnitude is compared and controlled using PI or IMC controller to get q-axis current. Similarly, rotor actual speed and optimal speed reference are controlled using PI or IMC to get d-axis reference current. They are compared with actual rotor d- and q-axis currents and controlled with tuned PI controllers to get the rotor injecting d- and q-axis voltages. The d and q voltages are converted into three-axis abc voltage by using phase locked loop (PLL) with inverse parks transformation and are given to a PWM pulse generator for getting pulses to RSC converter.

The d-axis decoupled voltage derivation block diagram is shown in Figure 2(b). The d- and q-axis stator flux and stator flux magnitude derivation block diagram is shown in Figure 2(c). The flux derivation technique helps in understanding the operation of DFIG during steady state and transient state. The accuracy of system performance during steady state depends on accuracy of wind speed measurement, action of pitch angle controller, instantaneous values of stator and rotor voltage, current, flux, and other parameters. The more accurate these measurements are, the more the real power can be extracted from DFIG wind turbine system. The equations from ((8a) and (8b)) to (11) play a vital role in understanding the behavior of DFIG during steady state and accuracy of RSC control action depends on control of d- and q-axis voltages.

3. Mathematical Analysis of RSC and GSC Converters for the Grid Connected DFIG during Transient State

3.1. Three-Phase Symmetrical Faults. The stator voltage will reach zero magnitude during severe three-phase symmetrical

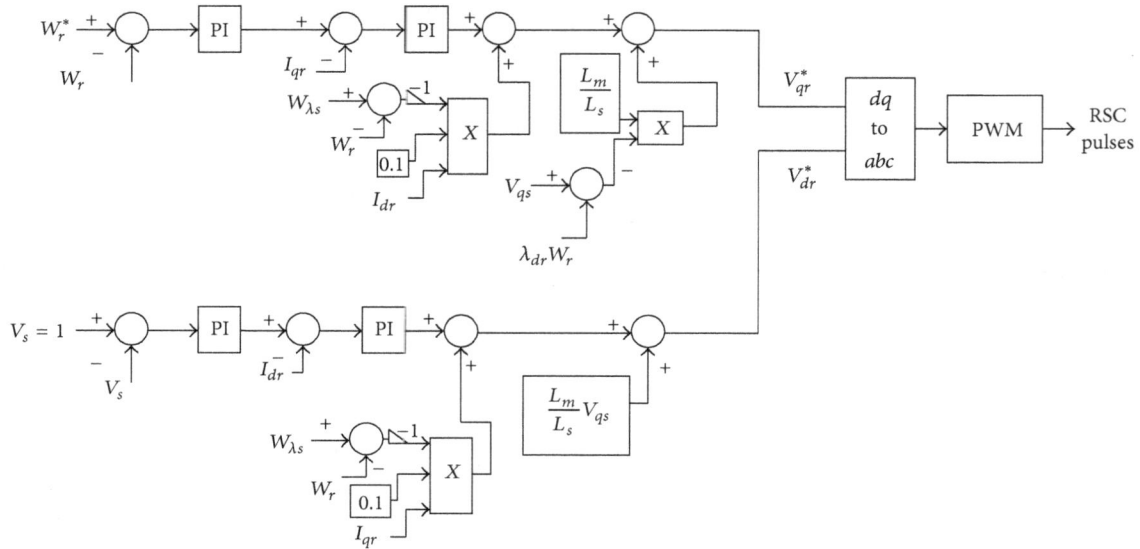

FIGURE 1: The RSC controller with EFOC technique design for grid connected DFIG.

fault of very low impedance and stator flux Φ_s gets reduced to zero magnitude. The decay in flux is not as rapid as in voltage and can be explained from the flux decay theorem from past observations and further can be explained as delay is due to inertial time lag $\tau_s = L_s/R_s$ affecting the rotor induced Electromotive Force (EMF) V_{0r}. The flux during fault is given by

$$\Phi_{sf}^s = \Phi_s^s e^{-t/\tau_s} \tag{10}$$

and $d\Phi_{sf}^s/dt$ is negative, indicating its decay. By substituting (10) into (1b),

$$V_{0r}^s = -\frac{L_m}{L_s}\left(\frac{1}{\tau_s} + j\omega\right)\Phi_s^s e^{-t/\tau_s}. \tag{11}$$

The above equation is converted into a rotor reference frame and neglecting $1/\tau_s$

$$V_{0r}^s = -\frac{L_m}{L_s}(j\omega)\Phi_s^s e^{-j\omega t}. \tag{12}$$

By substituting $\Phi_s^s = (V_s^s/j\omega_s)e^{-j\omega_s t}$ into (12),

$$V_{0r}^r = -\frac{L_m}{L_s}(1-s)V_s. \tag{13}$$

$|V_{0r}^r|$ is proportional to $(1-s)$.
Converting equation (1a) into the rotor reference frame

$$V_r^r = V_{0r}^r e^{-j\omega t} + R_r i_r^r + \sigma L_r \frac{di_r^r}{dt}. \tag{14}$$

Thus, rotor equivalent circuit derived from (12) is as shown in Figure 3 [11].

From equivalent circuit in Figure 3, the rotor voltage during fault is given by

$$V_r = i_r R_r + \sigma L_r \frac{di_r}{dt} + V_{or} \tag{15a}$$

or

$$V_r = i_r R_r + \sigma L_r \frac{di_r}{dt} + \frac{L_m}{L_s}\frac{d\Phi_s}{dt}. \tag{15b}$$

In (15b), the first two terms on RHS determine the voltage drop by rotor current due to passive elements and the last term determines the EMF induced by the stator flux.

A considerable decrease in prefault steady state voltage V_{0r}^r to a certain fault voltage during a three-phase fault is explained here. However, RSC converter is designed to meet V_r^r to match V_{0r}^r for rotor current control and the design has to be made for rating of only 35% of stator rated voltage. The voltage dip during fault can be adopted independently or in coordination by using two techniques as explained in Figure 3.

During fault, at first instant, Φ_s does not fall instantly (12) as shown in the flux and voltage trajectories [27]. If the machine is running at super synchronous speed with slip (s) near to -0.2 pu, during fault, rotor speed further increases based on the term $(1-s)$ as given by (12). The above speed change is uncontrollable for a generator having higher electrical and mechanical inertia constants. In order to control the rotor current change, V_r^r has to be increased. Based on the first reason, a voltage $V_{\Phi s}$ has to be injected in the feed forward path for improving the rotor dip to reach its near steady state value. By converting (12) into a synchronous reference frame and by considering direct alignment of Φ_{ds} with Φ_s, we get

$$V_{\Phi s} = -\frac{L_m}{L_s}\omega\Phi_{ds}. \tag{16}$$

The second technique for voltage increase requirement in a rotor is that dip can be compensated by replacing $s\omega_s$ with $(\omega_{\Phi s} - \omega)$ in cross coupling terms $s\omega_s L_r' i_{qr}$ and $s\omega_s L_r' i_{dr}$, respectively. The reduction in magnitude and frequency of

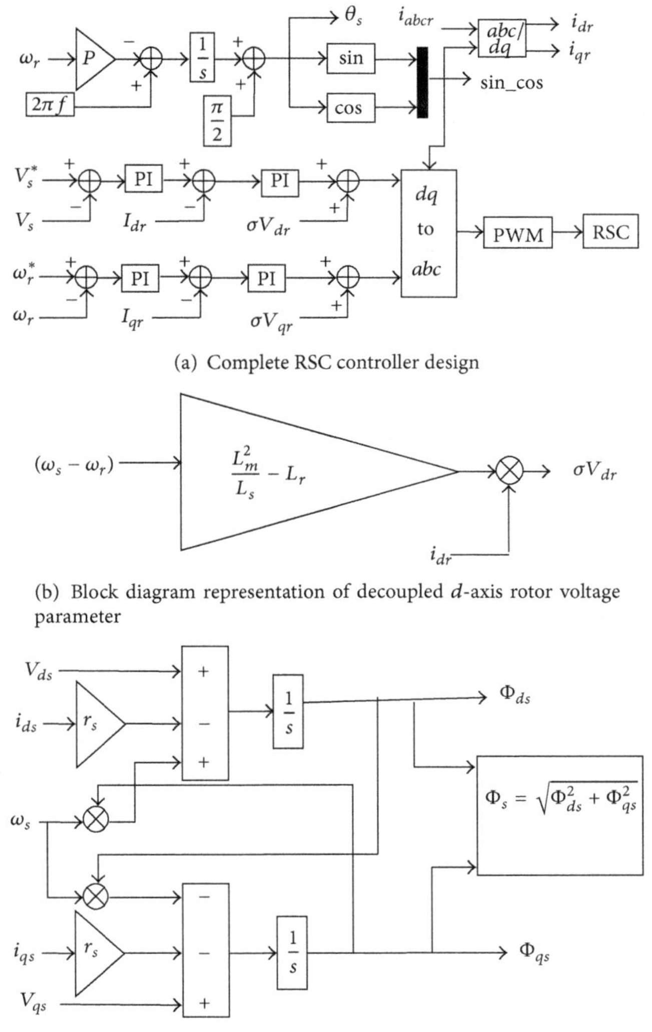

(a) Complete RSC controller design

(b) Block diagram representation of decoupled d-axis rotor voltage parameter

(c) Block diagram representation of stator flux calculation

FIGURE 2: Enhanced FOC control technique with the PI controller adopted for RSC.

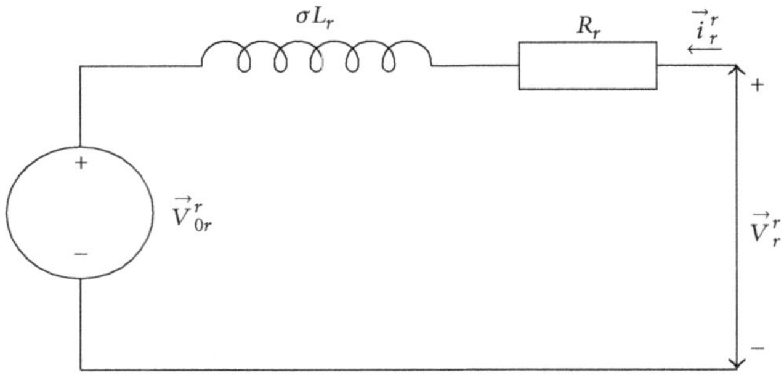

FIGURE 3: The rotor equivalent circuit.

When dynamic stability has to be improved, proposed technique controls the decrease in stator and rotor flux magnitude and also damps oscillations at the fault instances. To achieve better performance during transients, this paper proposes a strategy to change stator frequency reference to zero or other values depending on type and severity of disturbance. The accurate measurement of stator and rotor parameters like flux and current helps in achieving better performance during transients. The DC offset stator current reduction during transients and making the two-axis flux and voltage trajectory circular also improves the efficacy of the system performance during any faults. The equations (7) to (12) help in understanding DFIG behavior during transient conditions and accuracy of its working depends on measurement of rotor current and flux parameters.

3.2. EFOC Technique. The EFOC method of improving field flux oriented control technique helps in improving the performance of the RSC controller of DFIG during fault conditions as described in Figure 4. The DCOC observer carries two actions: the change in flux values of stationary frame stator references ($\Phi_{\alpha s}$, $\Phi_{\beta s}$) for tracking radius of trajectory and DCOC for offset change in stationary fluxes ($\Phi_{dc\alpha s}$, $\Phi_{dc\beta s}$) during fault conditions and controlling them.

There are two major controlling actions with proposed EFOC technique. The first action helps in gaining the trajectory of a circle point and helps in reaching its pre-fault state with the same radius and center of the circle. This helps in improving and maintaining the same rate of change in flux compensation even during fault without losing the stability. The second action helps in controlling and maintaining to nearly zero magnitude of EMF in DFIG with the action of (DC offset component) DCOC technique.

Based on the above two actions, if the former one is greater with change in trajectory which generally happens during disturbances from an external grid, stator synchronous frequency flux speed ($\omega_{\Phi s}$) changes to synchronous grid frequency flux (ω_s); otherwise, $\omega_{\Phi s}$ changes to fault angular frequency value and is injected to RSC voltage control loop as error compensator.

The general form of speed regulation is given by

$$T_e = J\frac{d\omega_r}{dt} + B\omega_r + T_l, \tag{19a}$$

$$T_e = (Js + B)\omega_r + T_l, \tag{19b}$$

flux Φ_s and alignment of flux with the stator voltage without the rate of change in flux angle $\theta_{\Phi s}$ indicate DC offset component in flux:

$$\frac{d\Phi_s}{dt} = \omega_{\Phi s} = 0 = \omega_f. \tag{17}$$

Here, ω_f is the speed of stator flux during fault and this value can be made to zero as offset.

The voltage injection components from equation (16) and (17) and compensating component as discussed above are estimated using enhanced flux oriented control (EFOC scheme whose flowchart is shown in Figure 4 and the determined values are incorporated in the RSC controller shown in Figure 1):

$$\frac{d\Phi_s}{dt} = \omega_{\Phi s} = \frac{V_{\beta s}\Phi_{\alpha s} - V_{\alpha s}\Phi_{\beta s}}{\Phi_{\alpha s}^2 + \Phi_{\beta s}^2} = \omega_f. \tag{18}$$

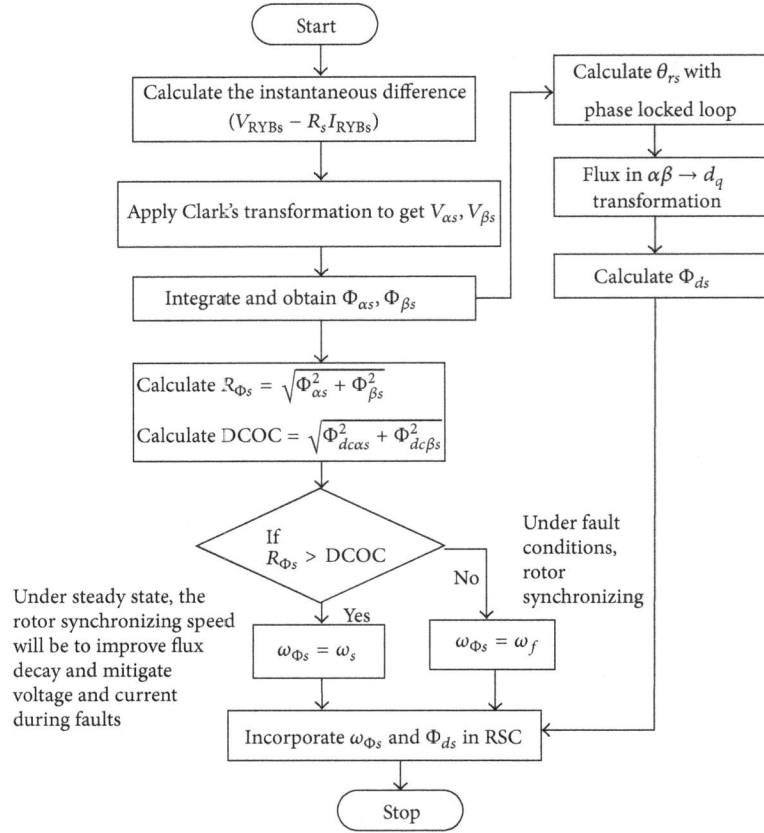

FIGURE 4: Scheme of enhanced flux oriented control, where DCOC = dc offset component of flux and $R_{\Phi s}$ = radius of flux trajectory.

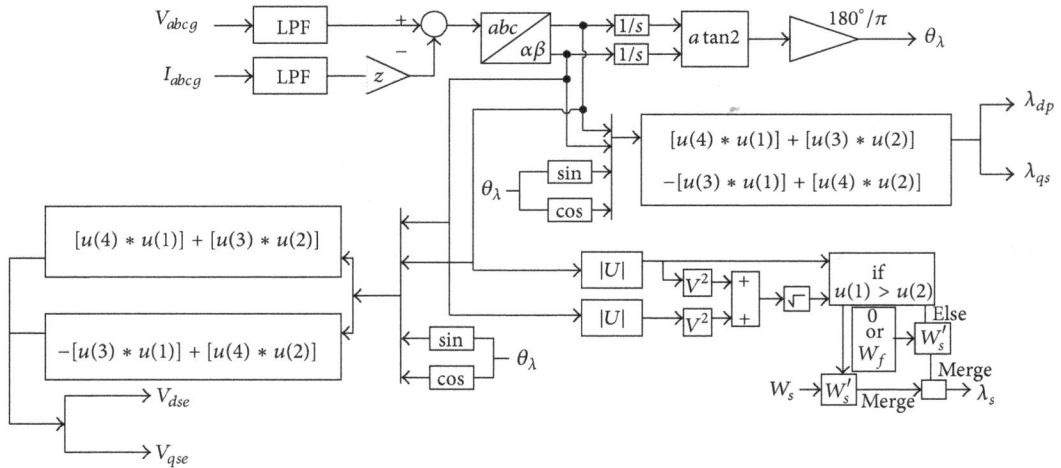

FIGURE 5: EFOC control loop design with DCOC and rotor flux trajectory control.

where T_e is electromagnetic torque, J is moment of inertia, B is friction coefficient, and T_l is considered to be disturbance. Multiplying both sides with ω_{error}, we get the equation as follows:

$$T_e\omega_{\text{error}} = (Js + B)\,\omega_r\omega_{\text{error}} + T_l\omega_{\text{error}}. \qquad (20)$$

Considering ω_r is constant and change in speed error ω_{error} is control variable, the above equation becomes

$$P_s^* = \left(K_{in}s + K_{pn}\right)\omega_{\text{error}} + P_l. \qquad (21)$$

As product of torque and speed is power, we will be getting stator reference power and disturbance power as shown in the following:

$$P_s^* - P_l = \left(K_{in}s + K_{pn} \right) \omega_{error}, \qquad (22)$$

where $K_{in} = J * \omega_r$ and $K_{pn} = B * \omega_r$.

Finally, direct axis reference voltage can be written by using (22) and from Figure 4 is

$$V_{rd}^* = -\left(\omega_{error} \right) \left(K_{pn} + \frac{K_{in}}{s} \right) + \left(P_s \right) \left(K_{pt} + \frac{K_{it}}{s} \right), \qquad (23)$$

$$V_{rq}^* = Q_{error} \left(K_{pQ} + \frac{K_{iQ}}{s} \right), \qquad (24)$$

$$V_{gd}^* = K_{gp} \left(i_{gd}^* - i_{gd} \right) + k_{gi} \int \left(i_{gd}^* - i_{gd} \right) dt \\ - \omega_o L_g i_{gd} + k_1 V_{sd}, \qquad (25)$$

$$V_{gq}^* = K_{gp} \left(i_{gq}^* - i_{gq} \right) + k_{gi} \int \left(i_{gq}^* - i_{gq} \right) dt + \omega_o L_g i_{gd} \\ + k_2 V_{sq}, \qquad (26)$$

$$i_{gq}^* = K_q sqrt \left(V_{dc}^{2*} - V_{dc}^2 \right) + k_{qi} \int \left(V_{dc}^* - V_{dc} \right) dt \\ + R_{dc} V_{dc}, \qquad (27)$$

$$i_{gd}^* = K_d sqrt \left(V_s^{2*} - V_s^2 \right) + k_{di} \int \left(V_s^* - V_s \right) dt. \qquad (28)$$

The rotating direct and quadrature reference voltages of rotor are converted into stationary abc frame parameters by using inverse parks transformation. Slip frequency is used to generate sinusoidal and cosine parameters for inverse parks transformation.

In general, during fault and after fault, the DC link voltage across the capacitor at the DFIG back-to-back converter terminal falls and rises and the STATCOM helps in improving the operation and assists in regaining its voltage value, respectively, to get ready for the operation during next fault. However, STATCOM provides efficient support to the grid-generator system under severe faults by fast action in controlling reactive power flow to grid by maintaining the DC link voltage at the capacitor terminal of DFIG converters constant particularly during transient state. Hence, it helps in improving the dynamic stability of the overall system.

3.3. Behavior of Mechanical and Electrical System with the Variation in Rotor Speed during Faults. The mechanical to electrical relationship is explained as follows. The rotor speed can be expressed as

$$\omega_r = (1 - s) \omega_s = p\eta\omega_{ot}, \qquad (29)$$

where s is slip of DFIG, p is pair of poles of DFIG, η is gear box ratio, and ω_{ot} is wind turbine speed. With the change in wind speed and depending on gears ratio and number of field poles, how the rotor speed varies is shown

in (27). When rotor speed varies, reference quadrature axis current changes; thereby, current flow in the rotor circuit varies. The stator output also varies with variation in wind turbine speed and DFIG output power. When slip varies, the voltage in rotor circuit also varies which can be explained as per (6) and (7). Further change in rotor voltage leads to change in rotor current; thereby, rotor power flow also varies. When a disturbance like symmetrical fault occurs, rotor speed increases so as to compensate the change in electrical power and the mechanical power. During faults, rotor of DFIG tries to accelerate and reaches a new operating point. Hence the rotor speed decreases with over voltage fault and decreases with under voltage fault. This statement can be understandable from the equal area criterion theory for electrical machines.

The mechanical turbine tip speed ratio (TSR) can be written in terms of radius of turbine wings (R), angular stator speed (ω_s), pole pairs, and gear box ratio as

$$\lambda = \frac{R\omega_s}{p\eta v_w} (1 - s). \qquad (30)$$

With the increase in stator or grid frequency, TSR increases and vice versa. Similarly, with increase in rotor speed or wind speed, TSR decreases and vice versa. Hence, when an electrical system gets disturbed, mechanical system also gets some turbulence and electrical to mechanical system is tightly interlinked. The steady state behavior of overall system must satisfy the following relation:

$$\Delta P = \frac{-P_{ot}}{(1 - s)} - P_{em} = 0. \qquad (31)$$

Under normal conditions, the change in turbine output has to be compensated by electrical power output from DFIG. Otherwise, slip gets changed and thereby rotor speed changes. Hence, with imbalance in mechanical to electrical power output ratios, the slip changes. With the change in coefficient of power Cp, the mechanical power varies. The mechanical power changes mostly when wind speed or air density around the turbine wings changes. The electrical power from DFIG changes when mechanical power changes or rotor speed changes or load demand from grid varies.

4. Result Analysis

In this paper, a general system was considered as shown in Figure 6. The DFIG is driven by a wind turbine and electric power from the generator is pumped to the grid for meeting different loads requirement. The RSC is designed with EFOC technique to enhance operation during grid fault and has better dynamic stability features than a conventional FOC. The role of RSC is to extract maximum power from wind turbine, so the generator is made to rotate at that optimal speed by adjusting gear wheels between generator and turbine shaft. Another role of RSC is to improve the reactive power requirement during any abnormal situations like undervoltage or overvoltage faults. The excess reactive power is supplied by RSC using the capacitor bank at RSC end and accuracy and fastness depend on the control strategy.

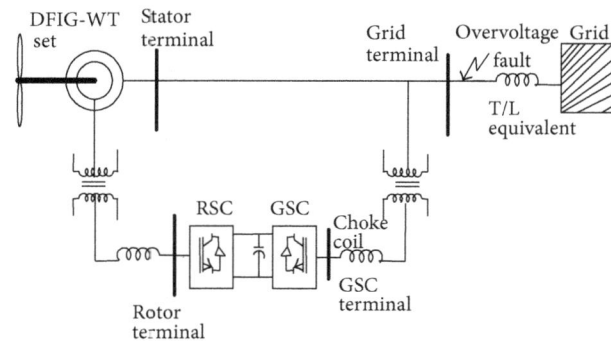

FIGURE 6: Grid connected DFIG showing the location of overvoltage fault.

The GSC of DFIG has two main functions. One function is to maintain nearly constant voltage profile at DC link capacitor so that voltage at point of common coupling (PCC) will also have the same value. The other function is to supply or absorb rapidly required reactive power. The GSC is also designed to bypass surge current to the converter terminal and store in capacitor bank and to reinject the excess power when fault gets relieved. In doing so, the fault current cannot reach stator and rotor and as a consequence GSC protects the two windings of the generator.

For the system in Figure 6, voltage swell is expected during the time 0.8 to 1.2 seconds. Swell in voltage occurs when generation is more than load or when a big load is switched off or during lightning. In this, grid voltage, electromagnetic torque, speed, stator and rotor two-axis fluxes and current, and three-phase stator and rotor current are analyzed and compared when grid voltage rises to 1.3, 1.62, and 2 times the rated voltage. The results with HVRT can be compared with [1, 2] for the performance when 1.3 times increase in grid voltage. The rated grid voltage is 270 volts, electromagnetic torque (EMT) is −200 Nm, 2660 rpm speed, 0.05 and −0.75 Weber of stator d- and q-axis flux. 0.51 and −0.55 Wb of rotor flux, −40 and −142 Amps of d and q axis stator current, 130 and 20 Amps of rotor two axis current and 110 Amps of stator and rotor current under steady state conditions respectively.

As shown in Figure 7(a)(i), the grid voltage has been increased from 270 volts to 350 volts, which are 1.3 times the rated voltage, and 437 volts' and 540 volts' rises during 0.8 to 1.2 seconds are shown in Figure 7(a)(ii) and (iii). For all the three rises in voltage, behavior of DFIG is analyzed. As per modern grid codes, DFIG needs to be in synchronism for 1.3 times' rise in voltage, but the proposed system is analyzed with twice the rise in voltage.

With increase in voltage to 1.3, 1.62, and 2 times, the changes in stator and rotor d- and q-axis fluxes are shown in Figure 7(b)(i), (ii), and (iii). A sudden increase in grid voltage leads to large inrush current into the stator and rotor initially. As a result, active power from rotor also increases and reaches the DC link capacitor. Since GSC is of very low rating (35% of rated machine rating), it is not having that ability to transfer current to grid [2, 16, 17]. Hence, in this case, DC capacitor rating if not high enough will get damaged and, overall, the system will lose ride through capability. In our proposed

theory, as explained in Section 3.1, the change in flux with DC offset components and change in speed is detected initially. The flux is controlled by selecting a new rotor speed and the flux further controls the change in voltage and current in rotor circuit by RSC control scheme as explained in flowchart. Thereby, damping of oscillations is eliminated and stability can be enhanced. The more voltage changes, the more magnitude of q-axis flux will be produced and d-axis flux will be reduced because flux linkage is proportional to stator and rotor voltages. When voltage is increased to 1.3 times in Figure 7(a)(i), q-axis stator flux changed from −0.75 to −0.9 Wb during fault and reaches again −0.75 Wb. The rotor q-axis flux is varied from prefault of −0.55 Wb to −0.9 Wb during fault at 0.8 s and reaches its prefault when fault was cleared at 1.2 s. In the same way, q-axis stator flux changed to −1.1 and −1.4 Wb with 1 and 2 oscillations and q-axis rotor flux change is −1 and −1.45 Wb for voltage rise of 1.62 and 2 times with much controlled oscillations as shown in Figure 7(b)(ii) and (iii). Even with voltage rise of two times also, system is highly stable with sustained oscillations in fluxes. The result from proposed scheme has better performance than the results obtained from [1–8] with additional FACTS devices.

With 1.3 times rise in voltage as shown in Figure 7(c)(i), torque rises from −200 Nm to −380 Nm when fault started and was maintained to −200 Nm during fault and reached normal value when fault was cleared. Speed of rotor decreased to 2400 rpm from 2660 rpm and was maintained at the same speed during fault and reached its prefault state when fault was cleared. From (9a) and (9b), when flux and current increase, torque will also increase. But with sustained increase in stator and rotor q-axis flux and decrease in d-axis flux with enhanced flux control scheme in RSC, helps in reaching its prefault state torque value. Since flux is having very less oscillations, torque is also having limited oscillations. With change in voltage to 1.62 and 2 times, surge in torque is nearly −580 Nm and −1000 Nm when fault was initiated as in Figure 7(c)(ii) and (iii). It is due to surge increase in magnitude of stator and rotor q-axis fluxes.

The d-axis stator surge current increased in magnitude from −40 A to −90 A, −150 A, and −220 A during voltage increase to 1.3, 1.62, and 2 times as shown in Figure 7(d)(i, ii, and iii) when fault started. The q-axis stator current increased from −140 A to −180, −200, and −220 A during 1.3, 1.62, and

Dynamic Stability Improvement of Grid Connected DFIG Using Enhanced Field Oriented Control Technique...

77

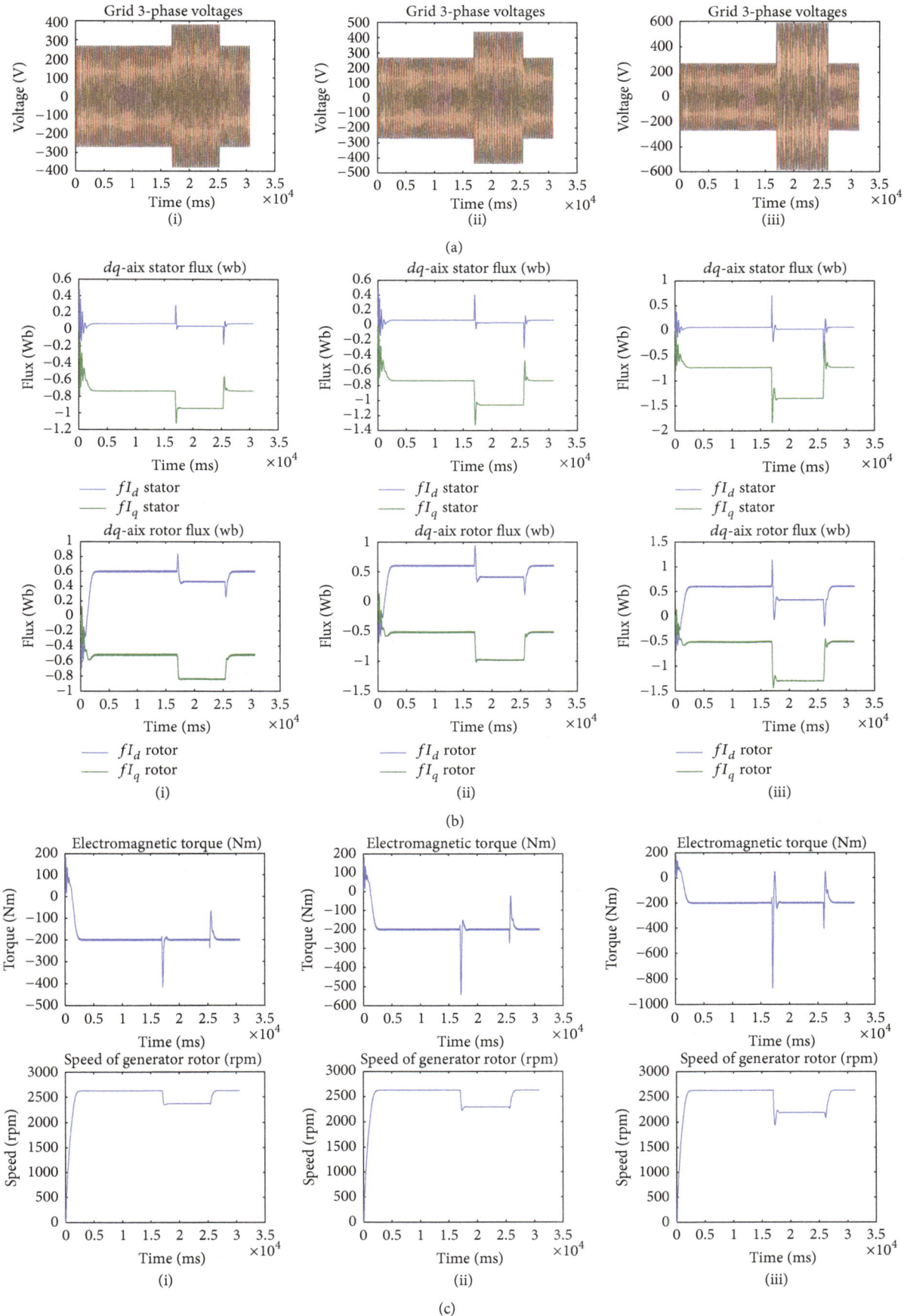

FIGURE 7: Continued.

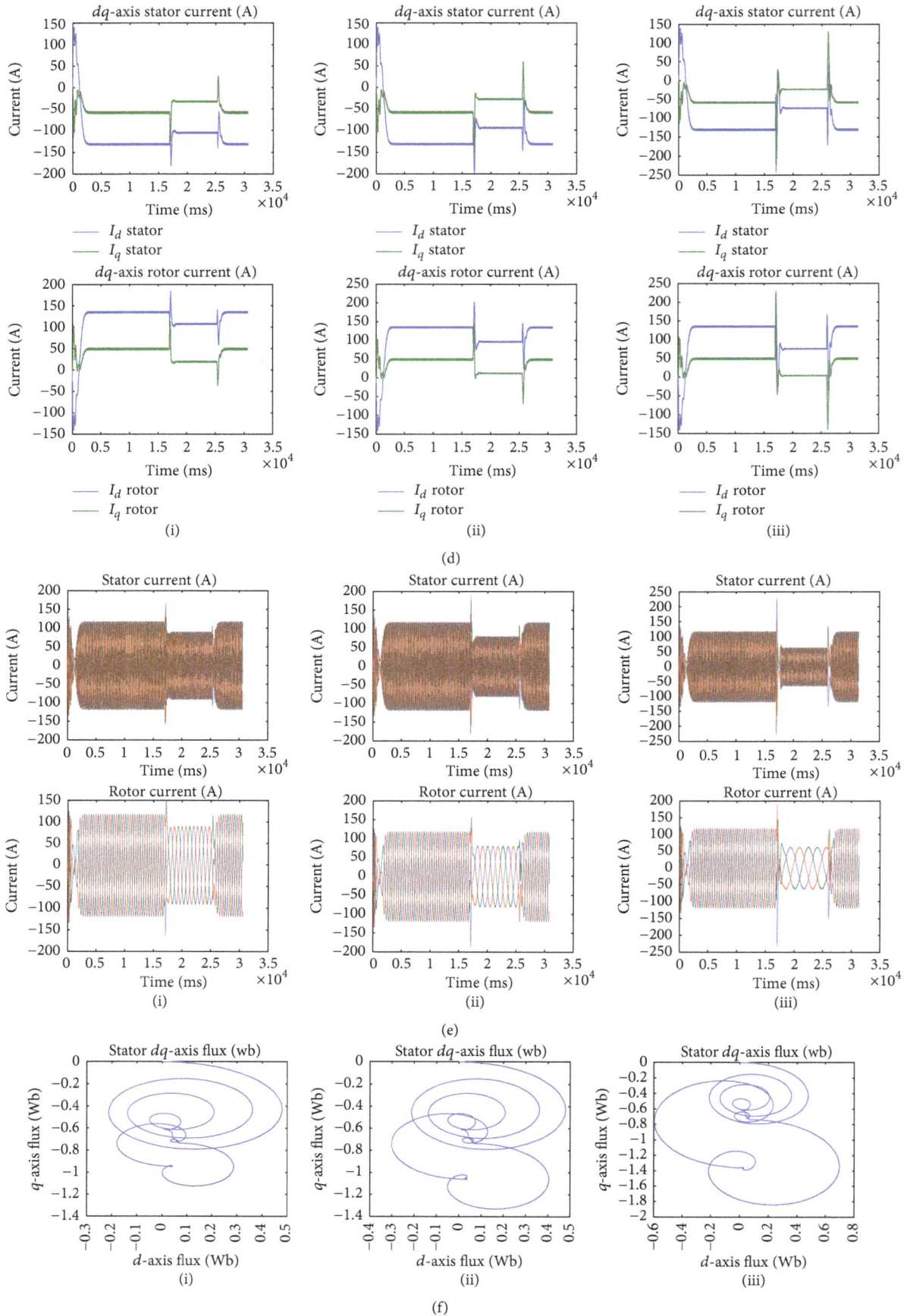

FIGURE 7: Continued.

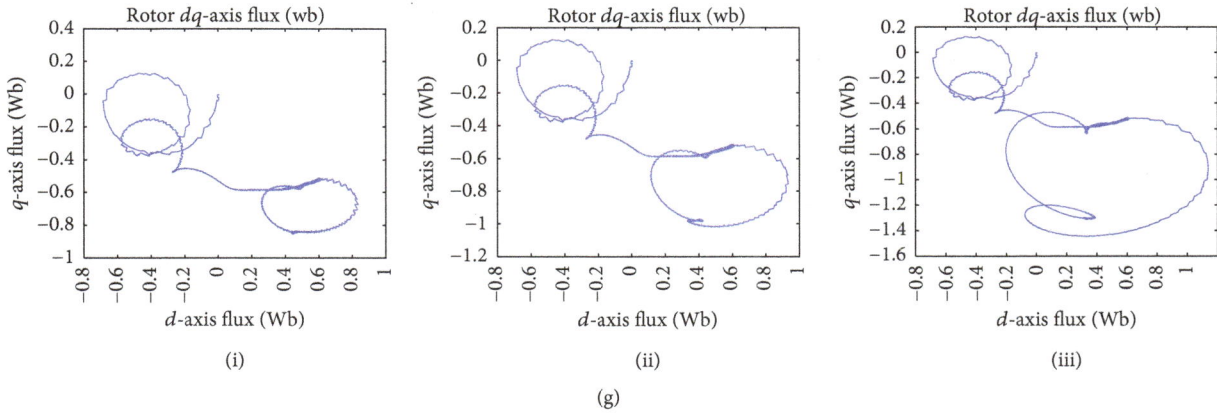

(i)

(ii)

(iii)

(g)

FIGURE 7: (a) Grid voltage at (i) 1.3 times the rated fault, (ii) 1.62 times the voltage rise, and (iii) 2 times the voltage rise. (b) dq-axis stator and rotor flux at (i) 1.3 times the rated fault, (ii) 1.62 times the voltage rise, and (iii) 2 times the voltage rise. (c) EM torque and speed at (i) 1.3 times the rated fault, (ii) 1.62 times the voltage rise, and (iii) 2 times the voltage rise. (d) dq-axis stator and rotor current at (i) 1.3 times the rated fault, (ii) 1.62 times the voltage rise, and (iii) 2 times the voltage rise. (e) Stator and rotor current at (i) 1.3 times the rated fault, (ii) 1.62 times the voltage rise, and (iii) 2 times the voltage rise. (f) dq-axis stator flux graph at (i) 1.3 times the rated fault, (ii) 1.62 times the voltage rise, and (iii) 2 times the voltage rise. (g) dq-axis rotor flux graph at (i) 1.3 times the rated fault, (ii) 1.62 times the voltage rise, and (iii) 2 times the voltage rise.

2 times' rise in grid voltage. But during fault, d-axis stator current and q-axis rotor current became zero when voltages swell occurred and q-axis stator current decreased to −105 A, −95 A, and −80 A when voltage increased to 1.3, 1.6, and 2 times and reached normal of −140 A when fault was cleared. Similarly, the rotor d-axis current decreased from 125 amps during steady state to 100 A, 80 A and 55 A when the voltage at grid rose from 1.3, 1.62 and 2 times the rated voltage because of an over-voltage fault. In the proposed control scheme, DC offset current components are eliminated by RSC control scheme and stator currents are controlled to some extent by GSC control scheme. The rotor voltage injection and current control action are as per (17) and ((19a), (19b)). The rapid and enhanced reactive power control action of proposed scheme helps in improving current profile of both stator and rotor. The three-phase stator and rotor current waveforms 1.3, 1.62, and 2 times' rises in grid voltage are shown in Figure 7(e)(i), (ii), and (iii).

The stator dq-axis flux patterns under 1.3, 1.62, and 2 times' rise in grid voltage are shown in Figure 7(f)(i), (ii), and (iii). The graph initially started at 0,0-axis and slowly increases in magnitude to the right hand side with positive d-axis and negative q-axis and in a clockwise spiral way reached a small circle during steady state. When there is a change in grid voltage, due to corresponding change in stator d- and q-axis flux linear graph as in Figure 7(b), the relative graph is in Figure 7(f). Due to overvoltage of 1.3 times, another circle away from small circle initiated and reaches the plane [0.25, −1.15] and then stabilises at [0.05, −0.95] coordinates during fault. When fault was cleared, the new coordinate is [−0.25, −0.55] and reached the same old small circle. The same explanation with change in coordinates holds well for 1.62 and 2 times' increase in grid voltage. However, pattern nearly remained the same with the change in coordinates.

The rotor flux path pattern for dq-axis during 1.3, 1.62, and 2 times' rise in grid voltage is shown in Figure 7(g)(i, ii, and iii). Initially, the dq-axis plane starts at [0, 0]. Slowly, with increase in time, the flux pattern in clockwise direction reaches the small upper circle shown in Figure 7(g)(i). It remained in this point till steady state is maintained. When an overvoltage occurred, the pattern changes its position at [−0.65, −0.35] at 1.2 s and later in short time reached [0.2, −0.9] at 1.22 s for 1.3 times' increase in voltage and continued to be in the big lower circle. For more severe fault, more magnitude of voltage is changed. Hence, flux d- and q-axis components also increase and the patterns for 1.62 and 2 times are shown in Figure 7(f)(ii and iii).

Table 1 gives a detailed picture of parameter variation under steady state, at the instant of fault with surge value and during the fault. Under normal conditions, grid voltage is 270 V, when voltage rised to 1.3, 1.6, and 2 times, the respective voltages are 350, 437, and 540 volts. The electromagnetic torque (EMT) during steady state is −220 Nm; at the instant when voltage rised to 1.3 times, a surge of −380 Nm was produced and, during the fault, it was again −200 Nm. In a similar way, the steady state stator d- and q-axis flux in webers are 0.08 and −0.75. At the instant of voltage rise to 1.3 times, stator d- and q-axis fluxes are 0.15 and −1 Wb and reach 0 and −0.8 Wb during the fault. The same explanation holds well for other parameters.

5. Conclusion

A conventional DFIG wind turbine system connected to grid was considered in the analysis. Voltage swell of 1.3, 1.62, and 2 times the rated voltage during 0.8 to 1.2 seconds was applied to grid and the behavior of DFIG system was studied. For a general system, with increase in grid voltage, stator and rotor

TABLE 1: The parameter variation before and during voltage swell.

Parameter under consideration	Normal system (at steady state)	Grid voltage 1.3 times' rise	Grid voltage 1.62 times' rise	Grid voltage 2 times' rise
Grid voltage (V)	270	350	437	540
EMT (Nm)	−200	Surge −380 During fault −200	Surge −580 During fault −200	Surge −1000 During fault −200
Speed (rpm)	2660	During fault −2400	During fault −2250	During fault −2100
(Φ_{ds}, Φ_{qs}) Wb	(0.05, −0.75)	Surge (0.15, −1) During fault (0, −0.8)	Surge (0.4, −1.2) During fault (0, −1.1)	Surge (0.6, −1.6) During fault (0, −1.4)
(Φ_{dr}, Φ_{qr}) Wb	(0.51, −0.55)	Surge (0.75, −0.92) During fault (0.48, −0.9)	Surge (1, −1.05) During fault (0.4, −1.02)	Surge (1.15, −1.5) During fault (0.25, −1.4)
(I_{ds}, I_{qs}) Amp	(−40, −140)	Surge (−80, −170) During fault (0, −110)	Surge (−140, −200) During fault (0, −100)	Surge (−210, −240) During fault (0, −80)
(I_{dr}, I_{qr}) Amp	(130, 20)	Surge (180, 90) During fault (110, −5)	Surge (220, −120) During fault (100, −1)	Surge (240, 145) During fault (90, −2)
(I_{st}, I_{rot}) 3-phase Amps	(110, 110)	Surge (120, 120) During fault (90, 100)	Surge (200, 150) During fault (80, 90)	Surge (210, 210) During fault (60, 70)

voltage levels will also increase. With increase in voltage, d-axis stator and rotor flux decrease and q-axis stator and rotor flux increase. The electromagnetic torque (EMT) of generator increases to small quantity and speed decreases with increase in grid voltage. d- and q-axis stator and rotor current decrease to a value depending on the swell in voltage. As per modern grid codes for the countries like Australia and Spain, the DFIG system must remain in synchronism with voltage up to 1.3 times occurring at grid. This is called high voltage fault ride through (HVRT).

In this paper, the performance of DFIG with EFOC technique during overvoltage was studied under three cases. In the first case, an overvoltage of 1.3 times the rated voltage at grid occurred between 0.8 and 1.2 seconds. In the other two cases with the same fault instant but with fault magnitude of 1.62 and 2 times occur at grid. Comparing the proposed enhanced field oriented control (EFOC) scheme with previous works, there is a significant reduction in the surges in electromagnetic torque at the instants of fault occurring or clearing. With 2 times the rated voltage, the torque magnitude is constant and the value is nearly same as the pre-fault condition. The oscillations at fault instants are completely eliminated with proposed EFOC. As a result of this, synchronism is maintained and hence overall stability was improved even for 2 times' increase in grid voltage. The DC offset component of flux is controlled by proposed system. Here, the flux in stator and rotor is maintained to a point so that sustained oscillations are damped with low magnitude and with quick settling time. In the conventional system with HVRT, stator and rotor current decay to a very small value during fault. It will take more time to reach its prefault condition after disturbance for DFIG system with conventional technique. But for our EFOC based DFIG

WECS, stator and rotor currents remain almost constant for 1.3 times' rise in grid voltage.

Thus, proposed system follows modern grid codes very strictly and can sustain to severe overvoltage issues easily. At present, meeting grid code for 1.3 times' swell is sufficient. With the proposed strategy, the system can easily withstand up to 2 times the swell fault without external devices or increased converter ratings. This paper describes the exact behavior of DFIG system during HVRT using analytics and MATLAB based simulation. The behavior of DFIG-grid connected system with EFOC technique is suitable for the system where grid voltage swell faults are most common. It does not require any additional reactive power sources like FACTS devices or energy storage devices. A conventional controller like PI with fast acting relay and converters are needed for RSC and GSC to enhance the system behavior during HVRT for proposed EFOC technique.

With the proposed control strategy, smooth transition in electromagnetic torque is achieved during symmetrical fault based transient state of drop in grid voltage and restoring is possible. The dynamic stability of DFIG was improved and thereby mitigation of generator stator and rotor voltages and current is superior with EFOC fuzzy technique. The output power from generator is better in damping the transient stator flux. This is possible by changing the reference flux reference value by choosing particular stator flux (λs) value. Otherwise, overcurrent in rotor winding makes the system performance and longevity degrade under these situations.

Hence, with the proposed EFOC fuzzy technique, a control over stator transient flux is possible so as to suppress rotor current surges and help in achieving better LVRT operating characteristics. EMT is smooth with suppressed oscillations and thereby prolongs the lifetime of the generator turbine system during voltage dip and recovery. The behavior during and after fault conditions is improved with the mitigation in stator and rotor current waveforms. The overall performance is improved and can sustain to severe faults with ensured stability.

Appendix

The parameters of DFIG used in simulation are as follows: Rated power = 1.5 MW, rated voltage = 690 V, stator resistance $R_s = 0.0049$ pu, rotor resistance $R_r = 0.0049$ pu, stator leakage inductance Lls = 0.093 pu, rotor leakage inductance Llr1 = 0.1 pu, inertia constant = 4.54 pu, number of poles = 4, mutual inductance $L_m = 3.39$ pu, DC link voltage = 415 V, DC link capacitance = 0.2 F, wind speed = 14 m/sec, grid voltage = 25 KV, grid frequency = 60 Hz, grid side filter: Rfg = 0.3 Ω, Lfg = 0.6 nH, rotor side filter: Rfr = 0.3 mΩ, Lfr = 0.6 nH, STATCOM: capacitance = 0.1 F, transformer-690/440 V, 50 kVA rating, and PWM frequency = 2 kHz.

Conflict of Interests

The authors declare that there is no conflict of interests regarding the publication of this paper.

References

[1] Y. Ling and X. Cai, "Rotor current dynamics of doubly fed induction generators during grid voltage dip and rise," *International Journal of Electrical Power & Energy Systems*, vol. 44, no. 1, pp. 17–24, 2013.

[2] M. Mohseni, M. A. S. Masoum, and S. M. Islam, "Low and high voltage ride-through of DFIG wind turbines using hybrid current controlled converters," *Electric Power Systems Research*, vol. 81, no. 7, pp. 1456–1465, 2011.

[3] C. Feltes, S. Engelhardt, J. Kretschmann, J. Fortmann, F. Koch, and I. Erlich, "High voltage ride-through of DFIG-based wind turbines," in *Proceedings of the IEEE Power and Energy Society General Meeting—Conversion and Delivery of Electrical Energy in the 21st Century*, pp. 1–8, IEEE, Pittsburgh, Pa, USA, July 2008.

[4] Y. Wang, Q. Wu, H. Xu, Q. Guo, and H. Sun, "Fast coordinated control of DFIG wind turbine generators for Low and High voltage ride-through," *Energies*, vol. 7, no. 7, pp. 4140–4156, 2014.

[5] R. Li, H. Geng, and G. Yang, "Asymmetrical high voltage ride through control strategy of grid-side converter for grid-connected renewable energy equipment," in *Proceedings of the International Electronics and Application Conference and Exposition (PEAC '14)*, pp. 496–501, Shanghai, China, November 2014.

[6] Z. Zheng, G. Yang, and H. Geng, "High voltage ride-through control strategy of grid-side converter for DFIG-based WECS," in *Proceedings of the 39th Annual Conference of the IEEE Industrial Electronics Society (IECON '13)*, pp. 5282–5287, IEEE, Vienna, Austria, November 2013.

[7] Y. M. Alharbi, A. M. S. Yunus, and A. Abu-Siada, "Application of STATCOM to improve the high-voltage-ride-through capability of wind turbine generator," in *Proceedings of the IEEE PES Innovative Smart Grid Technologies Asia (ISGT '11)*, pp. 1–5, IEEE, Perth, Wash, USA, November 2011.

[8] C. Wessels and F. W. Fuchs, "High voltage ride through with FACTS for DFIG based wind turbines," in *Proceedings of the 13th European Conference on Power Electronics and Applications (EPE '09)*, pp. 1–10, IEEE, Barcelona, Spain, September 2009.

[9] A. M. S. Yunus, A. Abu-Siada, and M. A. S. Masoum, "Application of SMES unit to improve the high-voltage-ride-through capability of DFIG-grid connected during voltage swell," in *Proceedings of the IEEE PES Innovative Smart Grid Technologies Asia (ISGT)*, pp. 1–6, Perth, Australia, November 2011.

[10] Z. Xie, Q. Shi, H. Song, X. Zhang, and S. Yang, "High voltage ride through control strategy of doubly fed induction wind generators based on active resistance," in *Proceedings of the IEEE 7th International Power Electronics and Motion Control Conference (IPEMC '12)*, vol. 3, pp. 2193–2196, IEEE, Harbin, China, June 2012.

[11] C. Feltes, S. Engelhardt, J. Kretschmann, J. Fortmann, F. Koch, and I. Erlich, "High voltage ride-through of DFIG-based wind turbines," in *Proceedings of the IEEE Power and Energy Society General Meeting: Conversion and Delivery of Electrical Energy in the 21st Century (PESGM '08)*, pp. 1–8, IEEE, Pittsburgh, Pa, USA, July 2008.

[12] J. Liang, D. F. Howard, J. A. Restrepo, and R. G. Harley, "Feedforward transient compensation control for DFIG wind turbines during both balanced and unbalanced grid disturbances," *IEEE Transactions on Industry Applications*, vol. 49, no. 3, pp. 1452–1463, 2013.

[13] J. Liang, W. Qiao, and R. G. Harley, "Feed-forward transient current control for low-voltage ride-through enhancement of DFIG wind turbines," *IEEE Transactions on Energy Conversion*, vol. 25, no. 3, pp. 836–843, 2010.

[14] T. D. Vrionis, X. I. Koutiva, and N. A. Vovos, "A genetic algorithm-based low voltage ride-through control strategy for grid connected doubly fed induction wind generators," *IEEE Transactions on Power Systems*, vol. 29, no. 3, pp. 1325–1334, 2014.

[15] J. P. da Costa, H. Pinheiro, T. Degner, and G. Arnold, "Robust controller for DFIGs of grid-connected wind turbines," *IEEE Transactions on Industrial Electronics*, vol. 58, no. 9, pp. 4023–4038, 2011.

[16] J. Vidal, G. Abad, J. Arza, and S. Aurtenechea, "Single-phase DC crowbar topologies for low voltage ride through fulfillment of high-power doubly fed induction generator-based wind turbines," *IEEE Transactions on Energy Conversion*, vol. 28, no. 3, pp. 768–781, 2013.

[17] C. Abbey and G. Joos, "Super-capacitor energy storage for wind energy applications," *IEEE Transactions on Industry Applications*, vol. 43, no. 3, pp. 769–776, 2007.

[18] W. Guo, L. Xiao, S. Dai et al., "LVRT capability enhancement of DFIG with switch-type fault current limiter," *IEEE Transactions on Industrial Electronics*, vol. 62, no. 1, pp. 332–342, 2015.

[19] W. Guo, L. Xiao, and S. Dai, "Enhancing low-voltage ride-through capability and smoothing output power of DFIG with a superconducting fault-current limiter-magnetic energy storage system," *IEEE Transactions on Energy Conversion*, vol. 27, no. 2, pp. 277–295, 2012.

[20] L. Wang and D.-N. Truong, "Stability enhancement of DFIG-based offshore wind farm fed to a multi-machine system using a STATCOM," *IEEE Transactions on Power Systems*, vol. 28, no. 3, pp. 2882–2889, 2013.

[21] W. Qiao, G. K. Venayagamoorthy, and R. G. Harley, "Real-time implementation of a STATCOM on a wind farm equipped with

doubly fed induction generators," *IEEE Transactions on Industry Applications*, vol. 45, no. 1, pp. 98–107, 2009.

[22] L. Wang and C.-T. Hsiung, "Dynamic stability improvement of an integrated grid-connected offshore wind farm and marine-current farm using a STATCOM," *IEEE Transactions on Power Systems*, vol. 26, no. 2, pp. 690–698, 2011.

[23] W. Qiao, R. G. Harley, and G. K. Venayagamoorthy, "Coordinated reactive power control of a large wind farm and a STATCOM using heuristic dynamic programming," *IEEE Transactions on Energy Conversion*, vol. 24, no. 2, pp. 493–503, 2009.

[24] S. Bozhko, G. Asher, R. Li, J. Clare, and L. Yao, "Large offshore DFIG-based wind farm with line-commutated HVDC connection to the main grid: engineering studies," *IEEE Transactions on Energy Conversion*, vol. 23, no. 1, pp. 119–127, 2008.

[25] S. V. Bozhko, R. V. Blasco-Giménez, R. Li, J. C. Clare, and G. M. Asher, "Control of offshore DFIG-based wind farm grid with line-commutated HVDC connection," *IEEE Transactions on Energy Conversion*, vol. 22, no. 1, pp. 71–78, 2007.

[26] L. Fan and Z. Miao, "Mitigating SSR using DFIG-based wind generation," *IEEE Transactions on Sustainable Energy*, vol. 3, no. 3, pp. 349–358, 2012.

[27] D. V. N. Ananth and G. V. Nagesh Kumar, "Fault ride-through enhancement using an enhanced field oriented control technique for converters of grid connected DFIG and STATCOM for different types of faults," *ISA Transactions*, 2015.

10

Optimal Operation Conditions for a Methane Fuelled SOFC and Microturbine Hybrid System

Vincenzo De Marco, Gaetano Florio, and Petronilla Fragiacomo

Department of Mechanical, Energy and Management Engineering, University of Calabria, Arcavacata, Rende, 87036 Cosenza, Italy

Correspondence should be addressed to Petronilla Fragiacomo; petronilla.fragiacomo@unical.it

Academic Editor: Wei-Hsin Chen

The study of a hybrid system obtained coupling a methane fuelled gas microturbine (MTG) and a solid oxide fuel cell (SOFC) was performed. The objective of this study is to evaluate the operation conditions as a function of the independent variables of the system, which are the current density and fuel utilization factor. Numerical simulations were carried out in developing a C++ computer code, in order to identify the preferable plant configuration and both the optimal methane flow and the current density. Operation conditions are able to ensure elasticity and the most suitable fuel utilization factor. To confirm the reliability of the models, results of the simulations were compared with reference results found in literature.

1. Introduction

The question of energy savings and optimization of resources is, nowadays, of primary importance, in both economic and environmental fields. In order to respect the Kyoto protocol it is important to invest in new forms of livelihood energy [1, 2]. The present work lies as part of these problems. An assigned hybrid system resulting from the coupling of a solid oxide fuel cell (SOFC) with a gas microturbine (MTG) was analyzed. For this purpose a C++ code was developed, so that it was possible to define the optimal conditions of operation of the system as a function of the current density dfc of the fuel cell and the fuel utilization (U_f). It has to be pointed out that only by developing such C++ code taking into account both chemical and physical phenomena (i.e., balance of power, chemical reactions, and change in the composition) was it possible to appreciate the difference in performance by comparing different plant schemes. Such calculations match with the current direction taken by the state of the art (i.e., the presence of an afterburner in place of a direct stream from the SOFC stack to the turbine). Although it is still difficult to find experimental data to validate the model completely, this was subjected to comparison with another model present in the literature. It is important to highlight that this kind

of approach can be easily adapted to other kinds of plants, only by changing the parameters utilized and by correctly defining the events considered. In this specific case, for the analysis of the gas turbine the data plate of the MTG Ansaldo 100 kW was taken as a reference, as well as in [3–7]. This type of system can reach values comparable to those of large size, from an efficiency point of view. This is why many researchers have been working for years in order to assess the most appropriate type of application [8–11]. Other types of similar systems, studied with the same approach, are those that focus more on evaluating the potential for residential cogeneration systems (SOFC-CHP) [12], those that evaluate coupling between SOFC and gas turbine of greater power (SOFC-GT) [13–15], or those that take into account molten carbonates fuel cell (MCFC) instead of SOFC [16–22]. In the same field specific SOFC-MTG can vary the nature of the fuel (which in this case is pure methane but which may be, e.g., synthesis gas or biomass) [14, 21–27] and there are also those that try, despite the fact that the state-of-the-art technology in question is not consolidated, a first economic evaluation [27, 28]. Even the same mathematical model can see different approaches. In this work, for example, a linear law is assumed as a function of temperature for the calculation of the current density of anodic and cathodic exchange, necessary for the calculation of activation losses, while in other works this is considered constant [23]. With regard to the current density limit, valid for the calculation of losses due to concentration, in this paper we consider the same for cathode and anode, while elsewhere, in spite of slight differences of the results, a diversification is operated [14]. Another approach, for the calculation of the cell voltage, is the use of a polynomial function of the current by which it calculates a reference voltage and the next calculation of the deviation from this voltage due to the temperature, the operating pressure, and the molars fractions of the different components [29]. In this case, from the definition of the limits of physical, chemical, and technological achievements of different components, the parameters "free" of the system and the range within which the values should vary were identified.

2. System Configurations

The system used as reference in this work is a hybrid system (MTG + SOFC) described in [3]. Here, in addition to the SOFC and MTG, there are a prereformer, an afterburner, and an ejector, in which methane is mixed with the gas recirculation. This type of system is capable of providing, at the design point, an electric power to the axis equal to 428 kW, divided into 319 kW produced by the fuel cell and 109 kW by gas turbine. The power spent by the two compressors (air and fuel) is, respectively, of 148 kW and 13 kW, for an efficiency of 0.61 and a TIT of 1240 K. Table 1 shows the main technical and thermodynamic data of the gas turbine and the cell and the meaning of both streams and blocks present in the following plant schemes. In this work, three different system configurations were analyzed, namely, the "base system," the "system with prereformer," and the "complete system." In both the second and the third configuration a component is added, with respect to the previous scheme. Compared to the system studied in [3], there is no ejector. The working pressure of the methane is assumed to be the same as that of the SOFC.

2.1. Base System. Figure 1 shows the diagram of the base system. The anode exhaust (in red) is divided into two parts: the left side exhaust is sent to the recirculation, while the right side flow enters the gas turbine. This stream finally unifies the cathode exhaust to make its entry in the expander. The exit gas from the turbine reaches the heat exchanger, where it provides thermal energy to preheat the air drawn from the MTG compressor. The "regenerated" air (in black) then enters the fuel cell from the cathode side. The recirculation is mixed with the pressurized methane (yellow) to enter then together the anode of the fuel cell (green).

2.2. System with Prereformer. The operation of the system with prereformer (Figure 2) differs, with respect to the hybrid system shown in Figure 1, only because of the presence of the prereformer. In the latter component, the mixture of methane and gas recirculation enters, so that a part of the methane is converted into hydrogen externally to the cell.

TABLE 1: Basic parameters of the components of the hybrid system and streams and blocks meaning in plant schemes.

Turbomachinery configuration	Radial
Compression ratio	3.9
Compressor isentropic efficiency	0.75
Expander isentropic efficiency	0.85
Air flow [kg/s]	0.808
MTG mechanical power [kW]	110
Rotation speed [rpm]	64000
Maximum turbine inlet temperature [K]	1250
Burner power [kW]	300
SOFC configuration	Tubular
Anode thickness [mm]	0.1
Cathode thickness [mm]	2.2
Interconnection thickness [mm]	0.085
Electrolyte thickness [mm]	0.04
Combustion efficiency	0.98
Heat exchanger efficiency	0.87
Current density (dfc) [A/m^2]	2000–3600
Fuel utilization (U_f)	0.7–0.9
Stream 1	Cold compressed air entering the heat exchanger
Stream 2	Hot compressed air entering the cathode
Stream 3	Uncompressed CH4
Stream 4	Compressed CH4
Stream 5	CH4 and anode outlet mixture entering the anode
Stream 6	Anode outlet
Stream 7	Anode outlet mixing with CH4
Stream 8	Anode outlet mixing with cathode outlet
Stream 9	Cathode outlet
Stream 10	Mixture entering the turbine
Stream 11	Turbine outlet entering the heat exchanger
Stream 12	Heat exchanger outlet
Stream 13	CH4 and anode outlet mixture entering the prereformer
Stream 14	Prereformer outlet entering the anode
Stream 15	Afterburner outlet entering the turbine
Stream 16	Mixer outlet
Block A	Air compressor
Block B	Turbine
Block C	Alternator
Block D	CH4 compressor
Block E	Heat exchanger
Block F	Inverter
Block G	Solid oxide fuel cell
Block H	Prereformer
Block I	Afterburner
Block J	Mixer

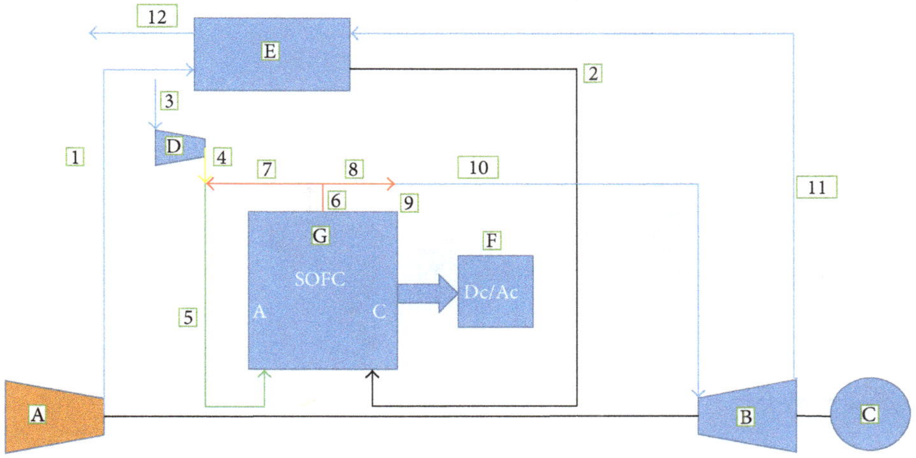

FIGURE 1: Technical scheme of the base system.

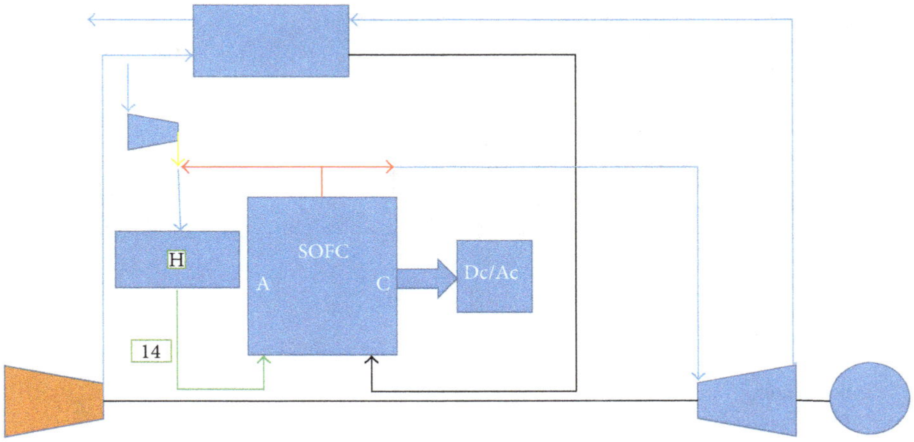

FIGURE 2: Technical scheme of the system with prereformer.

2.3. *Complete System*. In the case of complete system (Figure 3), the difference from the previous configuration is that the anode exhaust not sent to recirculation is sent downstream of SOFC, where an afterburner (or postcombustor) provides for oxidization of the hydrogen and carbon monoxide residues. The cathode exhaust provides the oxygen to the afterburner, while the products of the postcombustor constitute the working fluid of the expander of MTG. The rest of the system is entirely analogous to the previous configurations.

3. Mathematical Model

Here, the procedures used for the calculation of the main variables of the hybrid system as a function of U_f and dfc are described: recirculation flow, temperatures of mixing between recirculation and methane input anode cell operation, of afterburning, and of inlet air to the cathode, percentage of methane converted from the prereformer, and various performance parameters, that is, power output and

efficiency. *Calculation of flow recirculation*: we calculate the air recirculation rate according to

$$n_{\text{ric}} = \frac{n_{\text{H}_2\,\text{ric}}}{f_{\text{H}_2}}, \tag{1}$$

where the calculation of $n_{\text{H}_2\,\text{ric}}$ is done using

$$n_{\text{H}_2\,\text{ric}} = \frac{z/U_f - 3n_{\text{CH}_4}}{\left(1 + 0.3 f_{\text{CO}}/f_{\text{H}_2}\right)}, \tag{2}$$

where z is given by

$$z = \frac{\text{dfc} \cdot A_f}{2F}. \tag{3}$$

So, once U_f and dfc are assigned, the recirculation flow is therefore uniquely defined. *Calculation of the temperature of mixing between recirculation anodic and pressurized methane (T_{mix} block A in Figure 5)*: a mixture formed by gas recirculation and methane out from the compressor enters the fuel

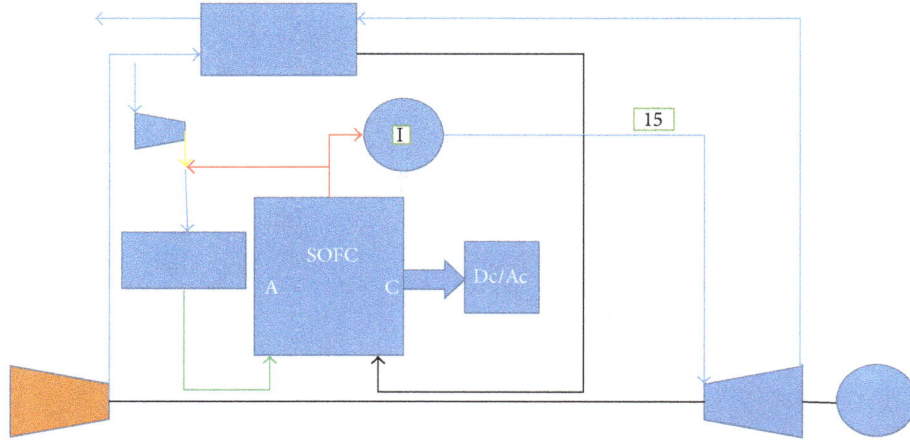

FIGURE 3: Technical scheme of complete system.

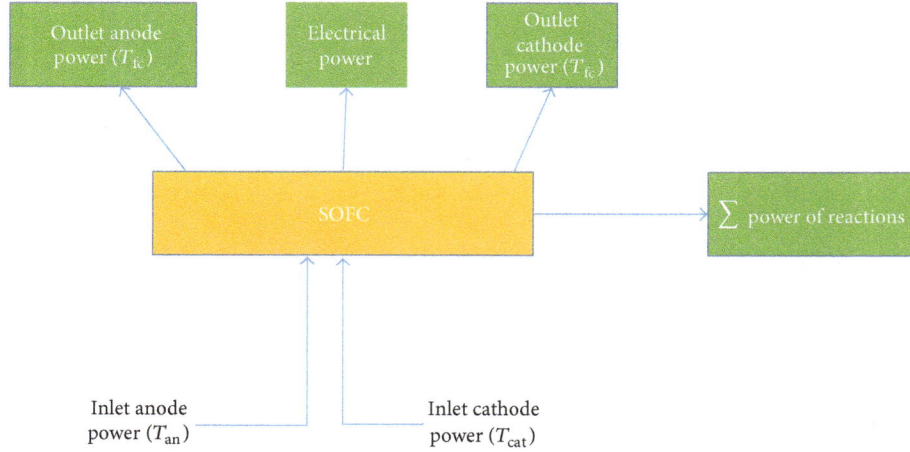

FIGURE 4: Flow input power and output from the fuel cell.

cell. The anode inlet temperature is calculated by attempts by the balance of thermal power expressed by

$$
\begin{aligned}
n_{CH_4} \cdot h_{CH_4}\,(600K) &+ n_{H_2O} \cdot h_{H_2O}\left(T_{fc}\right) + n_{H_2} \\
&\cdot h_{H_2}\left(T_{fc}\right) + n_{CO} \cdot h_{CO}\left(T_{fc}\right) + n_{CO_2} \cdot h_{CO_2}\left(T_{fc}\right) \\
&= n_{CH_4} \cdot h_{CH_4}\left(T_{mix}\right) + n_{H_2O} \cdot h_{H_2O}\left(T_{mix}\right) + n_{H_2} \\
&\cdot h_{H_2}\left(T_{mix}\right) + n_{CO} \cdot h_{CO}\left(T_{mix}\right) + n_{CO_2} \\
&\cdot h_{CO_2}\left(T_{mix}\right).
\end{aligned}
\tag{4}
$$

When (4) reaches convergence, the same T_{mix} represents the unknown searched. *Calculation of moles of methane converted from the prereformer (x, block A in Figure 5):* this parameter is obtained by attempts, starting from x of hypotheses to obtain the one that satisfies the power balance to prereformer. Inside the prereformer coupled reactions of steam reforming

reaction and that, coupled to it, of Water Gas Shift Reaction (WGSR) occur, expressed by

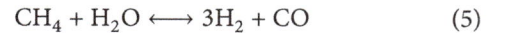

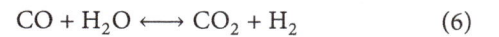

$$
CH_4 + H_2O \longleftrightarrow 3H_2 + CO \tag{5}
$$

$$
CO + H_2O \longleftrightarrow CO_2 + H_2 \tag{6}
$$

The ratio between (6) and (5) speed of reaction is unknown. We proceeded by calculating the speed of reactions for the operating temperatures of the SOFC and for different percentages of reforming, using as a parameter the convergence of equilibrium constants defined by

$$
K_{ref} = \frac{[H_2]^3 \cdot [CO]}{[CH_4] \cdot [H_2O]} P^2, \tag{7}
$$

$$
K_{shif} = \frac{[H_2] \cdot [CO_2]}{[CO] \cdot [H_2O]} \tag{8}
$$

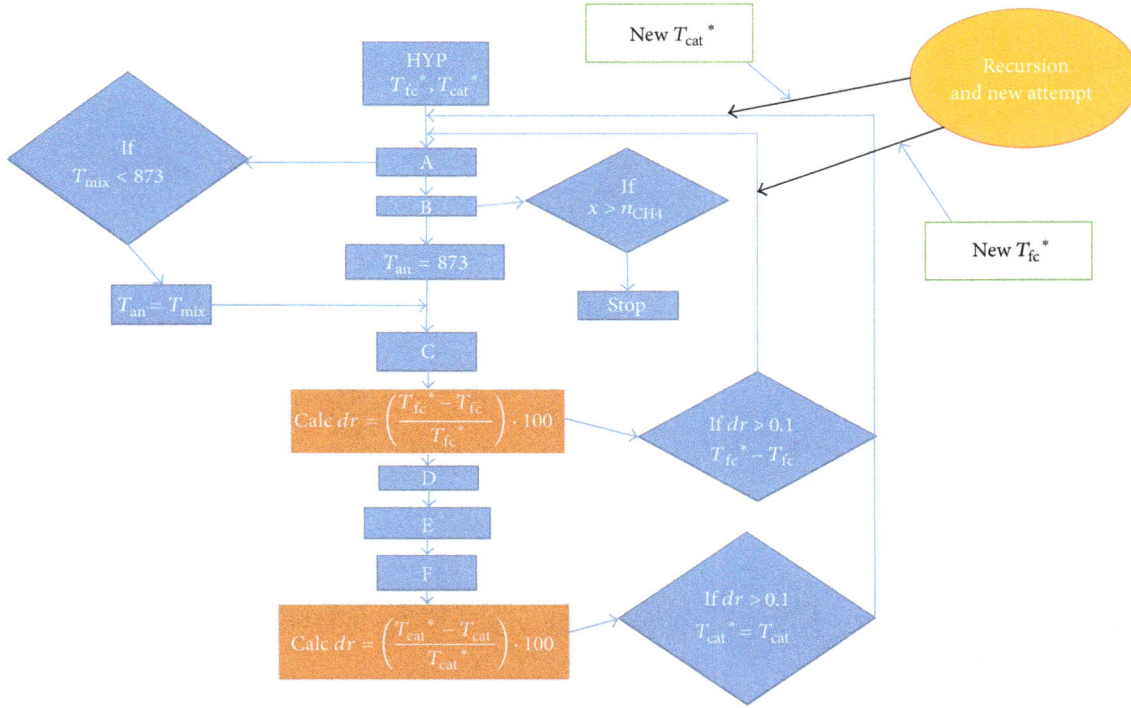

FIGURE 5: Flow diagram for the calculation of the parameters of the complete system.

and comparing (7) and (8) with the values of the equilibrium constants calculated as a function of the temperature according to

$$\log K = A \cdot T^4 + B \cdot T^3 + C \cdot T^2 + D \cdot T + E. \qquad (9)$$

Table 2 shows the values of the constants relating to (9) for the two reactions. The average ratio between speed of (6) and speed of (5) is equal on average to about 0.3. Equations (7) and (8) are taken from [31], while (9) is taken from [30]. Therefore, in the calculations of the mass balance and thermal power balance, we consider that, for each mole of CH_4 converted, 0.3 moles of H_2 is also generated from the conversion of CO produced by (6). In light of this approximation, the power balance to the prereformer appears to be expressed by

$$n_{CH_{4ip}} \cdot h_{CH_4}\left(T_{mix}\right) + n_{H_2O_{ip}} \cdot h_{H_2O}\left(T_{mix}\right) + n_{H_{2ip}}$$

$$\cdot \left(T_{mix}\right) + n_{CO_{ip}} \cdot h_{CO}\left(T_{mix}\right) + n_{CO_{2ip}}$$

$$\cdot h_{CO_2}\left(T_{mix}\right) - x \cdot \Delta h_{ref} - 0.3x \cdot \Delta h_{shif}$$

$$= \left(n_{CH_{4ip}} - x\right) \cdot h_{CH_4}(873) + \left(n_{H_2O_{ip}} - 1.3x\right) \qquad (10)$$

$$\cdot h_{H_2O}(873) + \left(n_{H_{2ip}} + 1.3x\right) \cdot (873)$$

$$+ \left(n_{CO_{ip}} + 0.7x\right) \cdot h_{CO}(873) + \left(n_{CO_{2ip}} + 0.3x\right)$$

$$\cdot h_{CO_2}(873).$$

TABLE 2: Values for the calculation of the equilibrium constant K [30].

	Reforming	Shifting
A	$2.6312 \cdot 10^{-11}$	$5.47 \cdot 10^{-12}$
B	$1.2406 \cdot 10^{-7}$	$-2.574 \cdot 10^{-8}$
C	$-2.2523 \cdot 10^{-4}$	$4.6374 \cdot 10^{-5}$
D	$5.12749 \cdot 10^{-1}$	$-3.915 \cdot 10^{-2}$
E	-66.139488	13.209723

Operating Temperature of the Cell (T_{fc}, Block C in Figure 5). An iterative method for the calculation of the operating temperature of the cell is used as well. We start from a temperature of attempt until the convergence of power balance is reached. This last is expressed by

$$n_{CH_4} \cdot h_{CH_4}\left(T_{an}\right) + n_{H_2O_{ifc}} \cdot h_{H_2O}\left(T_{an}\right) + n_{H_{2ifc}} \cdot \left(T_{an}\right)$$

$$+ n_{CO_{ifc}} \cdot h_{CO}\left(T_{an}\right) + n_{CO_{2ifc}} \cdot h_{CO_2}\left(T_{an}\right) + n_{O_{2ifc}}$$

$$\cdot h_{O_2}\left(T_{cat}\right) + n_{N_2} \cdot h_{N_2}\left(T_{cat}\right) - n_{CH_4} \cdot \Delta h_{ref}$$

$$- 0.3 n_{CH_4} \cdot \Delta h_{shif} - z \cdot \Delta h_{H_2O} \qquad (11)$$

$$= n_{H_2O_{ofc}} \cdot h_{H_2O}\left(T_{fc}\right) + n_{H_{2ofc}} \cdot \left(T_{fc}\right) + n_{CO_{ofc}}$$

$$\cdot h_{CO}\left(T_{fc}\right) + n_{CO_{2ofc}} \cdot h_{CO_2}\left(T_{fc}\right) + n_{O_{2ofc}}$$

$$\cdot h_{O_2}\left(T_{fc}\right) + n_{N_2} \cdot h_{N_2}\left(T_{fc}\right) + \dot{P}.$$

It is interesting to observe graphically the power flow of Figure 4, in which the contributions present in (11) are visible. It is now possible at this stage to calculate the power generated by the cell, through

$$\dot{P}_c = V \cdot I. \tag{12}$$

To calculate I the following equation is used:

$$I = \text{dfc} \cdot A_f \tag{13}$$

V is obtained by

$$V = V_0 - V_{\text{Nernst}} - V_{\text{att}} - V_{\text{ohm}} - V_{\text{conc}} \tag{14}$$

while (15) is used to calculate V_0:

$$V_0 = \frac{-\Delta G^0}{2F}. \tag{15}$$

V_{Nernst} is given by

$$V_{\text{Nernst}} = \frac{RT}{2F} \ln \left(\frac{f_{H_2O}}{f_{H_2} \cdot f_{O_2}^{0.5}} \cdot P^{0.5} \right). \tag{16}$$

V_{att} is provided by

$$V_{\text{act}} = V_{\text{act}_a} + V_{\text{act}_c}. \tag{17}$$

To calculate V_{act_a} we resort to

$$V_{\text{act}_a} = \left(\frac{RT}{F} \right) \sinh^{-1} \left(\frac{\text{dfc}}{2i0_a} \right). \tag{18}$$

Analogous calculation of V_{act_c} results by

$$V_{\text{act}_c} = \left(\frac{RT}{F} \right) \sinh^{-1} \left(\frac{\text{dfc}}{2i0_c} \right). \tag{19}$$

Once losses for activation have been defined, we calculate those for concentration V_{conc} by

$$V_{\text{conc}} = \left(\frac{RT}{F} \right) \ln \left(1 - \frac{\text{dfc}}{il} \right). \tag{20}$$

The voltage loss due to the ohmic resistance is obtained by

$$V_{\text{ohm}} = \text{dfc} \cdot \sum_{i=1}^{4} \frac{L_i}{\sigma_i}. \tag{21}$$

From (12) to (15) are taken from [3], whereas the equations in (16) to (21) are taken from [4]. *Calculation of afterburning temperature* (*block D in Figure 5*): once it has left the cell, the gas mixture reaches a postcombustor. Here, since the oxidation of both hydrogen and carbon monoxide still present in the anode exhaust, the temperature of the gas rises further. Then the following occur:

$$H_2 + \frac{1}{2}O_2 \longrightarrow H_2O \tag{22}$$

$$CO + \frac{1}{2}O_2 \longrightarrow CO_2 \tag{23}$$

By varying T_{pc} the balance of thermal power is solved, expressed by

$$
\begin{aligned}
&n_{H_{2\text{ipc}}} \cdot h_{H_2}(T_{\text{fc}}) + n_{H_2O_{\text{ipc}}} \cdot h_{H_2O}(T_{\text{fc}}) + n_{CO_{\text{ipc}}} \\
&\quad \cdot h_{CO}(T_{\text{fc}}) + n_{CO_{2\text{ipc}}} \cdot h_{CO_2}(T_{\text{fc}}) + n_{O_{2\text{ipc}}} \\
&\quad \cdot h_{O_2}(T_{\text{fc}}) + n_{N_{2\text{ipc}}} \cdot h_{N_2}(T_{\text{fc}}) \\
&\quad - \eta_{\text{comb}} \left[\left(n_{H_{2\text{ipc}}} \cdot \Delta h_{H_2O} \right) - \left(n_{CO_{\text{ipc}}} \cdot \Delta h_{CO_2} \right) \right] \\
&= n_{H_2O_{\text{opc}}} \cdot h_{H_2O}(T_{\text{pc}}) + n_{CO_{2\text{opc}}} \cdot h_{CO_2}(T_{\text{pc}}) \\
&\quad + n_{O_{2\text{opc}}} \cdot h_{O_2}(T_{\text{pc}}) + n_{N_{2\text{ipc}}} \cdot h_{N_2}(T_{\text{pc}}).
\end{aligned}
\tag{24}
$$

Once T_{pc} is known, which also corresponds to the TIT, since we know the isentropic efficiency of the expander MTG, we can easily calculate the temperature of the turbine outlet (T_{out}, *block E in Figure 5*) by

$$T_{\text{out}} = \text{TIT} - \eta_{\text{is}}(\text{TIT} - T_{\text{is}}). \tag{25}$$

Calculation of the inlet air temperature at the cathode (T_{cat}, block F in Figure 5): gas mixture, of known composition and temperature T_{out}, once expanded is sent to a countercurrent heat exchanger (or regenerator). Here, as the hot fluid and as the cold, respectively, the mixture under examination and the outlet air from the compressor (at a flow rate equal to 0.808 Kg/s and at a temperature of 404 K) enter. The balance equation of thermal power into the regenerator is expressed by

$$
\begin{aligned}
&n_{\text{air}} \cdot h_{\text{air}}(404\,\text{K}) + n_{H_2O} \cdot h_{H_2O}(T_{\text{out}}) + n_{H_2} \\
&\quad \cdot h_{H_2}(T_{\text{out}}) + n_{CO} \cdot h_{CO}(T_{\text{out}}) + n_{CO_2} \\
&\quad \cdot h_{CO_2}(T_{\text{out}}) + n_{O_2} \cdot h_{O_2}(T_{\text{out}}) + n_{N_2} \cdot h_{N_2}(T_{\text{out}}) \\
&= n_{\text{air}} \cdot h_{\text{air}}(T_{e1}) + n_{H_2O} \cdot h_{H_2O}(T_{e1}) + n_{H_2} \\
&\quad \cdot h_{H_2}(T_{e1}) + n_{CO} \cdot h_{CO}(T_{e1}) + n_{CO_2} \\
&\quad \cdot h_{CO_2}(T_{e1}) + n_{O_2} \cdot h_{O_2}(T_{e1}) + n_{N_2} \cdot h_{N_2}(T_{e1}).
\end{aligned}
\tag{26}
$$

It starts from T_{e1} attempted and the calculation is iterated until (26) is satisfied. T_{e1} represents the temperature at which the hot gases exiting the regenerator give part of their thermal power to cogeneration purposes. Once this first phase is completed, the calculation of T_{cat} through (27) is effected:

$$n_{\text{air}} \cdot h_{\text{air}}(T_{\text{cat}}) = n_{\text{air}} \cdot h_{\text{air}}(404\,\text{K}) + \dot{P}_{\text{term}}, \tag{27}$$

where

$$\dot{P}_{\text{term}} = 0.87 \cdot \dot{P}_{e1}. \tag{28}$$

Thus, after calculating T_{cat}, the circuit is completely defined. At the following iteration, this temperature T_{cat} is the input for the calculation of T_{fc}. The cycle continues until all the parameters arrive at convergence. Then, the evaluation of performance parameters is made, as follows.

Useful Power of the Cell Calculation ($\dot{P}_{uc}$). Consider

$$\dot{P}_{uc} = \eta_{inv} \cdot \dot{P}_{c}. \tag{29}$$

Cell Efficiency Calculation (η_c). Consider

$$\eta_c = \frac{\dot{P}_{uc}}{\dot{m}_{CH_4} \cdot LHV_{CH_4}}. \tag{30}$$

Gas Turbine Useful Power Calculation ($\dot{P}_{tg}$). Consider

$$\dot{P}_{tg} = \dot{P}_{ut} - \dot{P}_{ac}, \tag{31}$$

where

$$\dot{P}_{ut} = \dot{m}_t \cdot \Delta h. \tag{32}$$

Gas Turbine Efficiency Calculation (η_{tg}). Consider

$$\eta_{tg} = \frac{\dot{P}_{tg}}{\dot{m}_{CH_4} \cdot LHV_{CH_4}}. \tag{33}$$

The efficiency and useful power of the entire system are then calculated as the sum of efficiency and useful power of SOFC and MTG. We then calculate the cogeneration indices, that is, IRE,

$$IRE = 1 - \frac{\dot{P}_{tot}}{\dot{P}_{SI}/\eta_{SI} + \dot{P}_{cog}}, \tag{34}$$

and the thermal limit LT,

$$LT = \frac{\dot{P}_{cog} \cdot \eta_{term}}{\dot{P}_{tot}}. \tag{35}$$

It was assumed that the heat available downstream of the regenerator was transferable with an efficiency of 40% to a thermodynamic cycle downstream, to calculate the cogenerative values. The procedure described already is summarized in Figure 5.

4. Constraints Definition

"Setting" defines a given combination of parameters dfc and U_f with which it is possible to operate the hybrid system. Therefore, the set of all the possible settings by the range within which the parameters themselves can vary is defined (Table 1). By defining the constraints we proceed to identify the settings that are eligible for a given value of the flow of methane, so as to adequately assess the elasticity design connected to the same flow. *Steam to Carbon Ratio (STCR):* the lower limit of Steam to Carbon Ratio is the first restriction to be taken into account, defined as

$$STCR = \frac{n_{H_2O}}{n_{CO} + n_{CH_4}}. \tag{36}$$

The said parameter must remain above 2. In the event that this limit is not respected the humidification of the anode may not be satisfactory and it may cause cracking of both methane and carbon dioxide molecules, according to the reactions

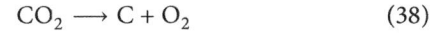

$$CH_4 \longrightarrow C + 2H_2 \tag{37}$$

$$CO_2 \longrightarrow C + O_2 \tag{38}$$

Consequently, we face the catalyst deactivation caused by the presence of carbonaceous deposits.

Constraint on Maximum Current Density. The methane is converted to hydrogen by (5) and (6). Given the assumptions previously made on the kinetics of these reactions, we have that the total conversion of one mole of methane per second gives rise to 3.3 moles of hydrogen per second. Simultaneously, according to (3), z moles of hydrogen per second is instead consumed. Thus, the consumption of hydrogen is directly proportional to the current density. Therefore, the settings that provide a value of z higher with respect to hydrogen product are considered ineligible.

Constraints on the Operating Temperature of the Fuel Cell (T_{fc}) and the Turbine Inlet Temperature (TIT). Constraints relating to temperature are the last to be taken into account. We excluded the settings that generate temperatures of the stack higher than typical operating temperatures of SOFCs and have turbine inlet temperatures above 1250 K (current technological limit of MTG). Thus, we summarize the conditions as follows:

(a) $873\,K < T_{fc} < 1200\,K$;

(b) $TIT < 1250\,K$.

5. Results

In this section, we proceed to the choice of the optimal configuration with which the hybrid system works and then to define the methane flow and the operative current density. The next step is sensitivity analysis of the main parameters at varying U_f, whereas at the end a first validation of the calculation model is operated. *Selection of the optimal configuration:* the optimum configuration is that of complete system. This choice stems from the following reasoning. According to (2), recirculation flow decreases at increasing U_f, and, the recirculation being at a temperature higher than compressed methane, this implies a lowering of T_{mix} which then propagates on all operating parameters of the plant, using as a parameter to control the fall percentage T_{fc} at varying U_f, defined by

fall of temperature [%]

$$= \frac{T_{fc}(U_f = i) - T_{fc}(U_f = i + 0,01)}{T_{fc}(U_f = i)}. \tag{39}$$

See Tables 3 and 4.

In the case of the base system, the fall of temperature is higher than 5%. This is considered excessive. By comparing

TABLE 3: Fall of temperature for increasing U_f for base system.

dfc [A/m^2]	U_f	T_{fc} [K]	Fall of temperature [%]
2500	0.81	963.9	
2500	0.82	913.7	5.21

TABLE 4: Fall of temperature for increasing U_f for system with prereformer.

dfc [A/m^2]	U_f	T_{fc} [K]	Fall of temperature [%]
2500	0.81	945.5	
2500	0.82	911.3	3.66

TABLE 5: Comparison between the gas turbine power plant with prereformer and complete system for dfc = 2500 A/m^2.

U_f	System with prereformer	Complete system
0.81	101.1	166.74
0.82	92.38	154.87
0.83	83.02	144.83

Tables 3 and 4 it is evident how, for homologous settings, in the case of system with prereformer the condition has improved. Having higher temperatures with lower values U_f implies that a significant part of the fuel is not properly used, a phenomenon that has an impact on the values of gas turbine power. Therefore, to remedy this gap is necessary to insert an afterburner downstream of the fuel cell, so that the configuration of the complete system becomes necessary. Table 5 shows how, for homologous conditions, the complete system ensures a significant increase of the gas turbine power.

Definition of Optimum Operating Conditions. The optimum operating conditions, that is, flow of methane and the current density to operate with, are chosen using design flexibility as the criterion. The model developed has been applied to the calculation of the conditions resulting from three different values of flow rate of methane, low ($\dot{m}_b$ = 0.012 kg/s), medium ($\dot{m}_m$ = 0.015 kg/s), and high ($\dot{m}_a$ = 0.018 kg/s). In the case of high flow rate of methane there is no setting compatible with all the constraints. In contrast, from a comparison between Tables 6(a) and 6(b), it is shown that the medium flow rate ensures greater design flexibility, thus resulting in a specific value (asterisks are the settings eligible). Table 6(b) shows how, for dfc = 2900 A/m^2, there is a greater choice of the possible settings that satisfy all the constraints outlined above, so that this value is identified as the operating current density and is used in the following sensitivity analysis.

Sensitivity of Operating Parameters and Performance at Varying U_f. A sensitivity analysis is performed to determine the effect of varying U_f on the operating parameters and performance. According to (2), the recirculation flow decreases at increasing U_f (first effect). Consequently, all operating temperatures of the plant should decrease. However, the decrease of recirculation flow implies a greater flow to the afterburner

TABLE 6: Plan of the possible settings for low and medium flow rate of methane.

(a)

U_f	dfc [A/m^2]									
	2300	2400	2500	2600	2700	2800	2900	3000	3100	3200
0.70										
0.71										
0.72										
0.73										
0.74		*								
0.75		*								
0.76		*								
0.77		*								
0.78		*								
0.79										
0.80										
0.81			*							
0.82			*							
0.83			*							
0.84			*							
0.85			*							
0.86			*							
0.87				*						
0.88				*						
0.89				*						
0.90				*						

(b)

U_f	dfc [A/m^2]									
	2300	2400	2500	2600	2700	2800	2900	3000	3100	3200
0.70						*				
0.71						*				
0.72						*				
0.73						*				
0.74						*				
0.75						*	*			
0.76							*			
0.77							*			
0.78							*			
0.79							*			
0.80							*			
0.81							*			
0.82							*	*		
0.83								*		
0.84								*		
0.85								*		
0.86								*		
0.87										
0.88									*	
0.89									*	
0.90									*	

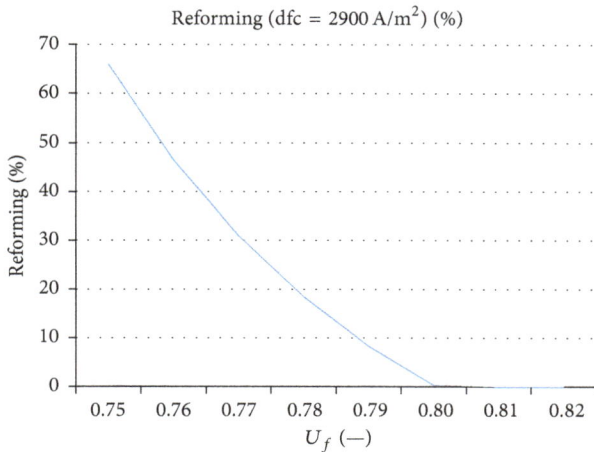

FIGURE 6: Percentage of prereforming at varying U_f.

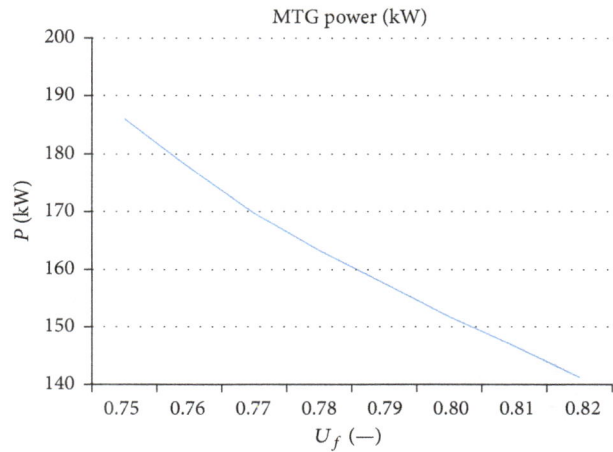

FIGURE 8: Turbogas power at varying U_f.

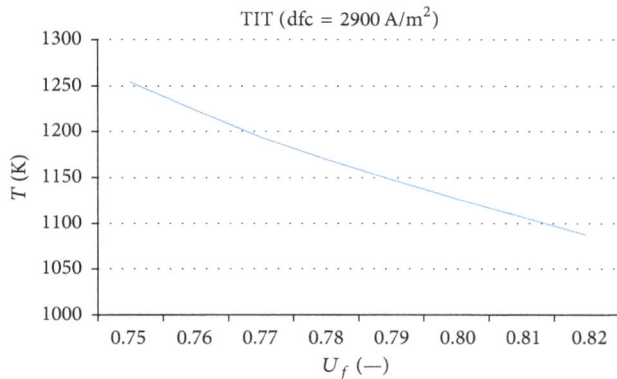

FIGURE 7: TIT at varying U_f.

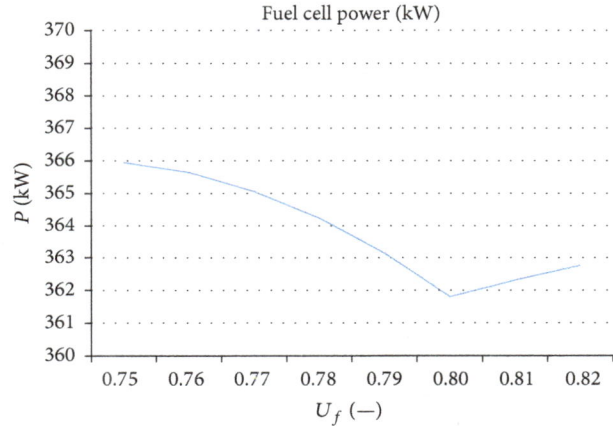

FIGURE 9: Fuel cell power at varying U_f.

as well (second effect), so that temperatures should increase. The first effect prevails on the second one. Therefore, the overall effect is a lowering of all operating temperatures of the hybrid system. Consequently, the temperature being lower, to keep the anode inlet temperature at the desired value, an inferior amount of methane flow has to be reformed before entering the cell. Thus, the percentage of reforming decreases, as Figure 6 shows. Figure 7, owing to the already described effects, shows how the temperature at the turbine inlet monotonically decreases and the turbogas power depending on the TIT (TIT decrease means a decreasing in Δh, thus a reduction in useful power, according to (31) and (32)); this means also a decreasing in terms of MTG power, as one can observe in Figure 8. Instead, a nonmonotonous trend is that concerning the power of the cell. In fact, this is affected, for low values of U_f, by a prereforming effect, which changes the composition in the anode input (reactions (5) and (6)). Therefore, according to (16), the composition change means that the percentage of reforming decreases, while Nernst-losses increase, causing an overall power decrease in the stack. Therefore, when it is no longer necessary to reform, the Nernst-loss decreases with decreasing temperature, so that the power of the cell starts growing (Figure 9). Finally,

it is interesting to note that, with increasing U_f, while overall performance parameters decrease, there is an increase in the index IRE (Figure 10), whereas the thermal limit remains nearly constant.

6. Discussion

6.1. First Validation of the Calculation Model. A first testing of the model calculation was carried out, both of a qualitative and of a quantitative nature. The "trend" of some fundamental parameters with respect to developments known from the literature was evaluated, and the results obtained here were compared with those calculated in [3].

6.2. Qualitative Validation. First, for purpose of qualitative model validation, the data obtained were compared, for the same U_f, for different values of dfc. As we expected, Table 7 shows that an increase of the current density causes an increase of the operating temperature of the hybrid system and, consequently, an increase in the percentage of methane on which it performs the prereforming. Table 8 shows that,

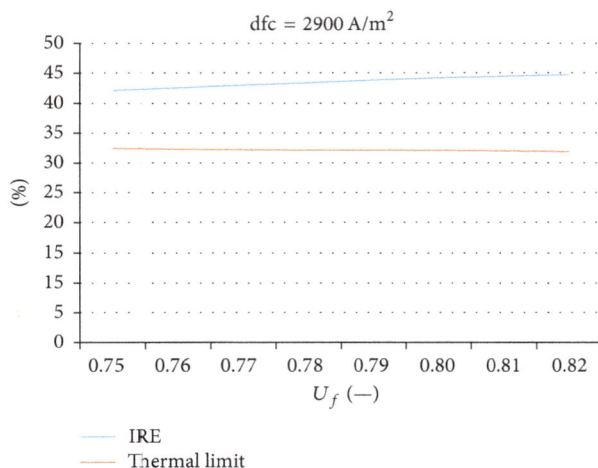

FIGURE 10: IRE (blue) and LT (red) at varying U_f.

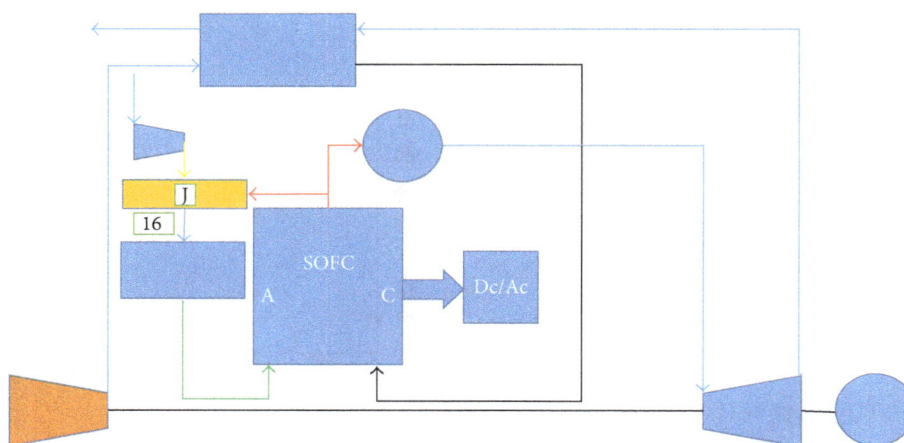

FIGURE 11: Hybrid system studied in [3].

with the increase of dfc, both the cell (despite an increase in voltage losses) and the gas turbine power rise, the second being directly dependent on the turbine inlet temperature.

6.3. Quantitative Validation. To end the first validation process, the model was applied to the system of Figure 15, studied in [3], and results were compared. In [3] the methane is compressed to 30 bars instead of the operating pressure of the MTG and then joined in a mixer and blend, with associated losses, from the anode recirculation. The mixer is the only difference compared to the complete system. The thermodynamic modeling of the mixer and of the ejector inside it would be very complex. In homologous conditions, the results turn out better for the complete system (consistent with the physical principles). Thus, one objective was to evaluate, in a first approximation, how the ejector affects the losses, using equivalent useful area as a parameter. This is defined as the percentage of usable area of Figure 3 hybrid system, compared with that of Figure 11 (without ejector), such that, in homologous operating conditions, both systems produce the same power. The results are as shown in Table 9.

It is seen that when the area is reduced up to 85% of the given "plate," the relative difference between the reference data and the data provided by the model remains around 1%, thus lending credibility to the mathematical model described in this paper.

7. Conclusions

The objective set at the beginning was to define the optimal conditions of operation of the hybrid system by developing a C++ code and to evaluate the suitability of this approach with the physical and chemical process present inside the SOFC-MTG plant. In the first instance we see that the optimal configuration of the hybrid system is that of the complete system. This ensures both a satisfactory temperature management and good values of gas turbine power. The flow rate of methane is excellent, given the guaranteed, high design flexibility, which is defined as $\dot{m}_m$, that is, 0.015 kg/s. For the said value of the flow rate of methane, current density that ensures the best compromise between performance and degrees of freedom to the designer (varying U_f eligible

TABLE 7: Operating parameters in equal value U_f for different dfc.

dfc [A/m^2]	U_f	%recirculation	T_{mix} [K]	%reforming	T_{an} [K]	T_{cat} [K]	T_{fc} [K]	TIT	T_{out} [K]
2800	0.75	83.43	823.3	0	823.3	619.71	851.73	1088.27	821.46
2900	0.75	86.17	1007.15	66.07	873	683.74	1050	1254.57	947.45
2900	0.82	77.24	831.16	0	831.16	620.1	874.53	1087.75	821.47
3000	0.82	81.69	995.97	45.59	873	676.38	1052	1235.74	933.23

TABLE 8: Performance parameters for different dfc.

dfc [A/m^2]	U_f	Voltage [V]	Voltage losses [V]	$\dot{P}_{uc}$ [kW]	$\dot{P}_{tg}$ [kW]	$\dot{P}_{tot}$ [kW]
2900	0.82	0.6788	0.5058	362.78	141.28	504.06
3000	0.82	0.6619	0.5227	365.96	180.78	546.74

TABLE 9: Comparison of the data obtained with the model and experimental data studied by evaluating an equivalent useful area equal to 85% of the effective area (dfc = 3200 A/m^2).

	Model data	Reference data	Relative difference [%]
Hybrid system power [kW]	431.47	428	0.80
Fuel cell power [kW]	320.52	319	0.47
Gas turbine power [kW]	110.95	109	1.76
Hybrid system efficiency [kW]	0.62	0.61	
Fuel cell efficiency [kW]	0.46	0.45	
Gas turbine efficiency [kW]	0.16	0.16	

between 0.75 and 0.82) is that of 2900 A/m^2. The last step is the choice of operating U_f, which may vary depending on the objective it set out: choosing a low U_f if there is directed towards energy optimization, U_f high if the goal is to maximize the cogeneration yield, and a medium U_f if seeking a compromise between the two requirements. Since systems of this type are still under study, of the 3 options described above, at the current state of the art, it seems to make sense to focus on energy optimization, and when consolidated on the market, there will be consideration later with the economic scenario of the moment. This factor is closely related to the evaluation of the investment from the perspective of cogeneration. The developed C++ code matches with both the state of the art and reference data taken from the literature, suggesting the suitability of this approach to evaluate and describe SOFC-MTG and other kinds of plants.

Nomenclature

A_f: Useful area of the fuel cell [m^2]

c_p: Specific heat at constant pressure [J/(mol·K)]

c_{p_m}: Average specific heat of the mixture in the course of expansion [J/(kg·K)]

c_v: Specific heat at constant volume [J/(mol·K)]

dfc: Current density with which it operates within the fuel cell [A/m^2]

F: Faraday constant, that is, 96485 [C/mol]

f_{CO}: Molar fraction of carbon monoxide, dimensionless

f_{H_2}: Molar fraction of hydrogen, dimensionless

h_{air}: Molar enthalpy of the air [J/mol]

h_{CH_4}: Molar enthalpy of methane [J/mol]

h_{CO}: Molar enthalpy of carbon monoxide [J/mol]

h_{CO_2}: Molar enthalpy of carbon dioxide [J/mol]

h_{H_2}: Molar enthalpy of hydrogen [J/mol]

h_{H_2O}: Molar enthalpy of the water vapor [J/mol]

h_{O_2}: Molar enthalpy of oxygen [J/mol]

h_{N_2}: Molar enthalpy of nitrogen [J/mol]

I: Operation current [A]

IRE: "Energy saving index," dimensionless

$i0_a$: Current density exchange anode side [A/m^2]

$i0_c$: Current density exchange cathode side [A/m^2]

il: Limit current density [A/m^2]

K_{ref}: Equilibrium constant of the reaction of steam reforming, dimensionless

K_{shif}: Equilibrium constant of the reaction of Water Gas Shift Reaction, dimensionless

$\dot{m}$: Air mass flow rate [kg/s]

$\dot{m}_{CH_4}$: Methane mass flow rate [kg/s]

$\dot{m}_t$: Mass flow rate in the expander [kg/s]

n_{CH_4}: Methane molar flow [mol/s]

n_{CO}: Carbon monoxide molar flow rate [mol/s]

n_{H_2}: Hydrogen molar flow rate [mol/s]

n_{H_2O}: Steam molar flow rate [mol/s]

$\dot{P}$: Electric power obtained through the electrochemical reaction of water formation [W]

$\dot{P}_{ac}$: Power absorbed by the compressor gas turbine system [W]

$\dot{P}_c$:   Power generated by the cell [W]

$\dot{P}_{cog}$:   Cogeneration power transmitted to the thermodynamic cycle placed downstream of the hybrid system [W]

$\dot{P}_{e1}$:   Thermal power transferred to air in the event that the regenerator has efficiency 1 [W]

$\dot{P}_{SI}$:   Hybrid system power [W]

$\dot{P}_{term}$:   Thermal power transferred to air [W]

$\dot{P}_{tg}$:   Gas turbine useful power [W]

$\dot{P}_{tot}$:   Total power supplied by the hybrid system [W]

P_{uc}:   Useful power generated by the cell [W]

$\dot{P}_{ut}$:   Gas turbine expander useful power [W]

R: Universal gas constant 8.314 [J/(mol·K)]

STCR: Steam to Carbon Ratio, dimensionless

T_{an}:   Anode inlet temperature [K]

T_{cat}:   Cathode inlet temperature [K]

T_{e1}:   Temperature efficiency 1 [K]

T_{fc}:   Operating temperature of the cell [K]

T_{is}:   Isentropic temperature of the turbine outlet [K]

T_{mix}:   Temperature mixing recirculation-methane [K]

T_{out}:   Turbine outlet temperature [K]

T_{pc}:   Afterburning temperature [K]

TIT: Turbine inlet temperature [K]

U_f: Fuel utilization factor, dimensionless

V: Cell operating voltage [V]

V_0:   Maximum voltage obtained in standard conditions, at a pressure of 1 atm and at a temperature of 25°C [V]

V_{att}:   Voltage activation losses [V]

V_{act_a}:   Voltage activation losses anode side [V]

V_{act_c}:   Voltage activation losses cathode side [V]

V_{conc}:   Voltage concentration losses [V]

V_{Nernst}:   Nernst-loss [V]

V_{ohm}:   Voltage ohmic losses [V]

z: Number of moles of hydrogen which react in a second inside the fuel cell [mol/s].

Greek Alphabet

β: Compression ratio, dimensionless

ΔG^0:   Variation in Gibbs free energy in formation water reaction −228600 [J/mol]

Δh_{CO_2}:   Standard enthalpy of formation of carbon monoxide oxidation reaction [J/mol]

Δh_{H_2O}:   Enthalpy of formation of electrochemical water formation reaction [J/mol]

Δh_{ref}:   Enthalpy of formation in reforming reaction [J/mol]

Δh_{shif}:   Enthalpy of formation in shifting reaction [J/mol]

η_c: Cell efficiency, dimensionless

η_{comb}:   Combustion efficiency, dimensionless

η_{inv}:   Inverter conversion efficiency, dimensionless

$\eta_{is,c}$:   Isentropic efficiency of the compressor, dimensionless

$\eta_{is,t}$:   Isentropic efficiency of the turbine, dimensionless

η_{SI}:   Hybrid system efficiency, dimensionless

η_{term}:   Thermal efficiency, dimensionless

η_{tg}:   Gas turbine efficiency, dimensionless

σ_a: Anode resistivity [Ω^{-1}·mm]

σ_c: Cathode resistivity [Ω^{-1}·mm]

σ_e: Electrolyte resistivity [Ω^{-1}·mm]

σ_i: Interconnection resistivity [Ω^{-1}·mm].

Subscripts

ifc: Fuel cell inlet

ip: Prereformer inlet

ipc: Afterburner inlet

it: Turbine inlet

ofc: Fuel cell outlet

opc: Afterburner outlet

ric: Recirculation.

Conflict of Interests

The authors declare that there is no conflict of interests regarding the publication of this paper.

References

[1] A. Demirbas, "Fuel cells as clean energy converters," *Energy Sources Part A: Recovery, Utilization, and Environmental Effects*, vol. 29, no. 2, pp. 185–191, 2007.

[2] Z. Ziaka and S. Vasileiadis, "Pretreated landfill gas conversion process via a catalytic membrane reactor for renewable combined fuel cell-power generation," *Journal of Renewable Energy*, vol. 2013, Article ID 209364, 8 pages, 2013.

[3] A. Pontecorvo, R. Tuccillo, and F. Bozza, *Studio di una microturbina a gas per sistemi cogenerativi ed ibridi [Ph.D. thesis]*, Università degli Studi di Napoli Federico II, Napoli, Italy, 2010.

[4] F. Bozza, M. C. Cameretti, and R. Tuccillo, "Adapting the micro-gas turbine operation to variable thermal and electrical requirements," ASME Paper 2003-GT-38652, 2003.

[5] F. Bozza and R. Tuccillo, "Transient operation analysis of a cogenerating micro-gas turbine," ASME Paper ESDA 2004-58079, 2004.

[6] M. C. Cameretti and R. Tuccillo, "Comparing different solutions for the micro-gas turbine combustor," ASME Paper 2004-GT-53286, 2004.

[7] R. Tuccillo, "Performance and transient behaviour of MTG based energy systems," Tech. Rep. RTO-MP-AVT-131 VKI/LS, Micro Gas Turbines, 2005.

[8] S. H. Chan, H. K. Ho, and Y. Tian, "Modelling of simple hybrid solid oxide fuel cell and gas turbine power plant," *Journal of Power Sources*, vol. 109, no. 1, pp. 111–120, 2002.

[9] S. K. Nayak and D. N. Gaonkar, "Modeling and performance analysis of microturbine generation system in grid connected/islanding operation," *Journal of Renewable Energy*, vol. 2, no. 4, pp. 750–757, 2012.

[10] C. Stiller, B. Thorud, and O. Bolland, "Safe dynamic operation of a simple SOFC/GT hybrid system," ASME Paper 2005-GT-68481, ASME, 2005.

[11] S. H. Chan, H. K. Ho, and Y. Tian, "Multi-level modeling of SOFC–gas turbine hybrid system," *International Journal of Hydrogen Energy*, vol. 28, no. 8, pp. 889–900, 2003.

[12] L. Barelli, G. Bidini, F. Gallorini, and P. A. Ottaviano, "Design optimization of a SOFC-based CHP system through dynamic analysis," *International Journal of Hydrogen Energy*, vol. 38, no. 1, pp. 354–369, 2013.

[13] H.-W. D. Chiang, C.-N. Hsu, W.-B. Huang, C.-H. Lee, W.-P. Huang, and W.-T. Hong, "Design and performance study of a solid oxide fuel cell and gas turbine hybrid system applied in combined cooling, heating, and power system," *Journal of Energy Engineering*, vol. 138, no. 4, pp. 205–214, 2012.

[14] L. Barelli, G. Bidini, and P. A. Ottaviano, "Part load operation of SOFC/GT hybrid systems: stationary analysis," *International Journal of Hydrogen Energy*, vol. 37, no. 21, pp. 16140–16150, 2012.

[15] P. Chinda and P. Brault, "The hybrid solid oxide fuel cell (SOFC) and gas turbine (GT) systems steady state modeling," *International Journal of Hydrogen Energy*, vol. 37, no. 11, pp. 9237–9248, 2012.

[16] X. Zhang, J. Guo, and J. Chen, "Influence of multiple irreversible losses on the performance of a molten carbonate fuel cell-gas turbine hybrid system," *International Journal of Hydrogen Energy*, vol. 37, no. 10, pp. 8664–8671, 2012.

[17] L. Leto, C. Dispenza, A. Moreno, and A. Calabrò, "Simulation model of a molten carbonate fuel cell-microturbine hybrid system," *Applied Thermal Engineering*, vol. 31, no. 6-7, pp. 1263–1271, 2011.

[18] O. Corigliano, G. Florio, and P. Fragiacomo, "A numerical simulation model of high temperature fuel cells fed by biogas," *Energy Sources Part A: Recovery, Utilization and Environmental Effects*, vol. 34, no. 2, pp. 101–110, 2011.

[19] G. De Lorenzo and P. Fragiacomo, "Technical analysis of an eco-friendly hybrid plant with a microgas turbine and an MCFC system," *Fuel Cells*, vol. 10, no. 1, pp. 194–208, 2010.

[20] G. De Lorenzo and P. Fragiacomo, "A methodology for improving the performance of molten carbonate fuel cell/gas turbine hybrid systems," *International Journal of Energy Research*, vol. 36, no. 1, pp. 96–110, 2012.

[21] S. Wongchanpai, H. I. Wai, M. Saito, and H. Yoshida, "Performance evaluation of a direct biogas solid oxide fuel cell—micro gas turbine (SOFC-MTG) hybrid combined heat and power (CHP) system," *Journal of Power Sources*, vol. 223, pp. 9–17, 2013.

[22] R. Toonssen, S. Sollai, P. V. Aravind, N. Woudstra, and A. H. M. Verkooijen, "Alternative system designs of biomass gasification SOFC/GT hybrid systems," *International Journal of Hydrogen Energy*, vol. 36, no. 16, pp. 10414–10425, 2011.

[23] Y. Zhao, J. Sadhukhan, A. Lanzini, N. Brandon, and N. Shah, "Optimal integration strategies for a syngas fuelled SOFC and gas turbine hybrid," *Journal of Power Sources*, vol. 196, no. 22, pp. 9516–9527, 2011.

[24] P. V. Aravind, C. Schilt, B. Türker, and T. Woudstra, "Thermodynamic model of a very high efficiency power plant based on a biomass gasifier, SOFCs, and a gas turbine," *International Journal of Renewable Energy Development*, vol. 1, no. 2, pp. 51–55, 2012.

[25] C. Bang-Møller and M. Rokni, "Thermodynamic performance study of biomass gasification, solid oxide fuel cell and micro gas turbine hybrid systems," *Energy Conversion and Management*, vol. 51, no. 11, pp. 2330–2339, 2010.

[26] C. Bao, N. Cai, and E. Croiset, "A multi-level simulation platform of natural gas internal reforming solid oxide fuel cell-gas turbine hybrid generation system—part II. Balancing units model library and system simulation," *Journal of Power Sources*, vol. 196, no. 20, pp. 8424–8434, 2011.

[27] S. Douvartzides and P. Tsiakaras, "Thermodynamic and economic analysis of a steam reformer-solid oxide fuel cell system fed by natural gas and ethanol," *Energy Sources*, vol. 24, no. 4, pp. 365–373, 2002.

[28] D. F. Cheddie and R. Murray, "Thermo-economic modeling of a solid oxide fuel cell/gas turbine power plant with semi-direct coupling and anode recycling," *International Journal of Hydrogen Energy*, vol. 35, no. 20, pp. 11208–11215, 2010.

[29] Y. Zhao, N. Shah, and N. Brandon, "Comparison between two optimization strategies for solid oxide fuel cell-gas turbine hybrid cycles," *International Journal of Hydrogen Energy*, vol. 36, no. 16, pp. 10235–10246, 2011.

[30] U. G. Bossel and B. C. H. Swiss, *Final Report on SOFC Data facts and Figures*, Federal Office of Energy, 1992.

[31] O. Levenspiel, *Ingegneria delle reazioni chimiche*, Casa Editrice Ambrosiana, Milano, Italy, 1972.

11

Inherent Difference in Saliency for Generators with Different PM Materials

Sandra Eriksson

Division for Electricity, Department of Engineering Sciences, Uppsala University, P.O. Box 534, 751 21 Uppsala, Sweden

Correspondence should be addressed to Sandra Eriksson; sandra.eriksson@angstrom.uu.se

Academic Editor: Shuhui Li

The inherent differences between salient and nonsalient electrical machines are evaluated for two permanent magnet generators with different configurations. The neodymium based (NdFeB) permanent magnets (PMs) in a generator are substituted with ferrite magnets and the characteristics of the NdFeB generator and the ferrite generator are compared through FEM simulations. The NdFeB generator is a nonsalient generator, whereas the ferrite machine is a salient-pole generator, with small saliency. The two generators have almost identical properties at rated load operation. However, at overload the behaviour differs between the two generators. The salient-pole, ferrite generator has lower maximum torque than the NdFeB generator and a larger voltage drop at high current. It is concluded that, for applications where overload capability is important, saliency must be considered and the generator design adapted according to the behaviour at overload operation. Furthermore, if the maximum torque is the design criteria, additional PM mass will be required for the salient-pole machine.

1. Introduction

The material most commonly used as permanent magnets (PMs) in electrical machines is neodymium-iron-boron, $Nd_2Fe_{14}B$ (shortened as NdFeB). During the last five years the price for NdFeB has increased and fluctuated extremely. In addition, 97% of all mining currently occurs in China and the export quotas for rare-earth metals from China are politically controlled [1]. Furthermore, the environmental aspects of rare-earth metal are an issue. Therefore, the idea to substitute NdFeB with ferrites has been suggested [2] and tested [3]. Several theoretical comparisons of generators with NdFeB and ferrites for wind power generators have previously been performed [2, 4–6]. In most of these studies [4–6] the better magnetic performance of NdFeB is used as an argument for its advantage. However, when the availability and unstable price development for NdFeB are considered, ferrites are considered superior, especially for applications where the weight increase is of less importance [2]. NdFeB generators have been extensively studied [7, 8]. In wind turbines, variable speed operation is increasingly popular. Therefore, the behaviour of a permanent magnet synchronous generator (PMSG) with NdFeB at variable speed and load has been

especially evaluated [9]. When substituting NdFeB with ferrites, very similar characteristics can be found at the nominal operating point [2]. However, the intrinsic difference between these two generator types was not considered in [2]; that is, the difference in saliency which may affect the variable load operation and especially overload behaviour.

A typical NdFeB generator is of round rotor type (also called cylindrical or nonsalient), with magnets surface mounted on a cylindrical iron core. Ferrites have much lower remanence than NdFeB. Therefore, the proposed ferrite generator has tangentially magnetized magnets mounted between magnetic pole shoes reinforcing the field and is thereby a salient-pole machine. The difference in saliency makes these two generator types behave differently at varying speed and load. Therefore, the whole operating range must be considered for a complete comparison between these two generators. In this paper, general machine characteristics are compared for the two generators from [2], over the whole operating range with focus on overload operation.

The NdFeB generator was designed to be used with a vertical axis wind turbine (VAWT) [10–12], even though the comparative study presented here is applicable to other types of PM machines. For this application, a high maximum

TABLE 1: Properties for different PM materials at 20°C. Data for NdFeB are taken from [2]. Data for ferrites are taken from a supplier (E-magnets UK). Costs are taken from [2] and have been updated by using the current metal price. The Y30 is used in the machine presented here. The Y40 is included for reference.

Property	NdFeB (N40H)	Ferrite (Y30)	Ferrite (Y40)
Remanence, Br (T)	1.29	0.38	0.45
Normal coercivity, Hc, (kA/m)	915	240	342
Intrinsic coercivity, Hci (kA/m)	1353	245	350
Energy product (kJ/m^3)	318	28.3	39.7
Density (kg/m^3)	7700	4700	5000
Energy density (J/kg)	41.3	6.0	7.9
Relative required mass	1	6.9	5.2
PM cost (EUR/kg)	29	1	1
Relative magnetic energy cost per generator	1	0.24	0.18

torque at overload operation is required, since the wind turbine is fully electrically regulated eliminating the need for a pitch-mechanism on the turbine blades [13]. The maximum torque capability of the generator is thereby an important design feature, and one of the objectives of this study was to investigate the difference in torque at overload operation. To ensure high overload capacity the generators were designed with unusually low load angle, which makes the generators quite unique and their overload behaviour especially interesting to evaluate. A low load angle enables the use of a simple, robust passive rectifier. The electrical system of the VAWT consists of a diode rectifier and an IGBT inverter so the generator is operated at unity power factor.

The saliency of PM machines has previously been discussed and considered concerning motor control [14, 15]. Here, saliency is considered in the area of substituting rare-earth metal based PMs with alternative materials, with the desire of keeping the same electromagnetic properties of the machine. In order to compare machines with different PM material a common reference point is needed. In this paper, it is evaluated whether this is at all possible, by investigating two PM machines with inherently different electromagnetic properties, that is, difference in saliency.

2. Materials and Methods

2.1. Magnetic Materials.
Most PMs used in electrical machines today are made of NdFeB. In Table 1 magnetic properties of NdFeB are compared with two different grades of ferrite magnets. New ferrite grades have been developed since the ferrite generator considered here was designed. The presented generator has a ferrite magnet of grade Y30, whereas the new ferrite grade Y40 has a 40% higher energy product. Therefore, a machine with Y40 magnets would have a lighter rotor with less magnet volume.

As seen in Table 1, the required mass for ferrite magnets is much higher than that for NdFeB magnets. However, the cost for the PMs, when substituting NdFeB with ferrite, is only one

fourth to one fifth; that is, if the larger mass can be accepted a large cost reduction can be expected.

2.2. Saliency of PM Machines.
Saliency is a measure of the reluctance difference between the rotor and the stator around the circumference of the rotor. A generator with tangentially magnetized ferrite magnets placed between magnetically conducting pole shoes is a type of salient-pole machine and has different reluctance and inductance on the pole (direct axis inductance L_d) and between two poles (quadrature axis inductance L_q). A generator with surface-mounted PMs on a cylinder is a type of round rotor, which is defined as a rotor with equal reluctance and inductance all around the rotor ($L_d = L_q$). For a round rotor generator, the power output, P_o, if the inner resistance is omitted, can be found from

$$P_o = \frac{3V_a E_a \sin\delta}{X_s},\tag{1}$$

where V_a is the terminal phase voltage, E_a is the no-load phase voltage, δ is the load angle, and X_s is the synchronous reactance, which is equal to the direct axis reactance. For a salient-pole machine, omitting the inner resistance, the power is

$$P_o = \frac{3V_a E_a \sin\delta}{X_d} + \frac{3(X_d - X_q)}{2X_d X_q}V_a^2 \sin(2\delta),\tag{2}$$

where X_d is the direct axis reactance and X_q is the quadrature axis reactance. The second term in (2) is called the reluctance term and it has larger influence if the difference between L_d and L_q is large. The power equations show that the power will vary differently with the loading of the machine depending on the saliency.

For a generator connected to a resistive load or a passive rectifier (diodes) the power factor is unity. For a machine run with unity power factor, the voltage will drop with increasing current. For a round rotor generator with unity power factor and omitted resistance, the output voltage is

$$V_a = E_a \cos\delta,\tag{3}$$

so the voltage drops with increasing load angle. The corresponding voltage drop for the salient-pole machine with unity power factor and omitted resistance is

$$V_a = E_a \cos\delta - I_d(X_d - X_q)\cos\delta,\tag{4}$$

where I_d is the direct axis current. The voltage drops faster for the salient-pole machine (as long as $(X_d - X_q) > 1$) as is seen when comparing (3) and (4).

Inserting (3) in (1) gives the power output for the round rotor generator with diode rectification:

$$P_{o,\text{diode}} = \frac{3E_a^2 \cos\delta \sin\delta}{X_s} = \frac{3E_a^2 \sin(2\delta)}{2X_s}.\tag{5}$$

Equation (5) shows that, for a diode rectified round rotor generator, the maximum power is half the value and occurs at half the load angle compared to a generator where the output voltage equals the internal voltage, which can be achieved by boosting the generator voltage with an active rectifier. For the latter type of operation, the maximum power occurs at about 90° load angle. For a diode rectified generator, the maximum power occurs around 45° due to the internal voltage drop. Similar behaviour is expected for the salient-pole generator when operated at unity power factor.

2.3. Simulations. The results presented here are from simulations using a two-dimensional model solved with the finite element method (FEM) [16]. Circuit theory, as (1), (2), (3), (4), and (5) are based on, have simplifications and therefore field based simulations are performed to consider all aspects of material and geometry in machine design. The simulation model is a combined field and circuit model solved through time-stepping. The generator characteristics were derived by stationary simulations during an iterative design process and the behaviour at different operating points was found from dynamic time-dependent simulations. The simulation model has been described more thoroughly and verified with experiments for a 12 kW generator and for the 225 kW NdFeB generator in [17] and [12], respectively. In the simulations the generator is connected to a purely resistive load; that is, the power factor is unity.

2.4. The Generators. The NdFeB based generator, G1-Nd, was tested and built in 2010 [12]. The ferrite generator, G2-Fe, with a different rotor but exactly the same stator, was designed in 2012 to have similar electromagnetic properties as G1-Nd [2]. The most important generator properties and a comparison at rated operation can be found in Table 2 and in [2]. Figure 1 shows the different geometry of the two generators as well as results from FEM simulations showing the magnetic flux density in the generators.

3. Results and Discussion

The torque at different load angle for the two machines is compared in Figure 2. At part load operation, the generators have similar behaviour and their rated properties are almost identical, as shown in Table 2. However, at overload operation differences occur. The NdFeB generator has higher overload capacity. The maximum torque for the ferrite machine is only 89% of the maximum torque of the NdFeB machine. The small saliency of the ferrite machine affects the shape of the torque curve and the overload capability even though the differences between the two machines are small; see (1) and (2). Equation (2) shows that the power corresponding to the reluctance torque (the second term) varies with the load angle. For the salient-pole generator considered here, the reluctance term has largest influence at load angles around 20 to 30 degrees and decreases at higher load angles. The reluctance torque will thereby decrease the maximum torque slightly and will make the power curve steeper at low load angles. In order to achieve as high maximum torque as the round rotor machine, the salient-pole machine should have been designed at a lower rated load angle, which would imply

TABLE 2: Electromagnetic characteristics for rated speed and power as well as some geometrical characteristics from stationary simulations for the two generators [2].

Characteristics	G1-Nd	G2-Fe
PM grade	N40H	Y30
Active power (kW)	225	225
Load L-L voltage (V) rms	793	793
Current (A) rms	164	164
Electrical frequency (Hz)	9.9	9.9
Rotational speed (r/min)	33	33
Torque (kNm)	65	65
Cogging (% of rated torque)	1.8	1.3
Electrical efficiency (%)	96.6	96.6
Stator copper losses (kW)	5.4	5.4
Stator iron losses (kW)	2.4	2.4
Load angle (°)	9.6	9.3
Power factor	1	1
Airgap force (MN/m^2)	0.18	0.20
Number of poles	36	36
Magnet dimensions (mm)	$19 * 120 * 848$	$100 * 350 * 809$
PM weight	536	4789
Total rotor weight, active[a] (t)	1.7	8.5
Total rotor weight (t)	3.3	approx. 10.8
Total generator weight, active[a] (t)	6.5	13.3
Total generator weight (t)	13	approx. 20.5

[a]Active material is material active in the magnetic circuit; see Figure 1.

TABLE 3: Comparison of overload operation for the two generators.

Characteristics	G1-Nd	G2-Fe
Max. torque (kNm)	200.4	178.0
Load angle at max torque (°)	46.2	44.1
Max. power (kW)	607.5	551.1
Load angle at max. power (°)	43.2	41.4
Voltage drop at 600 A (% of no load voltage)	28	37

using more permanent magnets. However, when designing the machine the objective was to have similar characteristics at rated power.

A comparison of the maximum torque and maximum power for the two machines and at what load angle they occur can be seen in Table 3. The salient-pole machine reaches the maximum values at about two degrees lower load angle than the round rotor machine. The voltage drop at increasing current can be seen in Figure 3 for the two machines. The voltage drops with increasing current since the power factor is unity. The voltage drops faster for the salient-pole machine since the voltage drop depends both on increasing load angle and on increasing direct axis current as discussed in Section 2.2.

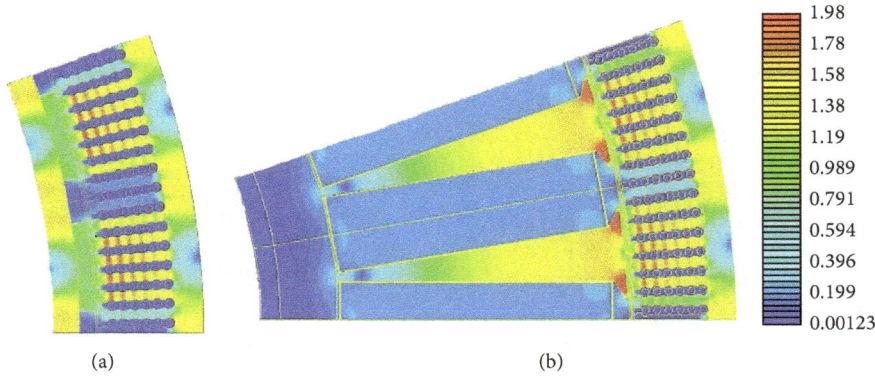

FIGURE 1: The magnetic flux density (in Tesla) in the two generators. (a) G1-Nd. (b) G2-Fe.

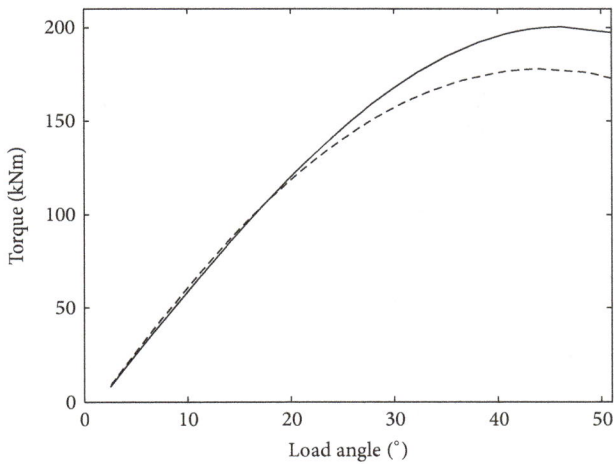

FIGURE 2: Torque as a function of load angle at rated speed and unity power factor for the round rotor generator G1-Nd (solid line) and the salient-pole generator G2-Fe (dashed line).

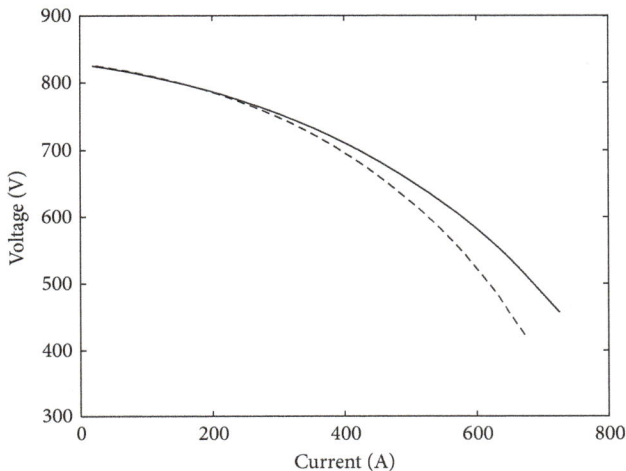

FIGURE 3: Voltage as a function of current at rated speed and unity power factor for the round rotor generator G1-Nd (solid line) and the salient-pole generator G2-Fe (dashed line).

This study has focused on FEM simulations of field equations which are more accurate than circuit equivalents of generators and since the inductances are difficult to estimate no exact estimation of inductances representing the machines has been done here. The salient-pole machine considered here has a small positive saliency; the saliency ratio L_d/L_q is between 1 and 1.5. A frequency response test in the FEM simulations of G1-Nd gives that L_d equals L_q, as expected for a round rotor, and is 8.1 mH, which is expected to be fairly accurate.

The larger the difference between L_d and L_q is, the more influence the reluctance torque has and the more the maximum torque point is moved to the left of the torque-load angle chart; see Figure 2. When the saliency is quite low, as here, the maxima are not shifted much. In this work, the generators have been designed to have certain rated operation values at, or close to, a particular load angle, in order to ensure a high maximum torque.

When a permanent magnet machine is operated at high current, there is a risk of permanently demagnetizing the permanent magnets. Therefore, the generator should be designed to withstand events such as short circuit currents, which usually are higher than the maximum current considered here. The aim of this study has been to show the differences between the two generators and not to evaluate the risk of permanent demagnetization. However, demagnetization during faults is not expected in any of these machines since the NdFeB magnets have high coercivity and the ferrite magnets are thick (in the direction of magnetization) and are protected by the rotor design.

4. Conclusions

In this paper, it has been shown that two generators with different saliency, designed at the same operational point for rated operation, have different overload characteristics. For machines rated at a low load angle, the difference in behaviour at part load and rated load can be neglected since the behaviour at low load angle is very similar. However, the operation at overload differs. If the maximum torque is the design criteria, the machines are not considered comparable, and a design for the ferrite generator with more magnets and lower rated load angle, as well as lower voltage drop, would have been chosen instead. For a more conventional generator design with a rated load angle at 20 to 30°, the differences would be more apparent and part load behaviour and maximum behaviour load might differ more.

It can be concluded that, for machines where overload capability is an important property, saliency needs to be considered. In addition, more magnetic material than expected when considering the energy product in Table 1 is needed for a ferrite machine, due to the intrinsic differences in saliency affecting machine properties. The small difference in saliency does not have any large effect on the generator rated operation, but it does have a small effect on the overload capability.

Conflict of Interests

The author declares that there is no conflict of interests regarding the publication of this paper.

Acknowledgments

The Swedish Research Council is acknowledged for contributions to Grant no. 2010-3950. Ångpanneföreningen's Foundation for Research and Development is acknowledged for its contributions. This work was conducted within the STandUp for ENERGY strategic research framework. Dr. A. Wolfbrandt and Professor U. Lundin are acknowledged for assistance with electromagnetic FEM simulations.

References

[1] C. Hurst, *China's Rare Earth Elements Industry: What Can the West Learn*, Institute for the Analysis of Global Security, Potomac, Md, USA, 2010, http://www.iags.org/rareearth0310hurst.pdf.

[2] S. Eriksson and H. Bernhoff, "Rotor design for PM generators reflecting the unstable neodymium price," in *Proceedings of the 20th International Conference on Electrical Machines (ICEM '12)*, pp. 1419–1423, September 2012.

[3] B. Ekergård, *Full Scale Applications of Permanent Magnet Electromagnetic Energy Converters*, Digital Comprehensive Summaries of Uppsala Dissertations from the Faculty of Science and Technology, Acta Universitatis Upsaliensis, Uppsala, Sweden, 2013.

[4] T. Sun, S.-O. Kwon, J.-J. Lee, and J.-P. Hong, "Investigation and comparison of system efficiency on the PMSM considering Nd-Fe-B magnet and ferrite magnet," in *Proceedings of the 31st International Telecommunications Energy Conference (INTELEC '09)*, pp. 1–6, October 2009.

[5] A. M. Mihai, S. Benelghali, L. Livadaru, A. Simion, and R. Outbib, "FEM analysis upon significance of different permanent magnet types used in a five-phase PM generator for gearless small-scale wind," in *Proceedings of the 20th International Conference on Electrical Machines (ICEM '12)*, pp. 267–273, September 2012.

[6] M. R. J. Dubois, *Optimized permanent magnet generator topologies for direct-drive wind turbines [Ph.D. thesis]*, Delft University, 2004.

[7] P. Lampola, *Directly driven, low-speed permanent-magnet generators for wind power applications [Ph.D. thesis]*, Department of Electrical Engineering, Helsinki University of Technology, 2000.

[8] A. Grauers, *Design of direct driven permanent magnet generators for wind turbines [Ph.D. thesis]*, Department of Electric Power Engineering, Chalmers University of Technology, 1996.

[9] S. Eriksson and H. Bernhoff, "Loss evaluation and design optimisation for direct driven permanent magnet synchronous generators for wind power," *Applied Energy*, vol. 88, no. 1, pp. 265–271, 2011.

[10] G. J. M. Darrieus, "Turbine having its rotating shaft transverse to the flow of the current," U.S. Patent 1.835.018, 1931.

[11] S. Eriksson, H. Bernhoff, and M. Leijon, "Evaluation of different turbine concepts for wind power," *Renewable and Sustainable Energy Reviews*, vol. 12, no. 5, pp. 1419–1434, 2008.

[12] S. Eriksson, H. Bernhoff, and M. Leijon, "A 225 kW direct driven PM generator adapted to a vertical axis wind turbine," *Advances in Power Electronics*, vol. 2011, Article ID 239061, 7 pages, 2011.

[13] J. Kjellin, S. Eriksson, and H. Bernhoff, "Electric control substituting pitch control for large wind turbines," *Journal of Wind Energy*, vol. 2013, Article ID 342061, 4 pages, 2013.

[14] N. R. Bianchi, S. Bolognani, and B. J. Chalmers, "Salient-rotor PM synchronous motors for an extended flux-weakening operation range," *IEEE Transactions on Industry Applications*, vol. 36, no. 4, pp. 1118–1125, 2000.

[15] R. H. Moncada, J. A. Tapia, and T. M. Jahns, "Analysis of negative-saliency permanent-magnet machines," *IEEE Transactions on Industrial Electronics*, vol. 57, no. 1, pp. 122–127, 2010.

[16] Anon. 1. Ace, "Modified Version 3.1, ABB common platform for field analysis and simulations," ABB Corporate Research Centre, Västerås, Sweden.

[17] S. Eriksson, A. Solum, M. Leijon, and H. Bernhoff, "Simulations and experiments on a 12 kW direct driven PM synchronous generator for wind power," *Renewable Energy*, vol. 33, no. 4, pp. 674–681, 2008.

12

Energy Efficient Hybrid Dual Axis Solar Tracking System

Rashid Ahammed Ferdaus,[1] **Mahir Asif Mohammed,**[1] **Sanzidur Rahman,**[1]
Sayedus Salehin,[2] **and Mohammad Abdul Mannan**[1]

[1] *Faculty of Engineering, American International University-Bangladesh, Road 14, Kemal Ataturk Avenue, Banani,*
  Dhaka 1213, Bangladesh
[2] *Department of Mechanical and Chemical Engineering, Islamic University of Technology (IUT),*
  Organisation of Islamic Cooperation (OIC), Board Bazar, Gazipur 1704, Bangladesh

Correspondence should be addressed to Rashid Ahammed Ferdaus; rashidferdaus@yahoo.com

Academic Editor: Jayanta Deb Mondol

This paper describes the design and implementation of an energy efficient solar tracking system from a normal mechanical single axis to a hybrid dual axis. For optimizing the solar tracking mechanism electromechanical systems were evolved through implementation of different evolutional algorithms and methodologies. To present the tracker, a hybrid dual-axis solar tracking system is designed, built, and tested based on both the solar map and light sensor based continuous tracking mechanism. These light sensors also compare the darkness and cloudy and sunny conditions assisting daily tracking. The designed tracker can track sun's apparent position at different months and seasons; thereby the electrical controlling device requires a real time clock device for guiding the tracking system in seeking solar position for the seasonal motion. So the combination of both of these tracking mechanisms made the designed tracker a hybrid one. The power gain and system power consumption are compared with a static and continuous dual axis solar tracking system. It is found that power gain of hybrid dual axis solar tracking system is almost equal to continuous dual axis solar tracking system, whereas the power saved in system operation by the hybrid tracker is 44.44% compared to the continuous tracking system.

1. Introduction

During the last few years the renewable energy sources like solar energy have gained much importance in all over the world. Different types of renewable or green energy resources like hydropower, wind power, and biomass energy are currently being utilized for the supply of energy demand. Among the conventional renewable energy sources, solar energy is the most essential and prerequisite resource of sustainable energy [1, 2].

Solar energy refers to the conversion of the sun's rays into useful forms of energy, such as electricity or heat. A photovoltaic cell, commonly called a solar cell or PV, is the technology used to convert solar energy directly into electrical power. The physics of the PV cell (solar cell) is very similar to the classical p-n junction diode. Sunlight is composed of photons or particles of solar energy. Semiconductor materials within the PV cell absorb sunlight which knocks electrons from their atoms, allowing electrons to flow through the material to produce electricity [3, 4]. Because of its cleanliness, ubiquity, abundance, and sustainability, solar energy has become well recognized and widely utilized [5].

Different researches estimate that covering 0.16% of the land on earth with 10% efficient solar conversion systems would provide 20 TW of power, nearly twice the world's consumption rate of fossil energy [6]. This proves the potential of solar energy which in turn points out the necessity of tracking mechanism in solar systems. The tracking mechanism is an electromechanical system that ensures solar radiation is always perpendicular to the surface of the photovoltaic cells (solar cells) which maximizes energy harnessing [7].

Over the years, researchers have developed smart solar trackers for maximizing the amount of energy generation. Before the introduction of solar tracking methods, static solar panels were positioned with a reasonable tilted angle based on the latitude of the location. In this competitive world of

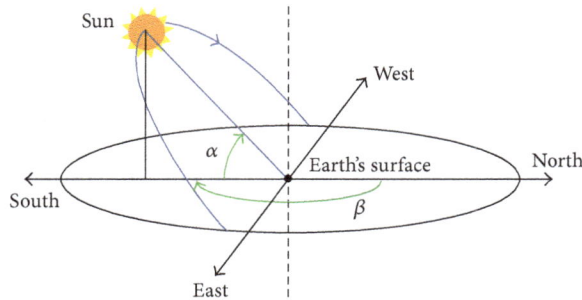

FIGURE 1: Illustration of the solar angles: (a) altitude angle, α; (b) azimuthal angle, β. The solar path corresponds to a day in the early fall or late winter seasons in the northern hemisphere, that is, just prior to the spring equinox or just after the fall equinox. Solar noon is the time of day when $\beta = 180$ degree, that is, the sun is directly at south and is halfway between sunrise and sunset.

advanced scientific discoveries, the introductions of automated systems improve existing power generation by 50% [8].

There are mainly two types of solar trackers on the basis of their movement degrees of freedoms. These are single axis solar tracker and dual axis solar tracker. Again these two systems are further classified on the basis of their tracking technologies. Active, passive, and chronological trackers are three of them [9, 10].

Previous researchers used single axis tracking system which follows only the sun's daily motion [11]. But the earth follows a complex motion that consists of the daily motion and the annual motion. The daily motion causes the sun to appear in the east to west direction over the earth whereas the annual motion causes the sun to tilt at a particular angle while moving along east to west direction [12].

Figure 1 shows the daily and annual motion of the sun. The sun's location in the sky relative to a location on the surface of the earth can be specified by two angles as shown in Figure 1. They are (1) the solar altitude angle (α) and (2) the solar azimuth angle (β). Angle α is the angle between the sun's position and the horizontal plane of the earth's surface while angle β specifies the angle between a vertical plane containing the solar disk and a line running due south [13].

Solar tracking is best achieved when the tilt angle of the solar tracking systems is synchronized with the seasonal changes of the sun's altitude. An ideal tracker would allow the solar modules to point towards the sun, compensating for both changes in the altitude angle of the sun (throughout the day) and latitudinal offset of the sun (during seasonal changes). So the maximum efficiency of the solar panel is not being used by single axis tracking system whereas double axis tracking would ensure a cosine effectiveness of one.

In active tracking or continuous tracking, the position of the sun in the sky during the day is continuously determined by sensors. The sensors will trigger the motor or actuator to move the mounting system so that the solar panels will always face the sun throughout the day. If the sunlight is not perpendicular to the tracker, then there will be a difference in light intensity on one light sensor compared to another. This difference can be used to determine in which direction

the tracker has to be tilted in order to be perpendicular to the sun. This method of sun tracking is reasonably accurate except on very cloudy days when it is hard for the sensors to determine the position of the sun in the sky [14].

Passive tracker, unlike an active tracker which determines the position of the sun in the sky, moves in response to an imbalance in pressure between two points at both ends of the tracker. The imbalance is caused by solar heat creating gas pressure on a "low boiling point compressed gas fluid, that is, driven to one side or the other" which then moves the structure. However, this method of sun tracking is not accurate [15, 16].

A chronological tracker is a time-based tracking system where the structure is moved at a fixed rate throughout the day as well for different months. Thus the motor or actuator is controlled to rotate at a slow average rate of one revolution per day (15° per hour). This method of sun tracking is more energy efficient [17].

To track the sun's movement accurately dual axis tracking system is necessary. The active/continuous tracking system tracks the sun for light intensity variation with precision. Hence, the power gain from this system is very high [18]. But to achieve this power gain the system uses two different motors continuously for two different axes. As a result it always consumes a certain amount of extra power compared to time-based tracking system. Therefore to reduce this power loss a combination of active and time-based tracking could be the suitable alternative to this system. Finally the motivation of the research was to design and implement a hybrid dual axis solar tracking system which reduces the motor power consumption while tracking accurately.

A simple energy efficient and rugged tracking model is presented in this paper in order to build a hybrid dual axis solar tracker. To track the sun's daily motion, that is, from east to west direction, a pair of light sensors are used and to track the seasonal motion of the sun real time clock (RTC) is used to create the accurate azimuth angle from some predetermined parameters. The light intensity is compared by microcontroller and it generates the suitable control signals to move the motors in proper direction. So a driver circuit is used to increase the voltage and current level for the operation of the motors. Two full geared stepper motors are used for rotating the solar module in two different axes.

A versatile mechanical system is introduced as a linear actuator to create proper tilt angle. In addition, this linear actuator has high weight lifting capability which is observed experimentally. It is found that energy efficient hybrid dual axis tracking yields almost same energy as continuous dual axis solar tracking system. It is also observed that in hybrid tracking system one motor can remain idle for one month and thus reduces more than 44% of power consumption as compared with continuous dual axis solar tracking system. This mechanism proved significant benefit of reducing energy consumption by hybrid tracker sacrificing a very little tracking loss. This paper also represents the comparative study between the continuous/active and hybrid solar tracking system.

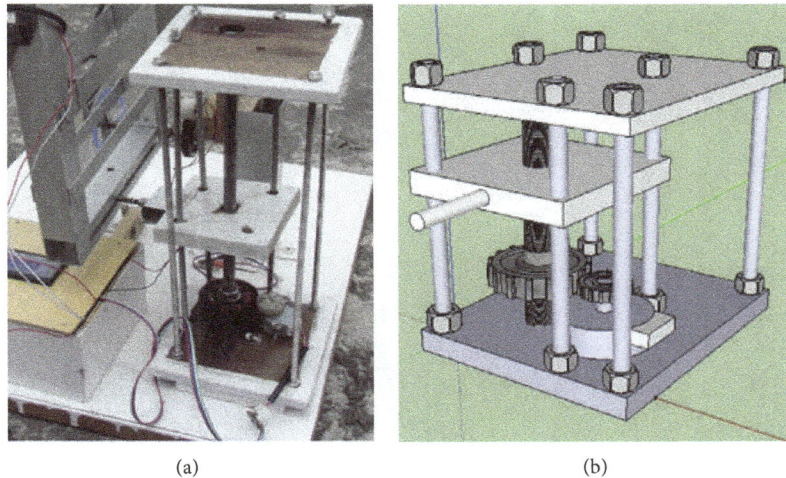

(a)   (b)

FIGURE 2: Design and implementation of linear actuator: (a) implementation and placement of linear actuator with the aluminium body of panel carrier; (b) hardware design of linear actuator in computer aided drafting tool.

2. Design and Implementation Process

The whole work involves the reading of different sensor values and then comparing them digitally to determine the exact position of the sun in east-west direction. Again the system is also given some predefined values based on the sun's geographical location in the north-south direction. Overall the entire system can intelligently track the sun's movement both in horizontal and vertical axis. In order to simplify the design and implementation process the whole system is divided into two parts.

These are as follows:

(A) mechanical system design;

(B) electrical circuit design.

(A) Mechanical System Design. Assembling the mechanical system was the most challenging part of this system because the objective was to make an energy efficient solar tracking system which demanded intelligent operations of the tracking motors. Generally one of these motors is used for daily tracking (east-west motion) and other for making a seasonal tracking (north-south motion). So the daily tracking motor operates continuously based on light sensors and the annual motion tracking motor operates only a few times over the year. So for design and implementation process the whole mechanical system is mainly divided into three parts as follows:

(1) linear actuator;

(2) panel carrier;

(3) panel carrier rotator.

(1) Linear Actuator. A linear actuator converts circular motion to a linear vertical motion in contrast to the circular motion of a conventional electric motor. The linear vertical motion is used for creating the seasonal angle of the sun. In this tracking system linear actuator consists of one stepper motor, screw thread, bolt, bearing, circular rod, and some pieces of wood. Figure 2 shows the mechanical design and structure of linear actuator. Experimentally it is found that this mechanical structure has a special feature of high weight lifting using a low power stepper motor.

Linear actuator gives the linear motion in vertical axis (upward and downward) and is connected to one end of panel carrier through a straight single rod hook. The rod hook is attached to the wooden frame. There are some bolts and these are tied with seven 15-inch long circular rods of 2 mm diameter. There is also a 13-inch long screw thread and its diameter is 6 mm. A bolt is attached in the middle of the wooden frame and this bolt is also tied with the screw thread. Four circular rods are also mortised through the wooden frame.

The wooden frame moves up and down along with the bolt and the single rod hook. It works in such a way that the wooden frame does not let the bolt move along with the thread screw rather when the thread screw moves then the four circular rods mortised into the wooden frame cause the bolt to move up or down. Now when the single rod hook moves upward or downward it moves along with the panel carrier. The two ends of screw thread are placed in two bearings which helps it to rotate smoothly. These bearings are mortised into the roof and floor. One gear is also placed at the bottom of the screw thread and this gear is connected to the stepper motor gear which is placed on the floor of the linear actuator body. The floor and roof of the linear actuator are made of wood which holds all the linear actuator instruments.

(2) Panel Carrier. Panel carrier is basically a rectangular frame made of aluminum which holds the solar panel with the help of a circular rod. One end of the horizontal base of the panel carrier is attached with the single rod hook of linear actuator and other with the panel carrier rotator. Figure 3 shows the design and implementation of panel carrier. A stepper motor

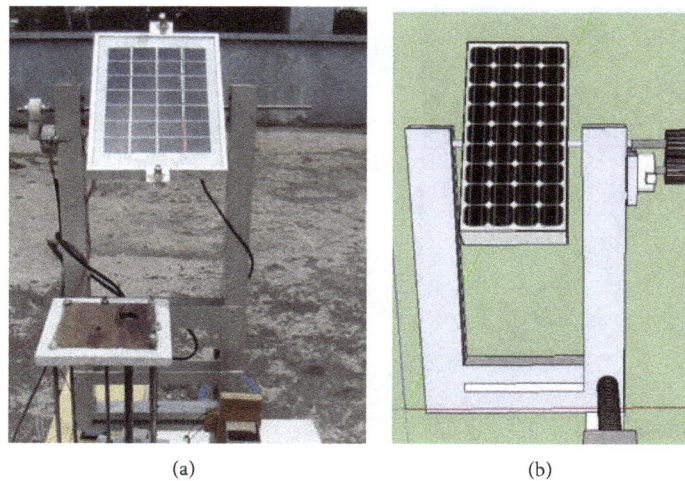

FIGURE 3: Design and implementation of panel carrier: (a) implementation and placement of panel carrier with the linear actuator and panel carrier rotator; (b) hardware design of Panel carrier in computer aided drafting tool.

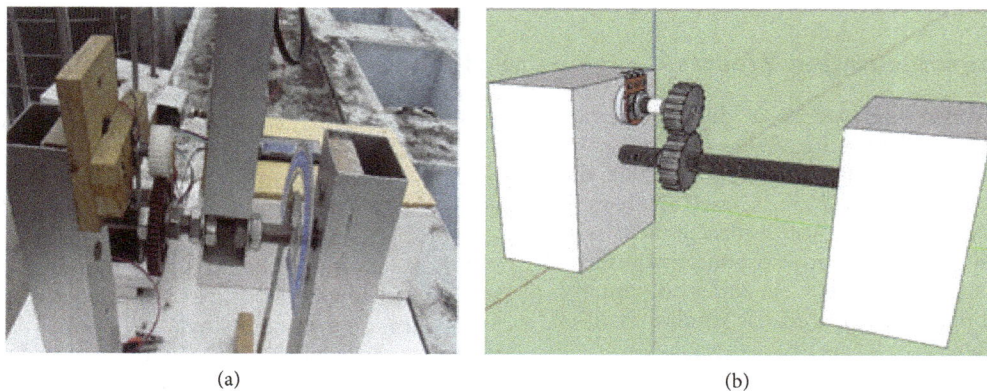

(a)   (b)

FIGURE 4: Design and implementation of panel carrier rotator: (a) implementation and placement of panel carrier rotator with the panel carrier and a wooden base; (b) hardware design of panel carrier rotator in computer aided drafting tool.

with a gear is placed on the body of the aluminium frame. When the stepper motor rotates along with its gear then the panel rotates from east to west by tracking sun's daily motion actively.

The light sensors are placed at the two ends of solar panel. Again the rectangular aluminium frame has a rectangular mortise in its horizontal base. Single circular rod hook from linear actuator goes through this mortise. Thus it helps to lift the panel carrier in a semi-circular path to get sun's tilt angle caused by seasonal/annual motion. While the linear actuator lifts one end of panel carrier the other end needs to be fixed with a panel carrier rotator to get the perfect circular motion.

(3) Panel Carrier Rotator. Panel carrier rotator is used to hold one end of the horizontal base of the solar panel carrier. One screw thread, gear, and position sensors are used in this panel carrier rotator to give a circular movement to the panel carrier. Its base is fixed on a wooden floor. Figure 4 shows the design and implementation of panel carrier rotator and Figure 5 shows the experimental setup of the hybrid dual axis solar tracker.

(B) Electrical Circuit Design. The whole electrical system is mainly divided into three units. These are sensor unit, control unit, and movement adjustment unit. Sensor unit senses three different parameters (light, time, and position) and converts it to appropriate electrical signals. Then the electrical signals from sensor unit are sent to control unit. Control unit determines the direction of the movement of the motors both in the horizontal and vertical axes. Finally the movement adjustment unit adjusts the position of the solar module by receiving signal from the control unit. This adjustment is done by using two geared unipolar stepper motors. Figure 6 shows the overall block diagram of the whole system.

(1) Sensor Unit. The sensor unit consists of three sensor circuits. These are as follows:

(a) light sensor;

(b) real time clock;

(c) position sensor.

FIGURE 5: Experimental setup of the Hybrid dual axis solar tracker.

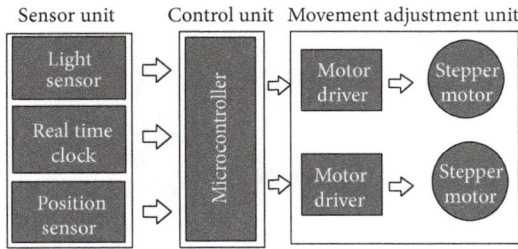

FIGURE 6: Block diagram of the electrical circuit.

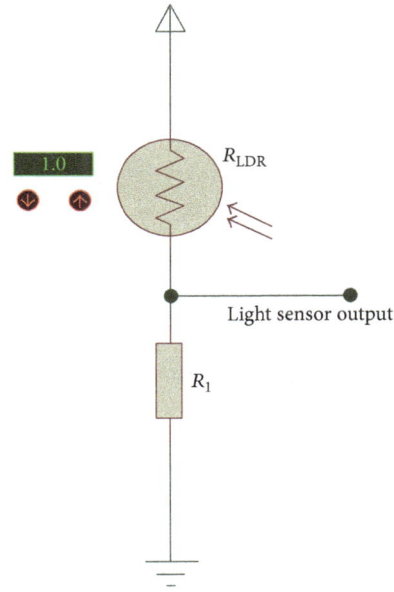

FIGURE 7: Basic LDR circuit.

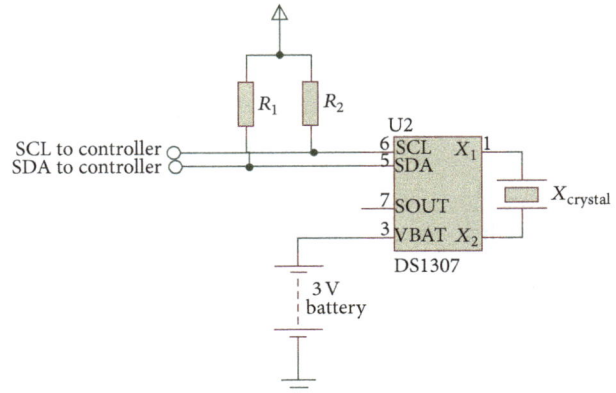

FIGURE 8: Real time clock circuit.

(a) Light Sensor. Light sensors are used for measuring light intensity and generating a corresponding analog voltage signal into the input of the analog to digital converter of the microcontroller. Since this is a hybrid dual axis solar tracking system so, to track the sun's daily motion continuously, that is, from east to west, a pair of light dependent resistors (LDR) is used as light sensors. On the other hand, the sun's annual motion, that is, from north to south, is tracked by the real time clock (RTC) device and position sensor.

A light dependent resistor (LDR) is a resistor whose resistance decreases with increasing incident light intensity. Figure 7 shows the basic LDR circuit and Table 1 shows the different specifications of LDR used in the tracking system [19]. The relationship between the resistance R_{LDR} (resistance of LDR) and light intensity (Lux) for a typical LDR is given in following equation [20]:

$$R_{LDR} = \left(\frac{500}{\text{Lux}}\right)\text{k}\Omega, \tag{1}$$

where R_{LDR} = Resistance of LDR.

(b) Real Time Clock. Real time clock is a clock device that keeps track of the current time. There are different types of real time clock (RTC) device; among them DS1307 is used here. This is a battery-backed real time clock (RTC), that is, connected to microcontroller via I2C bus to keep track of time even if it is reprogrammed or if the power is lost.

This device is suitable for data logging, clock-building, time stamping, timers, alarms, and so forth. Microcontroller takes the month and hour values from the RTC device to track the sun's annual motion and the darkness of night to take the solar panel at its initial position. Figure 8 shows the basic RTC circuit. In the figure U2 is the RTC chip. Address and data from RTC chip are transferred serially through an I2C, bidirectional bus. The two I2C signals are serial data (SDA) and serial clock (SCL) and these two signals are sent to controller with two pull-up resistors R_1 and R_2. Together, these signals make it possible to support serial transmission of 8 bit bytes of data-7 bit device addresses plus control bits-over the two-wire serial bus. $X_{crystal}$ is a 32.768 kHz quartz crystal used for required clock generation for RTC chip And B1 is a 3-volt battery used for power backup in case of power failure.

(c) Position Sensor. Position sensor detects the sun's annual motion. A variable resistor is used here as position sensor.

TABLE 1: Specification of LDR.

Dark resistance (MΩ)	Illuminated resistance (kΩ)	Sensitivity (Ω/lux)	Spectral application range (nm)	Rise time (ms)	Fall time (ms)
20	5–20	0.9	400–700	70	15

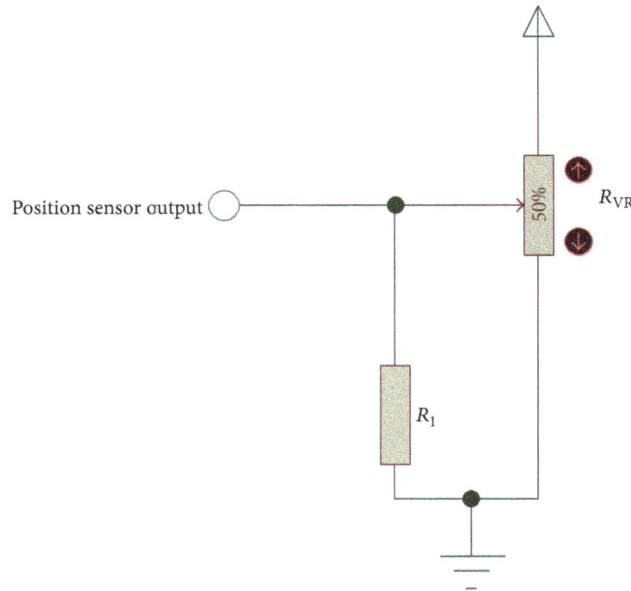

FIGURE 9: Position sensor circuit.

Figure 9 shows a variable resistor connected to another resistor R_1. So when the resistivity of variable resistor changes the position sensor output also changes. The output from this circuit goes to controller and different voltages in the output of position sensor circuit represent different latitude angle of the sun for its annual motion.

Position sensor is placed in the panel carrier rotator. When linear actuator moves linearly then panel carrier rotator rotates a semicircular path which causes the position sensor to change its voltage level. The panel carrier rotator rotates 50° degree in a semicircular path with respect to the horizontal axis as in the experimental location sun's latitude angle changes in between this 50°. The panel carrier rotator can rotate 75° in both sides which may also be applicable in other locations. In that case the sensor has to be calibrated accurately. For 12 months different 12 values of sun's latitude angle are predetermined and set in the microcontroller and with respect to these values microcontroller decides how much to move the linear actuator. Panel carrier rotator rotates due to the linear actuator's linearly upward and downward motion with the panel carrier. A gear is placed with the panel carrier rotator which also rotates with it. This gear rotation causes the variable resistor's gear to rotate and this is how the resistivity of the variable resistor changes. Thus the signal is changing from the position sensor.

(2) Control Unit. Microcontroller is the main control unit of this whole system. The output from the sensor unit comes to the input of the microcontroller which determines the direction of the movement of the motors both in the horizontal and vertical axes. For this research ATmega32 microcontroller is used. This is from the Atmel AVR family. Figure 10 shows the main flowchart of the microcontroller programming. Figure 11 shows another flowchart of the microcontroller programming which is a part of the main flowchart showed in Figure 10.

(3) Movement Adjustment Unit. Movement adjustment unit consists of two geared unipolar stepper motors along with their motor driver device. The output from microcontroller is sent to the motor driver which executes the proper sequence to turn the stepper motors in the required direction. To run the unipolar stepper motor in full drive or half drive mode ULN2803 is used as motor driver IC. This driver is an array of eight Darlington transistors. Darlington pair is a single transistor with a high current gain. Thus the current gain is required for motor drive and it reduces the circuit space and complexity. Figure 12 shows the unipolar stepper motor, motor driver device, and Darlington pair.

The two full geared stepper motors are used here for the accurate tracking of the sun. For our experimental purpose a small scale system was implemented for 3-watt solar panel. The specifications of solar panel and gear and stepper motors are listed as follows.

Specifications of solar panel as load of the motor:

mass of solar panel, $m = 0.75$ Kg;

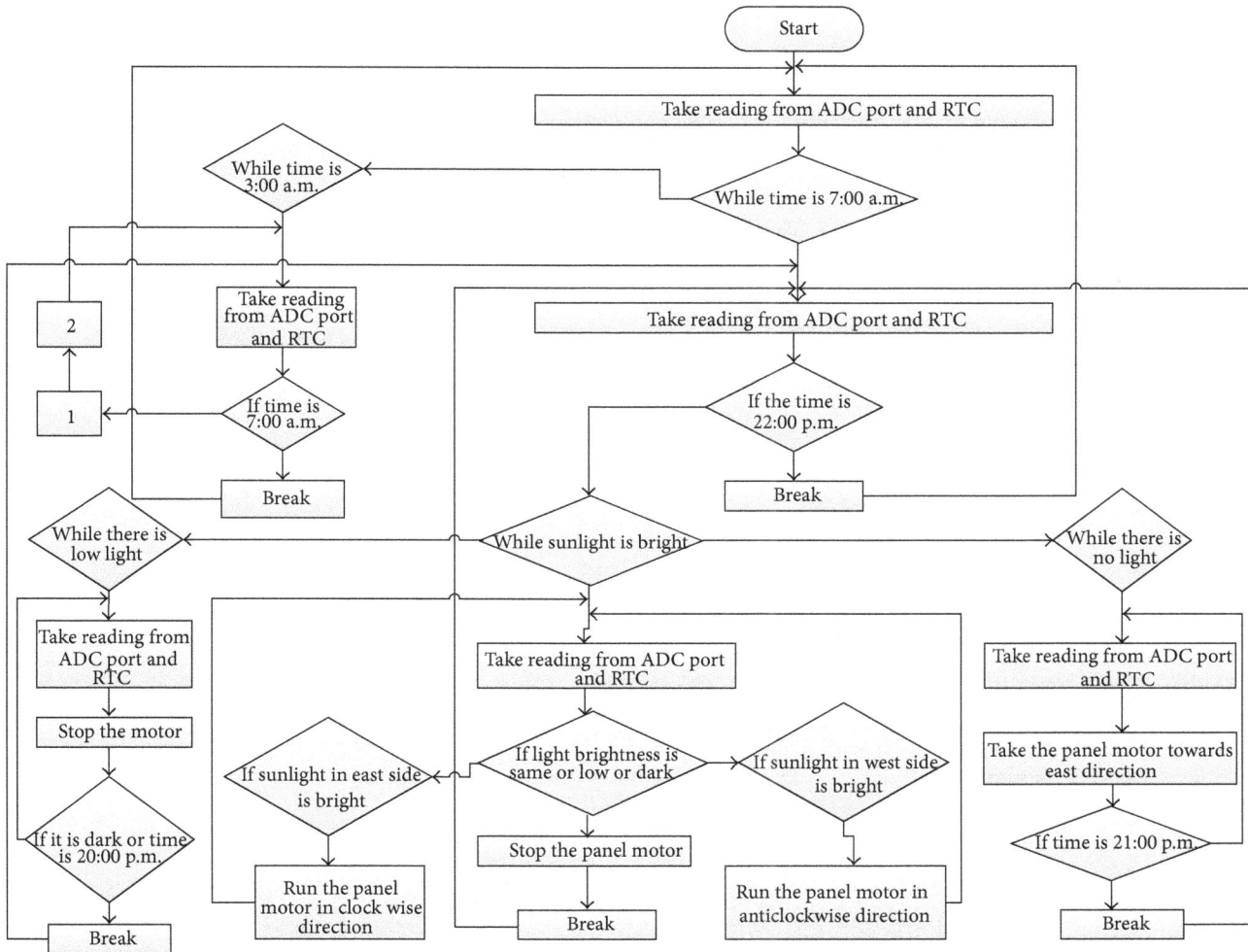

FIGURE 10: Main flowchart of microcontroller programming.

length of solar panel, $L = 0.165$ m;

width of solar panel, $D = 0.23$ m;

height of solar panel, $H = 0.015$ m;.

volume of solar panel, $v = L \times D \times H = 5.69 \times 10^{-4}$ m^3;

density, $\rho = (m/v) = 1318.10$ kg·m^{-3}.

Specifications of gears:

number of gear teeth, $N1 = N2 = 24$;

mass of gear, $m_{G1} = m_{G2} = 5 \times 10^{-3}$ kg;

diameter of gear, $D_{G1} = D_{G2} = 0.027$ m.

Specifications of motor:

inertia of motor, $J_m \approx 0$;

pull-out torque = 0.147 N·m;

pull-out frequency, $f = 100$ Hz;

friction coefficient, $\mu = 0.05$;

step angle, $\theta_s = 0.044°$;

angle coefficient, $n = 3.6°/\theta_s = 81.81$.

For the pull-out frequency of 100 Hz the required motor torque to rotate the panel is calculated as follows [21–23].

Calculation of moment of inertia is given as follows:

Inertia of the load is

$$J_L = \frac{\pi}{32} \times \rho \times L \times D^4 \times \left(\frac{N2}{N1}\right)^2 = 0.0597 \text{ kg} \cdot \text{m}^2, \quad (2)$$

inertia of gear 1 is

$$J_{G1} = \frac{1}{8} \times m_{G1} \times D_{G1}^2 \times \left(\frac{N2}{N1}\right)^2, \quad (3)$$

inertia of gear 2 is

$$J_{G2} = \frac{1}{8} \times m_{G2} \times D_{G2}^2, \quad (4)$$

number of gear teeth, $N1 = N2$ so,

$$J_{G1} = J_{G2} = \frac{1}{8} \times m_{G2} \times D_{G2}^2 = 4.56 \times 10^{-7} \text{ kg} \cdot \text{m}^2. \quad (5)$$

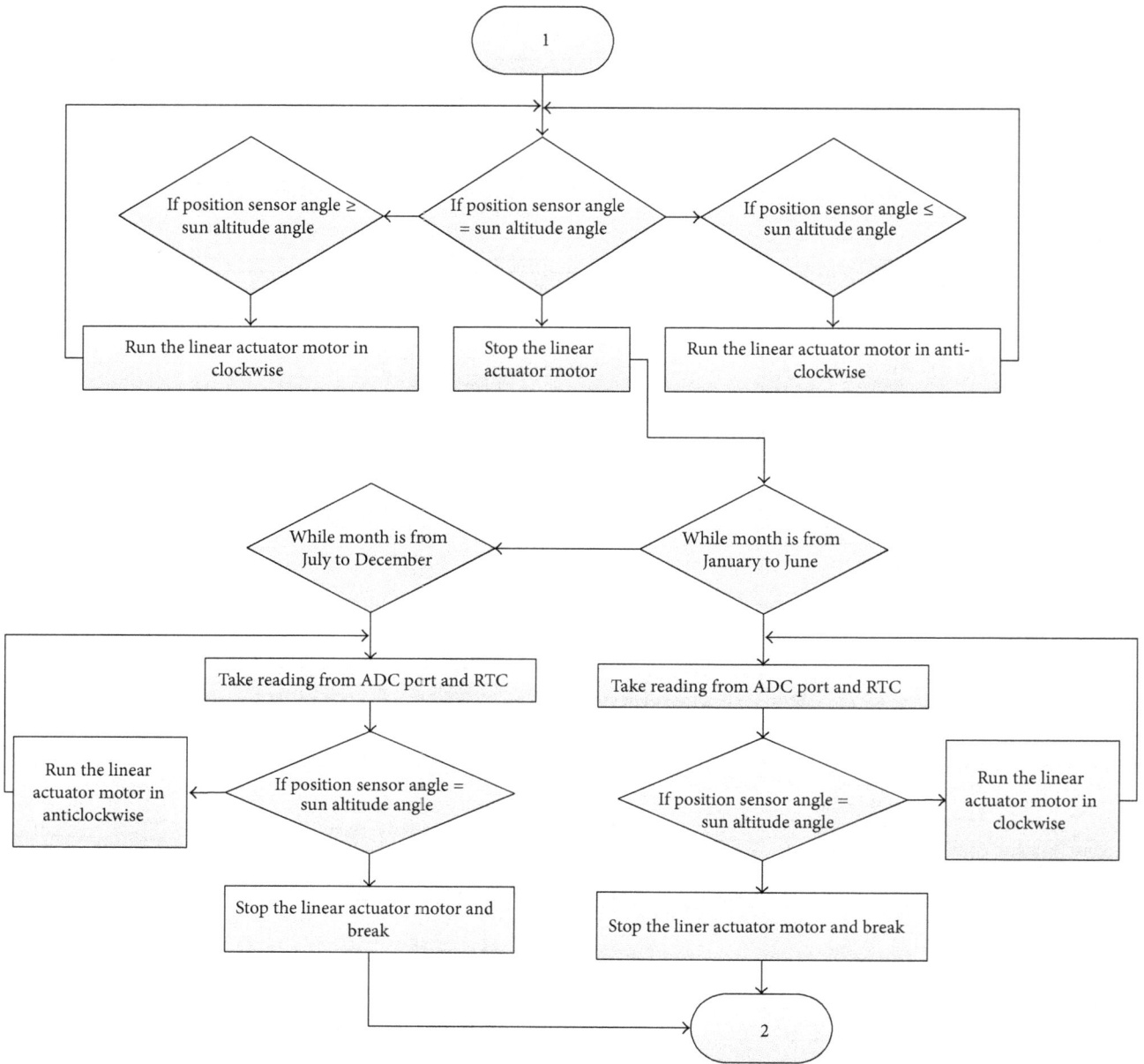

FIGURE 11: Continuation of main flowchart of microcontroller programming.

So, inertia of the system is

$$J_T = J_L + J_{G1} + J_{G2} + J_m = 0.0597 \, \text{kg} \cdot \text{m}^2. \qquad (6)$$

Calculation of acceleration torque is given as follows.
  Now, acceleration torque is

$$T_a = J_T \times \frac{\pi \times \theta_s}{180 \times n} \times f^2 = 5.60 \times 10^{-3} \, \text{N} \cdot \text{m}. \qquad (7)$$

Force to rotate the load is

$$F = m \times g \left(\sin \theta + \mu \times \cos \theta \right) = 0.3675 \, \text{N} \cdot \text{m}. \qquad (8)$$

Calculation of load torque is given as follows.
  Now, load torque is

$$T_L = \frac{F \times D}{2} + T_F = 0.0423 \, \text{N} \cdot \text{m}; \qquad (9)$$

here, load torque due to friction, $T_F \approx 0$.
  Calculation of required motor torque is given as follows.
  Total calculated torque is

$$T_T = T_a + T_L = 0.0479 \, \text{N} \cdot \text{m}. \qquad (10)$$

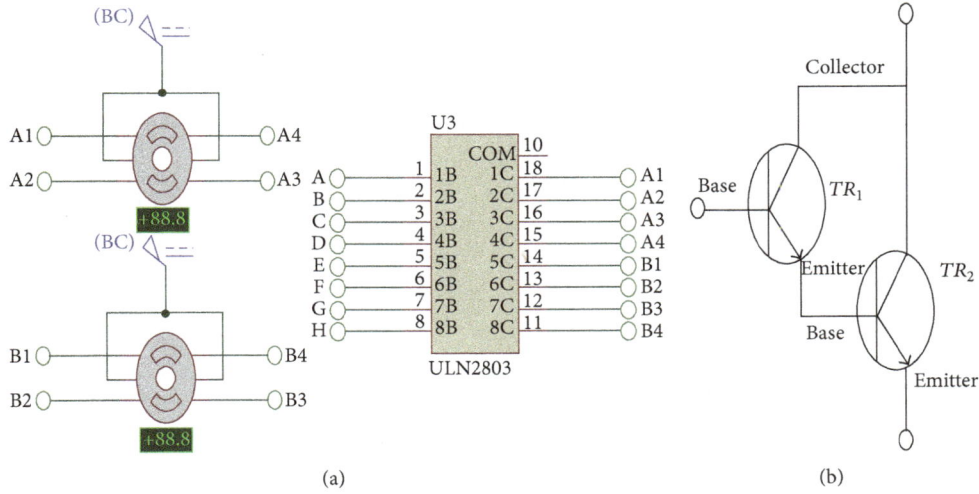

FIGURE 12: Movement adjustment unit: (a) the unipolar stepper motor and motor driver device (ULN2803) and (b) Darlington pair.

Required motor torque is

$$T_M = K_s \times T_T = 0.0958\,\text{N} \cdot \text{m}. \qquad (11)$$

Here, safety factor, $K_s = 2$.

$\therefore$ Required motor torque < pull-out torque.

So from the above comparison it is clear that pull-out torque of 0.147 N·m. of stepper motor is sufficient enough to rotate the solar panel of 0.75 kg.

3. Experimental Results and Data Analysis

(A) Comparative Study of Solar Panel Power Output. All the experiments have been conducted in Dhaka, Bangladesh ($23°42'0''$N $90°22'30''$E). Table 2 shows the current and voltage values received from the static panel, hybrid tracking system, and continuous tracking system for different times in a day. From Table 2 it is seen that at 8:00 am there is much improvement in current by both the tracking systems compared to the static panel. But as time goes on the difference in current among these three systems decreases up to around 11:00 am. After that when the sun rotates more towards west this difference increases again. The highest current of static panel, hybrid tracking system, and continuous tracking system is 0.47 amp, 0.48 amp, and 0.50 amp, respectively, at 12:30 pm. But in case of voltage the variation is less compared to current as the voltage has no direct relation with the sun light intensity. Figure 13 shows the comparison of current versus time curves for the static panel, hybrid tracking system, and continuous tracking system.

Table 3 shows the power values of the static panel and both the tracking systems. The power gain of tracking systems over static panel and between the two tracking systems for different times is also given in Table 3. The maximum power output of the static panel, hybrid tracking system, and continuous tracking system is found as 3.7036 watt, 3.7824 watt, and 3.94 watt, respectively at 12:30 pm. Much more power gain is achieved in the morning and afternoon because both the tracking systems can accurately track the

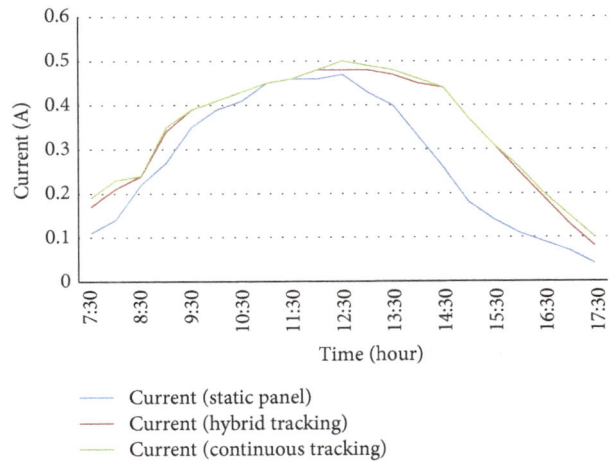

FIGURE 13: Comparison curve: comparison of current versus time curve for the static panel, hybrid tracking system, and continuous tracking system.

sun at these times while the static system cannot. For all these technologies power fall was very fast from 3:30 pm to 5:30 pm because of the low duration of day light.

The total power of static panel, hybrid tracking system, and the continuous tracking system throughout the day is 45.21 watt, 56.69 watt, and 58.24 watt, respectively. So the average power gain of hybrid tracking system over the static panel is 25.62%. Similarly the average power gain of continuous tracking system over the static panel is 28.10% and over the hybrid tracking system is 4.19%.

(B) Comparison of Stepper Motor Power Consumption. The power consumption by the stepper motors in both the solar tracking system is not same. Table 4 shows the comparison of stepper motors power consumption between the two tracking systems.

TABLE 2: Current and voltage values of static and tracking panel at different times in a day.

Time (hour)	Static panel		Hybrid tracking system		Continuous tracking system	
	Current (ampere)	Voltage (volt)	Current (ampere)	Voltage (volt)	Current (ampere)	Voltage (volt)
7:30	0.11	7.82	0.17	7.82	0.19	7.92
8:00	0.14	7.82	0.21	7.82	0.23	7.92
8:30	0.22	7.83	0.24	7.83	0.24	7.9
9:00	0.27	7.9	0.34	8	0.35	8
9:30	0.35	7.93	0.39	7.98	0.39	7.98
10:00	0.39	7.92	0.41	7.92	0.41	7.92
10:30	0.41	7.88	0.43	7.92	0.43	7.92
11:00	0.45	7.88	0.45	7.88	0.45	7.88
11:30	0.46	7.88	0.46	7.88	0.46	7.88
12:00	0.46	7.88	0.48	7.88	0.48	7.88
12:30	0.47	7.88	0.48	7.88	0.5	7.88
13:00	0.43	7.88	0.48	7.88	0.49	7.88
13:30	0.4	7.77	0.47	7.81	0.48	7.81
14:00	0.33	7.79	0.45	7.83	0.46	7.93
14:30	0.26	7.71	0.44	7.71	0.44	7.83
15:00	0.18	7.63	0.37	7.76	0.37	7.9
15:30	0.14	7.54	0.31	7.7	0.31	7.86
16:00	0.11	7.52	0.25	7.73	0.26	7.86
16:30	0.09	7.41	0.19	7.71	0.2	7.71
17:00	0.07	7.39	0.13	7.65	0.15	7.71
17:30	0.04	7.33	0.08	7.5	0.1	7.64

TABLE 3: Power values of static and tracking panel and the corresponding power gain by tracking panel over static panel at different times in a day.

Time (hour)	Static panel	Hybrid tracking panel	Continuous tracking panel	Power gain by hybrid tracking system over static panel	Power gain by continuous tracking system over static panel	Power gain by continuous tracking system over hybrid tracking system
	Power (watt)	Power (watt)	Power (watt)	%	%	%
7:30	0.8602	1.3294	1.5048	35.29412	42.83626	11.65603
8:00	1.0948	1.6422	1.8216	33.33333	39.89899	11.65603
8:30	1.7226	1.8792	1.896	8.333333	9.14557	9.848485
9:00	2.133	2.72	2.8	21.58088	23.82143	0.886076
9:30	2.7755	3.1122	3.1122	10.81871	10.81871	2.857143
10:00	3.0888	3.2472	3.2472	4.878049	4.878049	0
10:30	3.2308	3.4056	3.4056	5.132723	5.132723	0
11:00	3.546	3.546	3.546	0	0	0
11:30	3.6248	3.6248	3.6248	0	0	0
12:00	3.6248	3.7824	3.7824	4.166667	4.166667	0
12:30	3.7036	3.7824	3.94	2.083333	12.2449	0
13:00	3.3884	3.7824	3.8612	10.41667	12.2449	4
13:30	3.108	3.6707	3.7488	15.3295	17.09347	2.040816
14:00	2.5707	3.5235	3.6478	27.04129	29.52739	2.083333
14:30	2.0046	3.3924	3.4452	40.90909	41.8147	3.407533
15:00	1.3734	2.8712	2.923	52.16634	53.01403	1.532567
15:30	1.0556	2.387	2.4366	55.77713	56.67734	1.772152
16:00	0.8272	1.9325	2.0436	57.19534	59.52241	2.035623
16:30	0.6669	1.4649	1.542	54.47471	56.75097	5.436485
17:00	0.5173	0.9945	1.1565	47.98391	55.27021	5
17:30	0.2932	0.6	0.764	51.13333	61.62304	14.00778

TABLE 4: Comparison of stepper motor power consumption.

Hybrid tracking system		Continuous tracking system	
Power consumption for movement in east to west	Power consumption for movement in north to south	Power consumption for movement in east to west	Power consumption for movement in north to south
0.6 watt	Almost zero ≈ 0	0.6 watt	0.48 watt
Total = 0.6 watt		Total = 1.08 watt	
Power saved = 44.44%			

So power saved by hybrid tracking system over continuous tracking system is 44.44%.

(C) Data Analysis. So from all these data it is seen that the hybrid dual axis tracking system has average power generation of 56.69 watt whereas the continuous tracking system has 58.24 watt. Therefore continuous tracking system has only 4.2% average power gain over hybrid dual axis tracking system. On the other hand hybrid dual axis tracking system is saving 44.44% system power consumption compared to continuous tracking system. Though the continuous tracking system gives a slight improvement in power gain, due to its continuous tracking, it consumes much more power compared to the hybrid dual axis tracking system. Considering the case of 44.44% power saving by hybrid tracking system, it can be concluded that the hybrid dual axis tracking system can operate much more efficiently compared to the continuous tracking system while sacrificing little about 4.2% tracking accuracy.

4. Conclusion

The design, implementation, and testing of a hybrid dual axis solar tracking system is presented in the study. The Performance of the developed system was experimented and compared with both the static and continuous dual axis solar tracking system. This work demonstrates that hybrid dual axis solar tracking system can assure higher power generation compared to static panel as well as less power consumption compared to continuous dual axis solar tracking system. The result shows that the hybrid dual axis tracking system has 25.62% more average power gain over static system while it has 4.2% less average power gain compared to continuous tracking system. In hybrid dual axis solar tracking system one motor runs continuously to track continuous movement of sun due to daily motion and another motor runs once in a month to track suns seasonal motion. But in other trackers like in continuous solar tracker it needs to move both the motors continuously. Thus the hybrid system is saving motor power consumption while the power gain compared to other technology is almost marginal. So further comparative study about stepper motor power consumption shows that hybrid tracking system can save 44.44% power compared to continuous tracking system. This amount of power saving will have a significant effect in large systems like heliostat power plants where a lot of trackers are required and power saved by all the systems will show a big amount of power. Other than this the designed tracking system can also be implemented for the solar thermal systems. Finally the proposed design is achieved with low power consumption, high accuracy, and low cost.

Conflict of Interests

The authors declare that there is no conflict of interests regarding the publication of this paper.

Acknowledgment

The authors thank M.D. Ahasanul Kabir of American International University, Bangladesh (AIUB), for his help with the mechanical system implementation.

References

[1] G. Deb and A. B. Roy, "Use of solar tracking system for extracting solar energy," *International Journal of Computer and Electrical Engineering*, vol. 4, no. 1, pp. 42–46, 2012.

[2] T. Tudorache and L. Kreindler, "Design of a solar tracker system for PV power plants," *Acta Polytechnica Hungarica*, vol. 7, no. 1, pp. 23–39, 2010.

[3] C.-L. Shen and C.-T. Tsai, "Double-linear approximation algorithm to achieve maximum-power-point tracking for photovoltaic arrays," *Energies*, vol. 5, no. 6, pp. 1982–1997, 2012.

[4] K. Liu, "Dynamic characteristics and graphic monitoring design of photovoltaic energy conversion system," *WSEAS Transactions on Systems*, vol. 10, no. 8, pp. 239–248, 2011.

[5] T. Tudorache, C. D. Oancea, and L. Kreindler, "Performance evaluation of a solar tracking PV panel," *U.P.B. Scientific Bulletin, Series C: Electrical Engineering*, vol. 74, no. 1, pp. 3–10, 2012.

[6] H. Mousazadeh, A. Keyhani, A. Javadi, H. Mobli, K. Abrinia, and A. Sharifi, "A review of principle and sun-tracking methods for maximizing solar systems output," *Renewable and Sustainable Energy Reviews*, vol. 13, no. 8, pp. 1800–1818, 2009.

[7] M. Benghanem, "Optimization of tilt angle for solar panel: Case study for Madinah, Saudi Arabia," *Applied Energy*, vol. 88, no. 4, pp. 1427–1433, 2011.

[8] C. Praveen, "Design of automatic dual-axis solar tracker using microcontroller," in *Proceedings of the International Conference on Computing and Control Engineering (ICCCE '12)*, April 2012.

[9] D. F. Fam, S. P. Koh, S. K. Tiong, and K. H. Chong, "Qualitative analysis of stochastic operations in dual axis solar tracking environment," *Research Journal of Recent Sciences*, vol. 1, no. 9, pp. 74–78, 2012.

[10] A. M. Sharan and M. Prateek, "Automation of minimum torque-based accurate solar tracking systems using microprocessors,"

Journal of the Indian Institute of Science, vol. 86, no. 5, pp. 415–437, 2006.

[11] C. Alexandru and M. Comsit, *Virtual Prototyping of the Solar Tracking Systems*, Department of Product Design and Robotics, University Transilvania of Braşov, Brasov, Romania.

[12] A. Hsing, *Solar Panel Tracker*, Senior Project, Electrical Engineering Department, California Polytechnic State University, San Luis Obispo, Calif, USA, 2010.

[13] N. A. Kelly and T. L. Gibson, "Increasing the solar photovoltaic energy capture on sunny and cloudy days," *Solar Energy*, vol. 85, no. 1, pp. 111–125, 2011.

[14] M. B. Omar, *Low Cost Solar Tracker*, Faculty of Electrical & Electronics Engineering, Universiti Malaysia Pahang, 2009.

[15] A. Argeseanu, E. Ritchie, and K. Leban, "New low cost structure for dual axis mount solar tracking system using adaptive solar sensor," in *Proceedings of the 12th International Conference on Optimization of Electrical and Electronic Equipment (OPTIM '10)*, pp. 1109–1114, Braşov, Romania, May 2010.

[16] M. J. Clifford and D. Eastwood, "Design of a novel passive solar tracker," *Solar Energy*, vol. 77, no. 3, pp. 269–280, 2004.

[17] N. Barsoum, "Fabrication of dual-axis solar tracking controller project," *Intelligent Control and Automation*, vol. 2, no. 2, pp. 57–68, 2011.

[18] S. Rahman, R. A. Ferdaus, M. Abdul Manran, and M. A. Mohammed, "Design & implementation of a dual axis solar tracking system," *American Academic & Scholarly Research Journal*, vol. 5, no. 1, pp. 47–54, 2013.

[19] CdS Photoconductive Photocells, Advanced Photonix, http://www.cooking-hacks.com/skin/frontend/default/cooking/pdf/LDR-Datasheet.pdf.

[20] "Measure Light Intensity using Light Dependent Resistor (LDR)," http://www.emant.com/316002.page.

[21] Motor torque calculation, Leadshine technology, http://www.leadshine.com/Pdf/Calculation.pdf.

[22] "Selecting a stepping motor, Oriental motor," http://www.oriental-motor.co.uk/media/files/17112005105315.pdf.

[23] Technical reference, Oriental motor, http://www.orientalmotor.com/products/pdfs/2012-2013/G/usa_tech_calculation.pdf.

13

Assessment of Stand-Alone Residential Solar Photovoltaic Application in Sub-Saharan Africa: A Case Study of Gambia

Sambu Kanteh Sakiliba,[1] **Abubakar Sani Hassan,**[1] **Jianzhong Wu,**[1] **Edward Saja Sanneh,**[2] **and Sul Ademi**[3]

[1]*Institute of Energy, Cardiff University, Queen's Buildings, The Parade, Cardiff CF24 3AA, UK*
[2]*Ministry of Energy, Banjul, Gambia*
[3]*Institute for Energy and Environment, Department of Electronic & Electrical Engineering, University of Strathclyde, Technology and Innovation Centre, Level 4, 99 George Street, Glasgow G1 1RD, UK*

Correspondence should be addressed to Sambu Kanteh Sakiliba; sambuks@cardiff.ac.uk

Academic Editor: Yongsheng Chen

The focus of this paper is the design and implementation of solar PV deployment option, which is economical and easy to maintain for remote locations in less developed countries in Sub-Saharan Africa. The feasibility of stand-alone solar PV systems as a solution to the unstable electricity supply and as an alternative to the conventional resource, "diesel generators," is presented. Moreover, a design of a system is carried out, such that the electrical demand and site meteorological data of a typical household in the capital, Banjul, is simulated. Likewise, the life cycle cost analysis to assess the economic viability of the system, along with the solar home performance, is also presented. Such system will be beneficial to the inhabitants of Gambia by ensuring savings in fuel costs and by reducing carbon emissions produced by generators. The selection of appropriate-sized components is crucial, as they affect the lifetime, reliability, and initial costs. The design presented in this study represents a solution for domestic houses to adopt the system according to the location and environment, in order to meet electricity demand.

1. Introduction

In relation to the progress of a country, electricity is one of the elements required for agricultural, commercial, industrial, or residential development. In most countries of the world, areas with no electricity are less developed than those with electricity. The use of photovoltaics (PV) to produce electricity from sunlight would strongly benefit and improve the quality of life for those less developed countries such as Gambia. Closer studies would demonstrate that the energy sector in Sub-Saharan Africa can offer opportunities for implementing ambitious renewable energy (RE) programs. The decentralized approach based on power produced with locally available renewable energy resources is, for various reasons, gradually being recognized as a viable alternative in remote places of Gambia. This African country is making a considerable effort to provide electricity to rural and urban households generated from fossil fuel based resources, such as fuelwood and liquefied petroleum gas (LPG), although it has been a slow and ineffective process for many years [1]. Part of the population, including the capital Banjul, have no access to affordable energy resources with households living in situations where sometimes the electricity is unstable. Since 2007, power outages occur almost every day in Gambia, with an average duration of 6.86 hours [2].

Over the last few decades Gambia has witnessed a rapid increase in its population, infrastructure, and business expansion, triggering a rise in demand in the generation and transmission of electricity. Due to the poor state of transmission and distribution networks, the magnitude of the demand has exceedingly outstretched the available installed capacity, thus, culminating in the reoccurrence of frequent maintenance intervals and continuous load shedding [3]. Consequently, the historical and current billed electricity demand figures cannot reflect the actual requirements of the National Water and Electricity Company (NAWEC) customers, which has been under the management of the President's office since 2002. In Gambia regular power cuts have occurred since 1977 following the Sahelian drought; since then, the demand for electricity and water has increased. Over the past thirty years utility corporations have suffered mechanical breakdowns in the energy sector, with frequent power outages [2, 3]. Furthermore, businesses and hotels use generators; nevertheless, these are not financially sustainable due to the high cost of fuel [4]. The price of petrol and diesel in Gambia increased between February and April of 2013 from £0.72/l to £0.79/l for diesel and from £0.75 to £0.83/l for petrol, and it is expected to continue to rise in the future [5]. The daily average solar energy potential in the capital is 5.7 kWh/m^2 [6]; therefore, an alternative energy source such as solar PV systems could be a cost-effective alternative for households and allow their electricity requirements to be met, instead of using diesel generators.

In this paper, a method for the design of an alternative stand-alone solar PV system adoption option for Gambia was developed. The method was used to design and size a stand-alone system that will be economically competitive when compared to purchasing electricity from the utility grid.

The rest of the paper is organized as follows. Section 2 introduces Gambia as a country. Section 3 reviews the potential of the country's solar energy resource. Section 4 presents the typical building model for households in Banjul, the country's capital. Section 5 presents the model of the solar PV configuration of a typical solar PV system in Gambia [1]. Section 6 presents the methodology used for the stand-alone solar PV design, sizing, and simulation. Section 7 presents the results and discussion, and Section 8 represents the economic analysis of the system designed in Section 6.

2. Gambia

Gambia is situated on the Atlantic coast in West Africa surrounded by Senegal (Figure 1), and the capital is Banjul, located at latitude 13.2° north and longitude 16.6° of the equator. The country has a hot, tropical climate with a rainy season from June to November. The country's economy concentrates on tourism, farming, and fishing with a third of the population living under the international poverty line on £0.75/day [7].

2.1. Energy and Electricity Sector of Gambia. At the end of 2000, the energy consumption was 0.26 tonnes of oil per capita (TOE), which was supplied by 77% of firewood, 21% of petroleum products, and 2% of electricity (Figures 2 and 3). The principal consumers at this time were households and the transportation sector, with a percentage of 83% and 13%, respectively [2, 3]. NAWEC reported an electricity consumption of 80 GWh in 2002 with households being the major consumers facing extremely high electricity costs. As a result, many households struggled to pay electricity bills, due to the high price of the tariff. At that time, the tariff ranges were from £0.11/kWh to £0.15/kWh [3], and the capital, Banjul, was supplied by diesel generators with a total capacity of 44 MW. In Gambia, the electricity sector

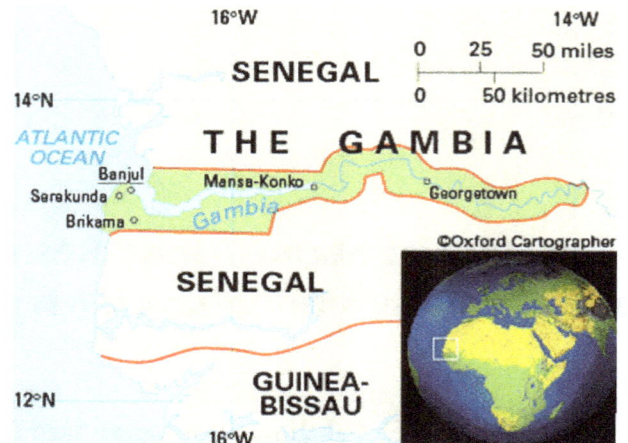

FIGURE 1: Map and location of Gambia [8].

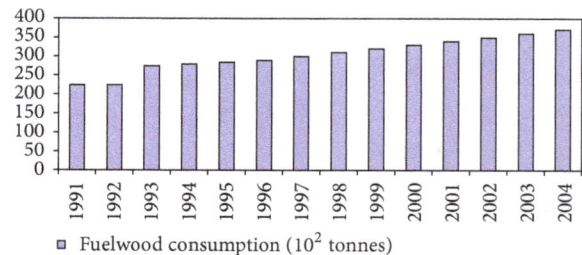

FIGURE 2: Fuelwood consumption from 1991 to 2004 [9].

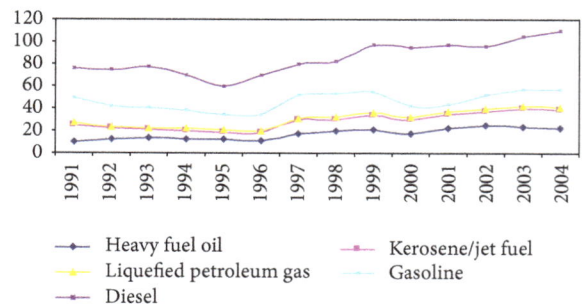

FIGURE 3: Petroleum consumption (10^2 tonnes) from 1991 to 2004 [9].

is presently facing serious issues and is unable to meet the demand. The transmission and distribution system is poor with technical and nontechnical losses requiring an improvement in efficiency, in order to reduce the high energy costs [6].

2.2. Household Demand. Most households in Gambia have no access to energy services. In 2005, 64% of urban users were connected to the grid [10], where six provincial centres were electrified with diesel fired isolated systems with a total capacity of 11 MW of electricity to supply, available for 15 hours on average [3, 10]. For lighting, other processes, for instance, kerosene or candles, can be used; nevertheless, these

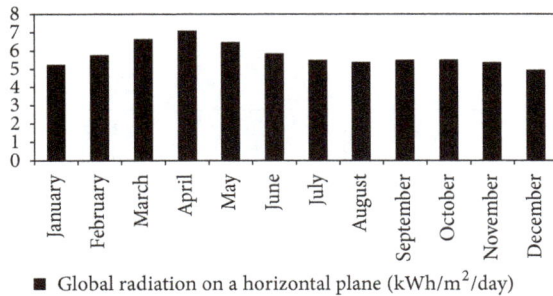

FIGURE 4: Monthly average daily global radiation.

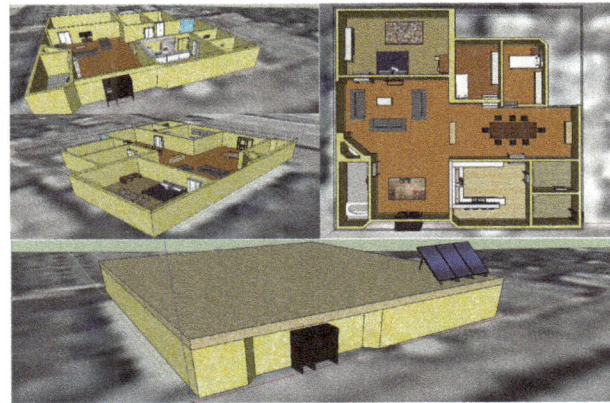

FIGURE 5: Three-dimensional model and representation of a typical urban house in Banjul.

methods are dangerous and reasonably inefficient. Therefore, stand-alone solar PV systems (SSPVS) in homes are perceived to be an excellent alternative to supply power to a population in need of solutions.

3. The Potential of Solar Energy

The energy sector in the country is facing difficulties leading to financial restrictions for the foreseeable future, due to its importation of fuel and a fragile supply. Gambia has considerable RE resource such as solar energy. The country benefits from high solar radiation all year round with an average of 4.5 to 6.7 kWh/m^2/day (Figure 4). Despite the rainy season (June to November) the country receives abundant amounts of solar radiation at about 5 kWh/m^2/day. At the moment, many SSPVS are in use across the country for many purposes such as electrification and water pumping in rural and urban areas. Most of the projects have been funded by the government and donors to provide energy in public services such as hospitals and schools and for street lighting.

4. Representation of a 3D Building Model

The goal of the 3D modelling as demonstrated in Figure 5 is to visualize the state of the art of a typical urban domestic dwelling in Banjul. The 3D model has also considered the domestic dwelling construction properties as well as the implemented renewable energy technology to provide electricity. The building was modelled using the graphical user interface Sketchup Pro 8 [11]. The domestic dwelling is south-oriented and is composed of a lounge, kitchen, dining room, three bedrooms, bathroom, electrical equipment room, and the SSPVS components room.

5. Household Solar PV System Configuration

The configuration for a SSPVS (Figure 6) was considered in this paper from a pilot study conducted in 2009 to evaluate the suitability of solar PV usage in urban communities [1]. The system consisted of three PV modules, inverter, battery bank, and load. The function of the PV panels is to convert sunlight into DC electrical power. The inverter is used to convert the DC electrical input power into AC output power to be supplied to the appliances. The battery bank stores

the excess DC power to be used when there is no sunshine. The controller monitors the electrical input generated from the solar panels, the amount entering the inverter, and the quantity of electrical energy for charging and discharging the battery bank.

6. Methodology

6.1. Stand-Alone Solar Photovoltaic System Design. The meteorological and environmental data was collected from the Photovoltaic Geographical Information System (PVGis) of the Joint Research Centre (JRC), Institute of Energy and Transports European Commission, to predict the performance of the SSPVS. Banjul's monthly average daily solar radiation incident on the horizontal surface is very high, especially in April, where radiation reaches 7.07 kWh/m^2/day on the horizontal (Figure 4). The load profile is assumed to run for 24 hours a day during whole year, as the average daily power outage in Gambia is 6.86 [2]. Total energy consumed was calculated from Table 1 as 3.332 kWh/day compensating previously the rated power (P_{rated}) of each appliance. In order to accomplish an effective power compensation (P_i), P_{rated} of each load has been divided by the adjustment factor. The adjustment factor is related to the efficiency of the inverter and reflects the actual power consumed from the battery bank to operate AC loads from the inverter [5]. For this application the adjustment factor is 0.90.

The design criteria for the SSPVS were considered using the minimum average solar radiation in a month to determine the size of the PV array and battery bank and by investigating the daily demands to meet the resulting sizes. Banjul is located at 13.2°N latitude and 16.6°E longitude. Therefore, the angle of inclination of the PV array measured from the horizontal (tilt angle) is maximized with an angle of 41° facing south, considering the less favourable season, winter:

$$\text{Tilt angle: } [(\text{Latitude} * 0.9) + \text{Season angle}], \quad (1)$$

where (i) latitude is 13.2°N and (ii) winter seasonal angle is 29°.

TABLE 1: Household load consumption data.

Appliances	Quantity	P_{rated} (kW)	$P_i = P_{rated}/0.90$ (kW)	Use (H/day)	Energy (kWh/day)
Lights	6	0.008	0.088	4	0.352
Fridge	1	0.300	0.333	6	1.998
Radios	1	0.008	0.088	1	0.088
Computers	1	0.250	0.277	2	0.554
TVs	1	0.020	0.022	4	0.088
DVD	1	0.010	0.011	2	0.022
Fans	1	0.042	0.046	5	0.230
Total		**0.678**	**0.865**	**24**	**3.332**

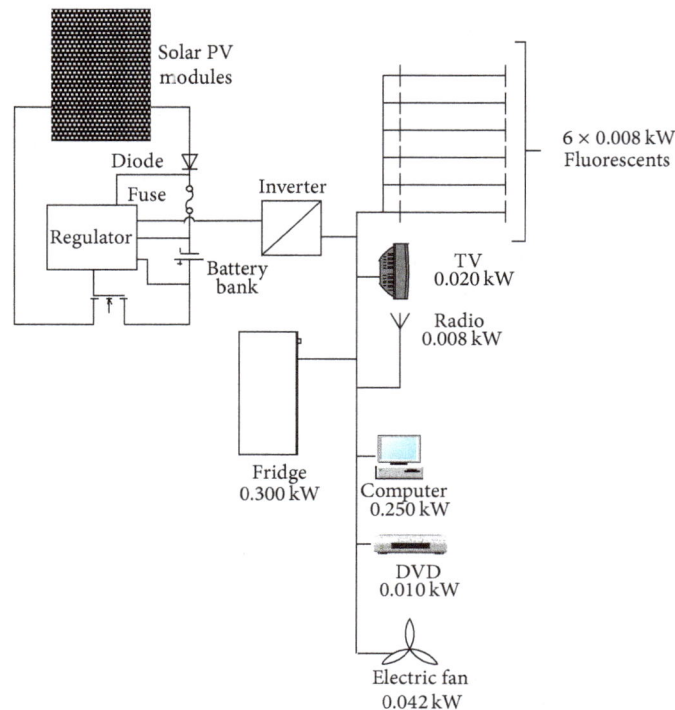

FIGURE 6: Ideal solar PV system for an urban Gambian home.

The PV array output is related to the intensity of light striking the panels, ambient and cell temperature, status of the loads, and characteristics of the PV array. The reason that the tilted angle was considered to be 41° and facing south is because the panels will be directly facing the sun at mid-day during those short winter days, hence, being more effective during the summer.

The lifespan of a battery depends on the depth of discharge (DoD); thus, in this case and in accordance with the selected battery, a DoD of 0.5 was considered. Battery sizing is based on the power supply during 3 autonomous days (N) [12]. The total power of the system is 0.865 kW (P_i) with a total energy consumed ($E_{d'}$) of 3.332 kWh consequently (Table 1); a DC system voltage of 24 V is suitable.

The current consumption by the load is 0.138 kAh, calculated by

$$I_d = \frac{E_{d'}}{\text{DC system voltage}}. \tag{2}$$

The selected battery is lead-acid vented tubular type, with a nominal voltage (V_{nb}) of 12 V, storage capacity (C_b) of 239 Ah, temperature derating factor (D.F) of 0.8 for a temperature of 25°C, and battery efficiency (μ) of 97% and without any maintenance required [12]. The battery bank is calculated below obtaining 2 batteries in series (B_s) and 3.46 ≈ 3.5 batteries in parallel (B_p):

$$B_s = \frac{\text{DC Voltage system}}{V_{nb}}$$

$$B_p = \left(\frac{((I_d * N)/\text{Dod})}{C_b} \right). \tag{3}$$

In consequence, the total number of batteries (B_t) required is 6.92 ≈ 7 from the product of (3):

$$B_t = B_s * B_p. \tag{4}$$

The maximum radiation time selected is 4.92 kWh/m²/day (M_{rt}) for the worst month of the year, which is December (Figure 4). Assuming a DC system voltage of 24 V, a power of 0.275 kW$_p$ (P_{panel}), short-circuit current of 5.80 A (I_{psc}), and rated voltage of 51.2 V (V_{rated}) from the panels and a 97% of battery efficiency (μ), the required output energy of the array (E_a) and output energy per module each day (E_{om}) calculated in this study are 3.435 kWh and 1.353 kWh, respectively, obtained by

$$E_a = \left(\frac{E_{d'}}{\mu} \right)$$

$$E_{om} = P_{panel} * M_{rt}. \tag{5}$$

The energy output of the modules (E_{out}) is related to the ambient temperature [13]. Energy temperature relationship influences the performance of the modules; hence, a D.F of 0.8 is justified for hot climates and critical applications such as Banjul [14]. Therefore, an energy output is obtained at an operating temperature of 1.082 kWh from the modules by

$$E_{out} = \text{D.F} * E_{om}. \tag{6}$$

In order to obtain the total number of modules (M_t), the PV panels in series (M_s) and parallel (M_p) are calculated priorly; therefore, the accomplished values are 0.48 and 6.34, respectively, by

$$M_s = \left(\frac{\text{DC system voltage}}{(V_{rated} * \mu)} \right)$$

$$M_p = \left(\frac{(E_a/E_{out})}{M_s} \right). \tag{7}$$

Thus, the total number of PV modules (M_t) is 3.04 ≈ 3, determined by

$$M_t = M_s * M_p. \tag{8}$$

The charge controller or regulator is required to charge the batteries and to maintain the long life of the battery bank. The regulators function is to carry 5.80 A of short-circuit current from the PV panels (I_{psc}). In this study, a maximum current charge controller (I_{pvc}) of 60 A is selected maintaining a constant DC voltage of 24 V. In order to confirm and ensure

that an appropriate regulator has been selected; the product of I_{psc}, M_p, and a factor of 125% [15] has to be equal to or not higher than I_{pvc} (9). For this case study, a total value of 45.96 was achieved, thus, performing under the limits of the charge controller carefully chosen (45.96 ≤ 60 A):

$$I_{psc} * M_p * 125\% \leq I_{pvc}. \tag{9}$$

The rated power of the inverter ($P_{inv\text{-}rated}$) is taken to be at least 20% higher than the appliances power (P_{rated}) [5]. The specifications for the required inverter are 2.4 kW (P_{inv}), 24 volts DC, frequency of 50 Hz, and a loss of coefficient produced by the inverter (k_c), running at optimum service taken to be 0.05. In addition, to make sure that a satisfactory inverter has been selected, $P_{inv\text{-}rated}$ (10) must be equal to or not higher than P_{inv}. When inverters are first turned on current flows are produced, exceeding the steady-state current value within a typical range from 3 to 6 times. This effect is known as In-Rush Current [16]; for this case it has been selected as 3:

$$P_{inv\text{-}rated} = \left[\left(\frac{P_{rated}}{(1 - k_c)} \right) * (\text{In-Rush Current}) \right]. \tag{10}$$

As a result, the power of the required inverter is higher than or equal to 2.141 kW ($P_{inv\text{-}rated}$), confirming the selected inverter as the appropriate to perform in the system (2.141 kW ≤ 2.4 kW). Table 2 represents a summary of the solar PV components selected in the design for this study.

6.2. System Simulation: PVSyst. The software PVSyst [17] was used in this study, in order to validate the design and data analysis of the PV system. The simulation dealt with the stand-alone system, including meteorological data, components of the system, and the solar energy tools. To simulate the system before, the daily energy consumption of an urban house was required and determined according to the minimum load requirements (Table 1).

The daily energy consumed was designed to be 3.332 kWh/day constantly, over the course of a year. Thus, the same parameters for the electrical demand have been introduced for the simulation.

6.2.1. Preliminary Design Input Data Procedure. The initial presizing was carried out, in order to give an estimation of the features of a PV system. The software carried out an evaluation of the systems yield in monthly values using the general parameters, without specifying the components of the system. The first step in the preliminary design was to specify the type of system (grid-connected, stand-alone, or pumping system). In this study, a stand-alone system was considered.

The next step involved choosing the site. Banjul was not available in the geographical site of the database software; hence, the location and meteorological data was imported from the RET Screen International software [18] with the purpose of obtaining the site parameters as shown in Table 3.

Once the meteorological data was imported monthly meteorological calculations were carried out, in order to determine the horizon and sun path diagrams for Banjul. The

TABLE 2: Solar PV system components.

Items	Quantity	Manufacturer model	Characteristics
PV panel	3	Auversun, AV275M96NB-5P	$P_{nom} = 0.275\,\text{kW}_p$ $G = 1000\,\text{W/m}^2$ $AM = 1.5$ $V_{rated} = 51.2\,\text{V}$
Charge controller	1	Morningstar, Tristar TS MPPT 60 A-24 V	$V = 24\,\text{V}$ $I_{max} = 60\,\text{A}$
Inverter	1	Sun Power, PVUP 3000	$P_{nom} = 2.4\,\text{kW}$ $V_{nom} = 230\,\text{V}$ Freq = 50 Hz Eff = 90%
Battery	7	Concorde, PVX-2580L	$C_{nom} = 239\,\text{Ah}$ $V_{nom} = 12\,\text{V}$ Eff = 97%

TABLE 3: Monthly meteorological data imported to the PVSyst [18].

Months	Global irradiation (kWh/m^2/month)	Diffuse irradiation (kWh/m^2/month)	Ambient temperature (C)	Wind velocity (m/s)
January	161.8	51.4	26.0	3.7
February	160.7	55.7	27.1	4.0
March	205.2	58.1	27.7	4.3
April	212.1	52.1	28.4	4.4
May	199.6	64.1	28.4	4.2
June	174.6	63.7	27.7	4.0
July	169.9	69.2	26.5	3.1
August	166.2	78.5	26.1	2.8
September	163.8	77.0	26.2	2.4
October	170.2	69.8	27.5	2.3
November	159.9	49.2	29.2	4.0
December	152.5	50.5	27.3	4.3
Year	2056.5	739.4	27.3	3.4

optimum angle of inclination for a fixed tilt collector in a stand-alone system maximizes the daily irradiation during the worst month. In this case, the worth month is December with the least irradiation of 4.92 kWh/m^2/day. Therefore, a tilt angle of 41° was also selected.

The final step was to verify and confirm the system parameters:

(i) Battery and system voltage, 24 V, was considered.

(ii) Number of days of autonomy (N) is the consecutive days without sunshine, to define the capacity of the battery bank. For this study, 3 days were considered.

(iii) Loss of Load Probability (LOLP) is the fraction of time where the electricity is not available during the time it is required. A LOLP of 5% was considered.

6.2.2. Preliminary Design Input Data Procedure. The aim of designing the system with PVSyst was performed using detailed monthly simulation. The simulation procedure was achieved as shown in Figure 7.

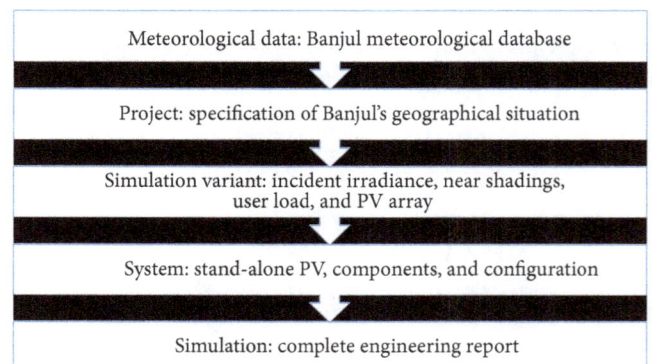

Meteorological data: Banjul meteorological database

Project: specification of Banjul's geographical situation

Simulation variant: incident irradiance, near shadings, user load, and PV array

System: stand-alone PV, components, and configuration

Simulation: complete engineering report

FIGURE 7: Outline of simulation process.

Initially, the input data included details of location, solar irradiance, and name of the project. Consequently, the feeding of the values was accomplished; the Albedo values were considered as a default of 0.20. Albedo values are

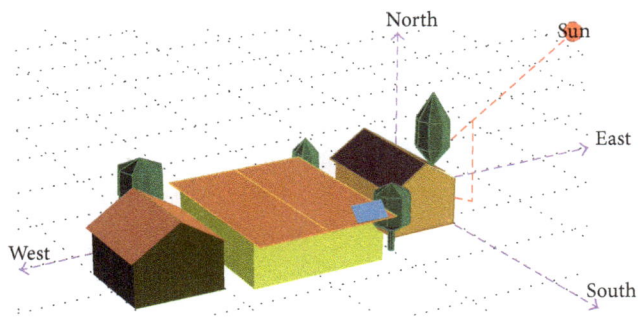

FIGURE 8: South facing PV array and its environment in Banjul.

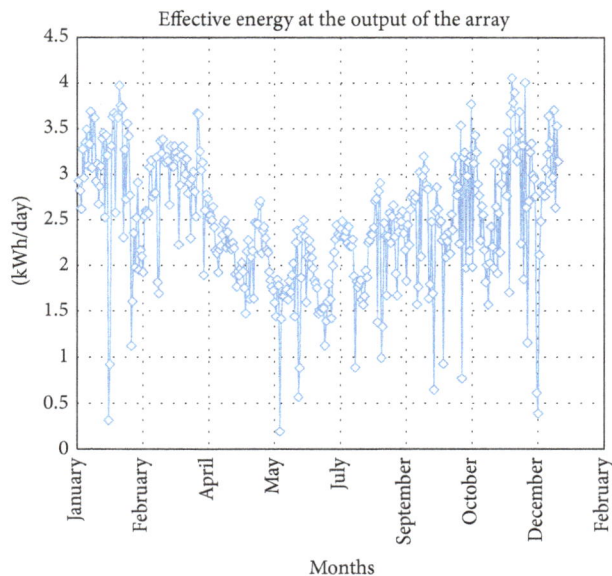

FIGURE 9: Daily array output energy.

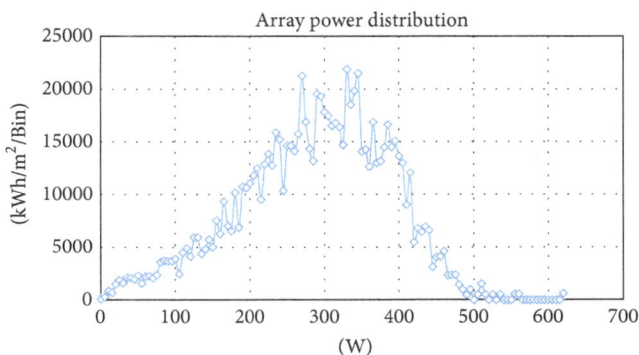

FIGURE 10: Array power distribution.

the reflecting power of the surface of a location, from the irradiance of the sunlight.

As it had been set in the preliminary design, the PV modules are on a fixed tilted plane, with an angle of 41°. The azimuth angle was 0° for the location of the sun on the east-west axis. A fixed tilted plane was designed for this study, since the system was optimized for lower monthly radiation. The properties of the system were defined as the third step. The definition of the parameters depends on the type of components necessary to achieve the load requirements for the households. The components were defined according to Table 2.

It was not possible to reconstruct the structure of the house with a computer aided design software such as Sketchup to introduce the 3D shading scenes. Therefore, the near shadings by trees and other buildings were not considered as part of the simulation. Figure 8 represents the reconstruction of the house modelled in PVSyst.

The simulation process involved the available variables in monthly tables and graphs. The data of interest was defined before the simulation, in order to be collected in hourly or daily values. The software offered three methods for the output of the data: (a) accumulating hourly values; (b) ASCII export files; and (c) special graphs.

After the simulation, as a final step, the results dialogue was made available, in order to obtain a printable report. The report included all the parameters used during the simulation, together with a description of the results.

7. Results and Discussions

The simulation engine used random daily data, in order to obtain real weather characteristics. The energy exported on a daily basis varies from day to day with different seasonal patterns. Figure 9 represents that the output energy is lower between June and October compared with the rest of the year, due to the losses produced by the rainfall during the wet season (June to October). In addition, the array output resulted in a reverse "Weibull curve" with high values of power generated between 250 and 350 W as shown in Figure 10. The energy production ceases above 550 W, due to the constant operation of the array at temperatures above the Standard Test Condition (STC).

In Banjul, the highest temperatures are in November with an average of 29°C; thus, this affects the operating temperatures of the PV array during the year as shown in Figure 11. It is ascertained that high temperatures will significantly affect the open-circuit voltage and, hence, the power that the PV array can deliver [19].

A yearly average performance ratio (PR) of 0.526 was observed in Figure 12, with the maximum in January. A low PR was obtained in the month of June due to the temperature derating through the hot and humid summer. Likewise, the solar fraction (SF), which is the fraction of the solar energy available (E_{sol}) and the energy needed for the user (E_{load}), was found to be 0.696 over the course of the year. PR values lower than 75% should be investigated according to [20, 21]. The analysis has shown that the performance depends on the components efficiency, their design, and load configuration. The PR alone cannot describe the operation of the SSPVS from a technical point of view; therefore, a more detailed analysis in terms of the system operation will require detailed and reliable monitoring devices, studies on the evolution of

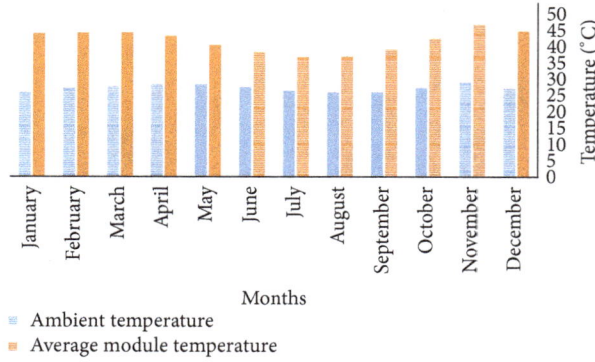

FIGURE 11: Ambient temperature versus average module temperature.

- Ambient temperature
- Average module temperature

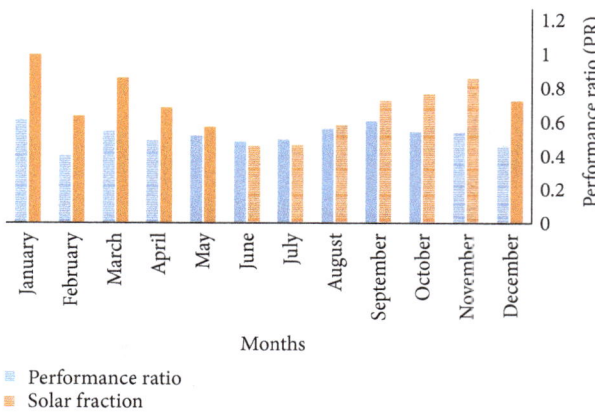

FIGURE 12: Performance ratio and solar fraction.

- Performance ratio
- Solar fraction

TABLE 4: Costs of all items [16, 17].

Items	Cost (Pound Sterling)	Cost (Gambian Dalasi)
PV panel	£0.41/W_p	GMD27.48/W_p
Battery	£0.53/Ah	GMD35.54/Ah
Charge controller	£2.65/A	GMD177.72/A
Inverter	£0.18/W	GMD12.16/W
Installation	10% of PV panel cost	
Maintenance/year	2% of PV panel cost	

TABLE 5: Total cost of items [16, 17, 19].

Items	Cost (C_{it})
Cost of the PV panel (C_{PVp})	$(3 * 0.275 \, kW_p) * £0.41 = £338.25$
Initial cost of the batteries (C_b)	$(7 * 239 \, Ah) * £0.53 = £886$
Cost of the charge controller (C_c)	$£2.65 * 60 \, A = £159$
Cost of the inverter (C_{inv})	$£0.18 * 2.4 \, kW = £432$
Installation costs (C_{inst})	$10\% * £338.25 = £33.82$
Maintenance cost per year (M/yr)	$2\% * £338.25 = £6.76$

they do not require maintenance; consequently, they have a life span of roughly 10 years. Therefore, a group of 7 batteries has to be purchased after 10 and 20 years. According to the "Gambia Mineral & Mining Sector Investment and Business Guide," the country's inflation rate (i) is 6%, with a discount rate (d) of 10% [22]. Consequently, the total cost of all items (C_{it}) based on the data collected in Table 4 can be calculated as shown in Table 5 [23].

The 1st battery group (C_{b1}) purchased after 10 years (N_1) can be calculated to be £612.21, and consequently a 2nd (C_{b2}) group of batteries to be purchased after 20 years (N_2) can be calculated to be £422.70 using the following equation:

$$C_{b1,b2,...,bn} = \left[C_b * \left(\frac{(1+i)}{(1+d)} \right)^{N_{1,2,...,n}} \right]. \quad (11)$$

The total annual maintenance costs ($C_{M/yr}$) over the life time of 20 years (N) can be calculated to be £93.69 ≈ £93.70 using the maintenance cost per year (M/yr) with the following equation [24]:

$$C_{M/yr} = (M/yr) * \left(\frac{1+i}{1+d} \right)$$
$$* \left[\frac{1 - ((1+i)/(1+d))^N}{1 - ((1+i)/(1+d))} \right]. \quad (12)$$

Therefore, L_{cc} of the system is calculated to be £2977.43 from the following equation:

$$L_{cc} = \left(C_{PVp} + C_b + C_{b1} + C_{b2} + C_c + C_{inv} + C_{inst} + C_{M/yr} \right). \quad (13)$$

the user behavior over time, or the use of advanced simulation tools to evaluate the influence of component sizes.

Figure 13 illustrates the SSPVS energy losses, where a 7% loss is produced from converting the horizontal irradiation to the global irradiation incident. Furthermore, the large loss of 9.2% is caused by the temperature derating effect (hot and humid summer), with converter losses at 4.9% and 0% shading losses. The rest of the losses are attributable to the Incidence Angle Modifier (IAM), module array mismatch losses at 1.9%, model quality losses at 2%, and cable losses at 3.1%. The irradiance level losses of 3.4% are due to the low irradiance and high STC irradiance levels.

8. Economic Analysis

8.1. Life Cycle Cost Analysis. In this section, the estimation of the life cycle cost (L_{cc}) is discussed. A high initial cost is essential, but the advantages are that the maintenance cost is low, and there are no fuel costs as using diesel generators. The expenditure for the system, including the purchasing and replacement costs, is represented with the corresponding currency rates in Table 4.

The life time (N) of all the items, excluding the battery bank, is considered to be approximately 20 years. The batteries selected for this project are lead-acid vented tubular and

2096 kWh/m²   Horizontal global irradiation

−7.0%   Global incident in coll. plane

−3.4%   IAM factor on global

1884 kWh/m² * 5 m² coll.   Effective irradiance on collectors

Efficiency at STC = 15.81%   PV conversion

1579 kWh   Array nominal energy (at STC effic.)

−3.7%   PV loss due to irradiance level

−9.2%   PV loss due to temperature

−2.0%   Module quality loss

−1.3%   Module array mismatch loss

−3.1%   Ohmic wiring loss

−23.5%   Loss with respect to the MPP running

Missing energy   Unused energy (full battery) loss

899 kWh   −0.0%   Effective energy at the output of the array

43.6%   Direct use   Stored   Battery storage

368.9 kWh   16.4%   83.6%   Battery stored energy balance

−0.7%   Battery efficiency loss

−5.8%   Gassing current (electrolyte dissociation)

−0.4%   Battery self-discharge current

−0.3%   Energy supplied to the user

847 kWh   Energy need of the user (load)

1215 kWh

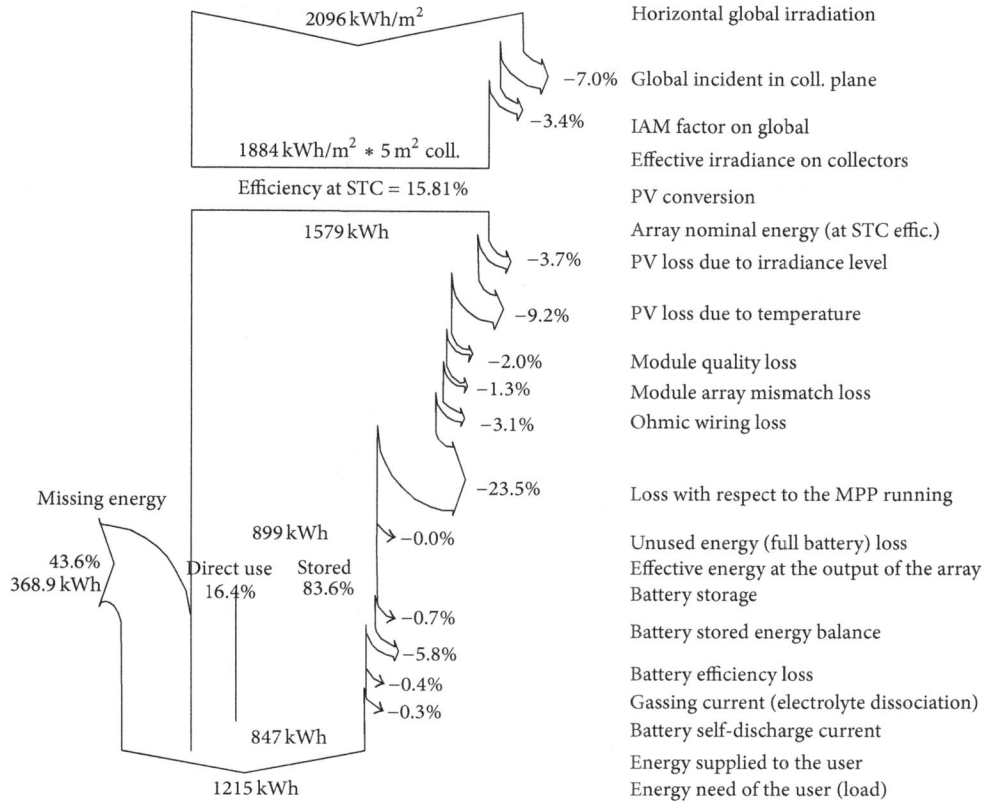

FIGURE 13: Energy losses diagram.

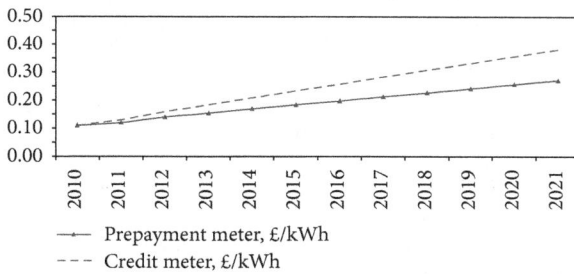

FIGURE 14: Projected NAWEC utility prices (2012–2021).

In order to calculate the unit electrical cost of 1 kWh, it is necessary to calculate L_{cc} of the system on an annual basis. Hence, the annual life cycle cost (AL_{cc}) of the SSPVS can be calculated to be £206.67, from [24]

$$AL_{cc} = L_{cc} * \left[\frac{1 - ((1 + i) / (1 + d))}{1 - ((1 + i) / (1 + d))^N} \right]. \quad (14)$$

Finally, once annual life cycle cost (AL_{cc}) is calculated, the unit electrical cost can be calculated to be £0.169/kWh ≈ 0.170/kWh using

$$\text{Unit electrical cost} = \frac{AL_{cc}}{E_{d'} * 365}. \quad (15)$$

Solar PV system installers are encouraged to sell the electricity of the PV system at a price not lower than £0.170/kWh, in order to maximize their profit. In 2010, NAWEC was offering a domestic utility tariff of £0.110/kWh for prepayment and credit meters; however, in a short period there was a price rice of 3% and two years later up to 5% [25]. Figure 14 represents a projected graph of the domestic utility prices from 2010 to 2021 if the energy prices continue rising. The results show that, in 2021, the prepayment meter and credit meter might increase for about 0.27£/kWh and 0.38£/kWh, respectively.

9. Conclusion

Gambia is currently in a challenging situation, with a large number of its citizens unable to access electricity in their daily lives. Network infrastructures are still underdeveloped and there is a general lack of detailed data on their deployment and, in addition, the expansion plans. A complete computer simulation was undertaken to study the performance of an SSPVS in a domestic house in Banjul. The purpose of the simulation was described with a methodology in detail and involved research for data collection and an investigation into the location. The manual system sizing and design made by the simulation engine had very accurate results, as the quantity of components required to supply power from the SSPVS has all matched, except the number of batteries (7 for the manual sizing and 8 from PVSyst).

The simulations have demonstrated the efficiency of the model, performance of the components, and production of the system's results, although the software lacks the facility of importing 3D models as presented in Section 4 using Sketchup Pro 8, which limits the capability of PVSyst. A model import function would permit the importation of proposed buildings into PVSyst for a better optimisation and modelling.

The life cycle cost analysis analysis noted that in Gambia the price of using SSPVS is lower compared to the unit electricity price. This price is predicted to diminish in the future, if the initial cost of the PV modules decreases. Meanwhile, if the unit cost of electricity grows to be three times its current value, due to an increase in fuel prices [3], the demand for solar PV systems for use in domestic housing will increase. The results of the L_{cc} study demonstrated that electrifying a domestic house is favourable and suitable with regard to long term investment, as the prices of the PV system components continue to reduce. Likewise, they represent a vital, efficient, and economical alternative resource to diesel generators [3, 4].

Conflict of Interests

The authors declare that there is no conflict of interests regarding the publication of this paper.

Acknowledgments

The authors would like to thank Toshiba Research Europe (TRE), the Engineering and Physical Sciences Research Council (EPSRC), and Cardiff University for their technical and financial support.

References

[1] E. S. Sanneh and A. H. Hu, "Lighting rural and peri-urban homes of the gambia using solar photo-voltaics (PV)," *Open Renewable Energy Journal*, vol. 2, no. 1, pp. 1–13, 2009.

[2] The Encyclopedia of the Nations, "Average Duration of Power Outages (hours)—Enterprise Survey Indicators, Country Comparison, Nations Statistics," November 2014, http://www.nationsencyclopedia.com/WorldStats/ESI-average-duration-power-outages.html.

[3] Lahmeyer International GmbH, *Renewable Energy Master Plan for the Gambia*, Lahmeyer International GmbH, 2006.

[4] Access Gambia, *Gambia Electrical Blackouts*, 2014, http://www.accessgambia.com/information/power-outages.html.

[5] J. V. Roger and A. Messenger, *Photovoltaic Systems Engineering*, vol. 40, CRC Press, Boca Raton, Fla, USA, 2nd edition, 2001.

[6] Lahmeyer International GmbH, *Feasibility Study Solar Home System Program, the Gambia*, Energy Division, Office of the President, Banjul, Gambia, 2006.

[7] United Nations Development Programme (UNDP), "Human development reports," Tech. Rep., 2008, http://hdr.undp.org/en/media/HDI_2008_EN_Tables.pdf.

[8] Brufut Eduction, "The Gambia—The Brufut Education Project," 2007, http://brufuteducationproject.com/the-gambia/.

[9] Lahmeyer International GmbH, *Energy Study Draft RE Master Plan—Module I. Energy Demand Assessment and Projection*, Lahmeyer International GmbH, Bad Vilbel, Germany, 2005.

[10] S. Kinteh, "Report of a National Household Energy," 2005.

[11] Last Software, "SketchUp Pro, SketchUp," October 2015, http://www.sketchup.com/products/sketchup-pro.

[12] Concorde Battery Corporation, "Battery Sizing Tips for Stand Alone PV Systems," 2009, http://www.concordebattery.com/.

[13] A. D. Jones and C. P. Underwood, "A thermal model for photovoltaic systems," *Solar Energy*, vol. 70, no. 4, pp. 349–359, 2001.

[14] M. Arif and M. E. Khan, "Design and life cycle cost analysis of a SAPV system to electrify a rural area household in India," *Current World Environment*, vol. 5, no. 1, pp. 101–106, 2010.

[15] National Electrical Code Committee, "Conductor sizing and over currents protection," in *National Electrical Code*, vol. 14, pp. 154–175, National Electrical Code Committee, McIntosh, Ala, USA, 2014.

[16] Internationational Electrotechnical Commission, *Internationational Electrotechnical Commission*, IEC 61000-4-30, Electromagnetic Compatibility (EMC), 2003.

[17] Rets. I. Government of Canada, Natural Resources Canada, Energy Sector, CANMET Energy Technology Centre—Varennes, "RETScreen International Climate data", http://www.retscreen.net/ang/d_data_w.php.

[18] N. R. C. E. S. C. E. T. C.–V. Rets. I.Government of Canada, "RETScreen International Climate data".

[19] S. Dubey, J. N. Sarvaiya, and B. Seshadri, "Temperature dependent photovoltaic (PV) efficiency and its effect on pv production in the world—a review," *Energy Procedia*, vol. 33, pp. 311–321, 2013.

[20] D. Mayer and M. Heidenreich, "Performance analysis of stand alone PV systems from a rational use of energy point of view," in *Proceedings of the 3rd World Conference on Photovoltaic Energy Conversion*, vol. 3, pp. 2155–2158, IEEE, Osaka, Japan, May 2003.

[21] W. G. J. H. M. Van Sark, N. H. Reich, B. Müller, A. Armbruster, K. Kiefer, and C. Reise, "Review of PV performance ratio development," in *Proceedings of the World Renewable Energy Forum (WREF '12)*, vol. 6, pp. 4795–4800, American Solar Energy Society, Denver, Colo, USA, May 2012.

[22] USA International Business Publications, *Gambia Mineral & Mining Sector Investment and Business Guide*, World Stra, 2007.

[23] P. Birajdar, S. Bammani, A. Shete, R. Bhandari, and S. Metan, "Assessing the technical and economic feasibility of a stand-alone PV system for rural electrification: a case study," *International Journal of Engineering and Research Applications*, vol. 3, no. 4, pp. 2525–2529, 2013.

[24] T. Markvart, *Solar Electricity*, John Wiley & Sons, 2000.

[25] Nawec, "National Water and Electricity Company—Utility Tariff," 2012, http://www.nawec.gm/index.php/faqs/tariff.

14

Prospect of *Pongamia pinnata* (Karanja) in Bangladesh: A Sustainable Source of Liquid Fuel

P. K. Halder,[1] N. Paul,[2] and M. R. A. Beg[3]

[1]*Jessore University of Science & Technology, Jessore 7408, Bangladesh*
[2]*Bangladesh University of Engineering & Technology, Dhaka 1000, Bangladesh*
[3]*Rajshahi University of Engineering & Technology, Rajshahi 6204, Bangladesh*

Correspondence should be addressed to P. K. Halder; pobitra.halder@gmail.com

Academic Editor: Adnan Parlak

Energy is the basic requirement for the existence of human being in today's digital world. Indigenous energy of Bangladesh (especially natural gas and diesel) is basically used in power generation and depleting hastily to meet the increasing power demand. Therefore, special emphasis has been given to produce alternative liquid fuel worldwide to overcome the crisis of diesel. *Pongamia pinnata* (karanja) may be an emerging option for providing biooil for biodiesel production. Although karanja biooil has been used as a source of traditional medicines in Bangladesh, it can also be used for rural illumination. This paper outlines the medical and energy aspects of *Pongamia pinnata*. It has been assessed that Bangladesh can utilize about 128.95 PJ through Pongamia cultivation in unused lands. The paper reviews the potentiality of *Pongamia pinnata* as a source of biodiesel and its benefits in Bangladesh. The paper also revives that, about 0.52 million tons of biodiesel can be produced only utilizing the unused lands per year in sustainable basis as it reduces CO_2, CO, HC, and NO_x emission compared to pure diesel.

1. Introduction

Pongamia pinnata (L.) Pierre (family: Leguminosae) is an important nonedible minor oilseed tree [1] that grows in the semiarid regions. It is probably originated from India and grows naturally in India, Bangladesh, Pakistan, Malaysia, Thailand, Vietnam, Australia, Florida, and Sri Lanka and also in northeastern Australia, Fiji, Japan, and the Philippines [2]. In the USA *Pongamia pinnata* was introduced into Hawaii in the 1960s by Hillebrand [3]. In Bangladesh it is popularly known as Koroch. It is an adaptable tree for tropical and subtropical regions which requires excellent drainage and a sunny location. In India, billions of karanja trees exist where karanja trees are cultivated commercially and seed is collected from December to April. However, in Bangladesh it is not cultivated commercially yet. In India, one person can collect 180 kg of seeds in 8 hours of a day where the collection cost is INR 4 per kg [4].

1.1. Classification

> Kingdom: Plantae
>
> Division: Magnoliophyta
>
> Class: Magnoliopsida
>
> Order: Fabales
>
> Family: Leguminosae
>
> Genus: *Pongamia*
>
> Species: *pinnata*

Source: [5]

1.2. Botanic and Chemical Characteristics. *Pongamia pinnata* (chromosome number: 22) is a very fast-growing medium size plant with an average height of 30–40 feet and spreads canopy for casting moderate shade. *Pongamia pinnata* has a

FIGURE 1: Botanical feature of *Pongamia pinnata*: (a) karanja tree with green leaf, (b) flower morphology of karanja showing standard petal; about 25–35% of flowers set to seed, (c) elliptical karanja seed pods containing seed inside them, (d) karanja seeds which weigh about 2-3 g/seed and contain about 30–40% oil, and (e) karanja pod shells which weigh about 2 g/shell.

varied habitat distribution and can grow in a wide range of conditions. It can grow in various types of soil like salty, alkaline, hefty clay, sandy, stony, and waterlogged soils and also shows high tolerance against drought bearing temperature up to 50°C. The trunk is usually short with a diameter of more than 1.64 feet. The leaves are comprised of 5–7 leaflets 5–10 cm long and 4–6 cm wide which are arranged in 2-3 pairs. On the other hand, the bark is thin and gray to grayish-brown in color with yellow on the inside where the tap root is thick and long [6]. Pea-shaped flowers are generally 15–18 mm long and pink, light purple, or white in color [5]. The elliptical pods consist of single seed inside the thick walled pod shell which are 3–6 cm long and 2-3 cm wide as shown in Figure 1. The pods are dried in sun and the seeds are extracted by thrashing. Seeds are light brown in color with a length of 1.0–1.5 cm [5]. About 9–90 kg of seed pods can be obtained from one tree which yields up to 40% oil per seed and around 50% of this oil is C 18 : 1, which is considered as suitable for biodiesel production [7]. Another study shows that about 8–24 kg of kernels is obtained from one tree which yields 30–40% oil [8, 9]. The seeds naturally exist for about 6 months [10]. The air dried kernels consist of 19% moisture, 27.5% oil, 17.4% protein, 6.6% starch, 7.3% crude fiber, and 2.3% ash [9, 10].

1.3. Cultivation of Pongamia pinnata in Bangladesh. *Pongamia pinnata* is one of the few nitrogen-fixing trees which are predominantly cultivated easily through seeds. The genetic diversity has been conserved through storage of seeds which is the most common conventional and economical method [11, 12]. The growth of *Pongamia pinnata* is seen from sea level to an altitude of around 1200 m and an optimal annual rainfall of 500 to 2500 mm. The trees are naturally distributed along the coasts and riverbanks in lands and are native to the Asian subcontinent. Furthermore, these are also cultivated along roadsides, canal banks, and open farmlands. About $60 \times 60 \times 60$ cm^3 pits are appropriate for planting where the spacing between rows should be 5 m and plant to plant distance is recommended to be 4 m [13]. Generally, three irrigations may be given in a year for better growth and development of the plants.

A simple and reliable method selection is the primary step for the successful propagation of *Pongamia pinnata* tree. However, coppicing and pollarding are considered as fruitful ways of agroforestry management practices for *Pongamia pinnata* [14]. The successful propagation methods of *Pongamia pinnata* are comprised of through seeds, through cuttings, and through layering and drafting. It can be easily propagated through seeds by direct sowing both in the nursery bed and in the polybags. However, seeds can be effectually used for mass propagation of *Pongamia pinnata* [15, 16]. A study shows the direct relationship between seed size and germination efficiency [17]. It can also be propagated through semihard wood stem cutting. Moreover, air layering and cleft drafting is the other process for successful propagation. The unused and marginal lands of Bangladesh can be brought under *Pongamia pinnata* cultivation to meet the need of liquid fuel.

2. Versatile Applications of *Pongamia pinnata*

Historically, all the parts of *Pongamia pinnata* like flower, seed, leaf, root, and so forth have been utilized as a source of traditional medicines, animal fodder, green manure, timber, fish poison, and fuel in India, Bangladesh, and other neighbouring regions.

2.1. Pongamia pinnata Wood. Traditionally, *Pongamia pinnata* wood is used as fuel in rural areas in Bangladesh. It has no distinct heartwood and varies from white to yellowish-grey color with a calorific value of 19.32 MJ/kg. The wood is considered as low quality timber due to its softness, tendency to split during sowing, and vulnerability to insect attack [6]. Therefore, the wood is used for stove top fuels, poles and ornamental carvings [18], cabinet making, cart wheels, posts [19], agricultural implements, tool handles, and some usual activities [20]. The ash produced from burning wood is used for dyeing [21].

2.2. Pongamia pinnata Oil. Oil is considered the most noteworthy product obtained from the *Pongamia pinnata* seeds. It is a thick, yellow or reddish-brown oil which has a calorific value of 40.756 MJ/kg, extracted through expeller, solvent extraction, and so forth. The oil is nonedible, acrimonious in taste, and offensive in smell and is used for commercial processes maybe as medicine and lamp fuel and for the production of biodiesel. Furthermore, it is used as fuel for cooking, as a lubricant, as water-paint binder, in leather dressing, and in soap-making, candles, and tanning industries [22]. Crude karanja oil (CKO) has also the application in body oils, salves, lotions, hair tonics, shampoos, and pesticides [23].

2.3. Pongamia pinnata as Fodder and Feed. The *Pongamia pinnata* leaves contain 43% dry matter, 18% crude protein, 62% neutral detergent fiber, and in vitro dry matter digestibility of 50% and are eaten by cattle and readily consumed by goats. The trees have a significant value in arid regions, however the use is not common. The deoiled cakes could be used as poultry feed and cattle feed [24].

2.4. Pongamia pinnata as a Medicine. Even though all parts of the plant are noxious, the flowers and fruits along with the seeds are used in many traditional medicines. Flowers are used to treat bleeding hemorrhoids whereas fruits aid treatment of abdominal tumors, ulcers, and hemorrhoids. Seed powder reduces fever and helps in treating bronchitis and whooping cough. On the other hand, leaves juices aid in treatment of leprosy, gonorrhea, diarrhea, flatulence, coughs, and colds. Besides, bark relieves coughs and colds and mental disorder. Root is used as a toothbrush for oral hygiene while root juice is used to clean ulcers [25]. *Pongamia pinnata* oil is styptic, anthelmintic, and good in leprosy, piles, ulcers, chronic fever, liver pain [26], and rheumatism arthritis scabies [27].

2.5. Seed Cake as Fertilizer. Seed cake, a byproduct of oil extraction, is bitter and unfit for animal feed. It is rich in protein and nitrogen and is used as green manure to fertilize the land. It is also used as a pesticide, especially against nematodes. Besides, the seed cake can be used for biogas production.

2.6. Soil Erosion. *Pongamia pinnata* trees are usually planted along the highways, roads, and canals to stop soil erosion. The plants develop a lateral network of roots for controlling soil erosion and binding sand dunes [28]. Thus, karanja plantation can reduce soil erosion with many other benefits as described.

3. *Pongamia pinnata*: A Viable Alternative to Liquid Fuel

Diesel and kerosene are considered as major liquid fuel and account for 90% of the country's total fuel. Thus *Pongamia pinnata*, a source of nonedible vegetable oil, is considered as the most significant to use as an alternative liquid fuel due to its nonfood use and less expense for production.

3.1. Alternative Oil for Kerosene. Kerosene, the most common liquid fuel, is traditionally used in rural illumination and for cooking purpose in Bangladesh. In the country, only about 49% of total population has access to electricity and has only few facilities to use LPG for cooking purpose. Therefore, population of remote and coastal areas use kerosene lamp and hurricane for lighting and the poor people of those villages use biomass burning stove for cooking and heating purposes. On the contrary, population of urban and semiurban areas use kerosene stove for cooking where LPG and biomass are not available. In the fiscal year 2012, the country has consumed about 508.5 million liters of kerosene as shown in Figure 2 [29]. The country imported about 124.28 million liters of kerosene in 2010 which was 24.41% of total consumed kerosene account as 509.24 million liters. Bangladesh Petroleum Corporation (BPC) has to pay USD 134.53 per barrel to import kerosene and jet fuel. BPC has planned to import 0.267 million tons of kerosene, jet fuel, and octane during the fiscal year 2013-2014. BPC is incurring loss of BDT 14–16 per liter for kerosene though the price is hiking.

Increasing price (68 BDT/liter) of kerosene is trending in reduction of kerosene consumption and encouraging finding out alternative liquid fuel option to kerosene. Furthermore, the reserve in the country is very limited. Hence, the consumption rate of kerosene is reduced by 37.42% from fiscal year 2001 to fiscal year 2012 though the demand is high. Recently, several techniques have been adapted to produce

TABLE 1: Comparison of *Pongamia pinnata* oil with other biomass derived oil and kerosene.

Variable	Pongamia seed [31]	*Jatropha curcas* [137]	Plum seed [138]	Mahogany seed [139]	Coconut seed [140]	Kerosene [30, 141]
Kinematic viscosity (cSt)	29.65[a]	52.76[a]	1.14[b]	3.8[b]	1.99[b]	2.71[c]
Density (kg/m^3)	912	932	940	1525	1095.5	780–810
Flash point (°C)	241	240	112	60	>100	37–65
HHV (MJ/kg)	40.756	39.774	22.4	32.4	21.40	46.2

[a]Value is at 30°C; [b]value is at 26°C; [c]value is at 20°C.

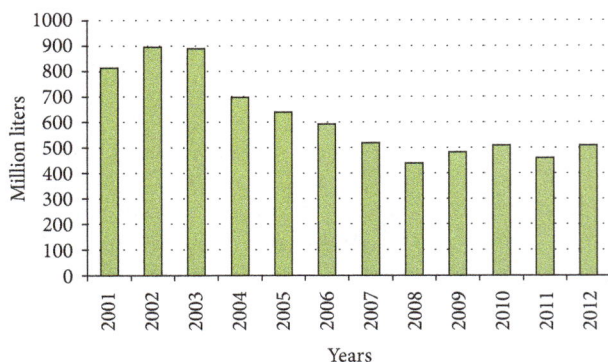

FIGURE 2: Kerosene consumption pattern in Bangladesh [29].

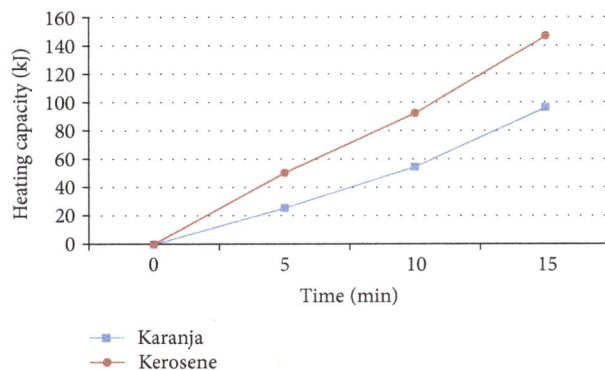

FIGURE 3: Performance characteristics of karanja and kerosene in gravity stove [31].

liquid fuels from nonedible seeds in renewable basis. Table 1 shows the comparison of some seed oils to kerosene. Kerosene has a calorific value of 46.2 MJ/kg with a density of 780–810 kg/m^3 [30] and a maximum distillation temperature of 205°C at the 10% recovery point. Among the seed oils *Pongamia pinnata* and *Jatropha curcas* have the competitive calorific value though the density is higher than kerosene. *Pongamia pinnata* has a calorific value of 40.756 MJ/kg [31] which is higher than other seed oils and slightly lower than kerosene.

Nonedible seed oil or straight vegetable oil (SVO) is rare to use in lighting and cooking due to its high density and viscosity. Hence, use of biooil in traditional kerosene stove for cooking and in wick-fed lamp for lighting shows the poor result due to its low capillary action. A recent study shows the effective way to use the crude karanja oil (CKO) for cooking purpose [31]. Use of karanja oil in gravity stove where fuel is feed under gravity shows the promising result as indicated in Figure 3 [31]. The oil also shows attractive heating performance in it though the method is not commercially well-known. Therefore, karanja oil may be an effective option for an alternative to kerosene.

3.2. Pongamia pinnata Biodiesel as an Alternative to Diesel.

Diesel is an indispensable fuel generally used in industrial and agricultural goods transports, in vehicles, and in diesel tractors and pumps for irrigation. Bangladesh is a developing country with a total population of 150 million. The population of the country is growing rapidly which is creating the petrodiesel utilization sector.

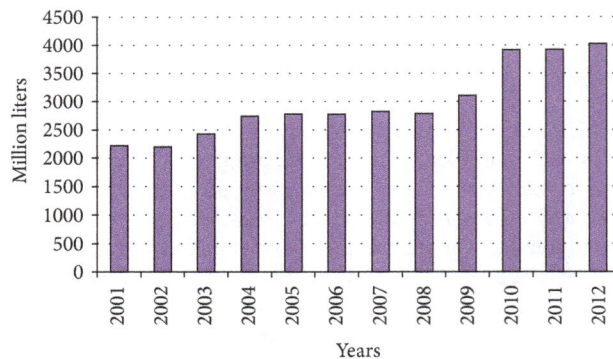

FIGURE 4: Diesel consumption pattern in Bangladesh [29].

In Bangladesh demand of diesel fuel is increasing day by day; hence it is necessary to find renewable alternative to diesel immediately. Not only Bangladesh's but also the world's diesel requirement is growing firstly. Bangladesh consumed about 4021.65 million liters of diesel in 2012 where the neighboring country India consumed about 66 million tons in 2011-12 [32]. Figure 4 shows the increasing trend of diesel consumption pattern in Bangladesh. The country has planned to import about 3 million tons of diesel where the country's average import is about 2.4 million tons [33]. The import price of diesel fuel is USD 133 per barrel and the government is incurring loss of BDT 13–15 per liter for diesel though the price has been increased to BDT 68 per liter.

TABLE 2: Comparison of Pongamia biodiesel to diesel and other fuels.

Analysis	Pongamia biodiesel [142]	Diesel [143]	Heavy fuel [144]	Fossil fuel [142]
Kinematic viscosity (cSt)	5.867	2.61	200	2–5
Density (kg/m^3)	870	827.1	980	820–860
Flash point (°C)	186	53	90–180	35 min
HHV (MJ/kg)	39-40	45.18	42-43	44.03

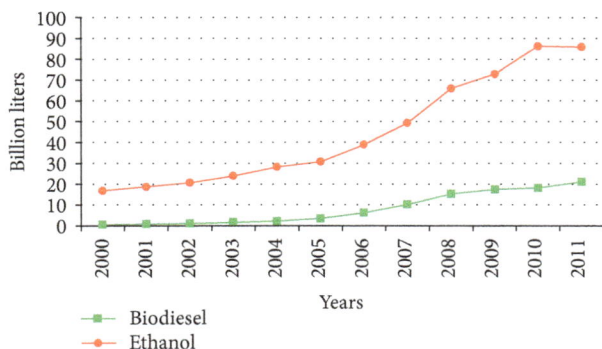

FIGURE 5: Worldwide biofuel production scenario [35].

Considering the increasing prices and environmental aspects of fossil fuels especially diesel fuel, interests have been revived around the world to find renewable substitute for fossil fuels. Biodiesel obtained from vegetable oils is considered the most suitable alternative to diesel around the world [34]. World's biofuel production has been growing rapidly to meet the increasing demand of petrodiesel and has reached 107.5 billion liters in 2011 comprising 21.4 billion liters of biodiesel and 86.1 billion liters of ethanol as shown in Figure 5 [35]. The USA, Germany, Brazil, Argentina, and France are the world's top biodiesel producer countries where the USA has increased the biodiesel production mainly from soybeans by 159% to nearly 3.2 billion liters in 2011 [35]. However, Indonesia, Malaysia, Thailand, the Philippines, and India are the largest biofuel producing countries in Asia [36].

Various vegetable oils like coconut, jatropha, karanja, rapeseed, peanut, sunflower, and soybean have been used to produce biofuel for the last few years [37]. In the year 1910, Dr. Rudolf Diesel first introduced peanut oil as fuel in compression ignition (CI) engine [38]. However, high viscosity, low volatility, and polyunsaturated character of SVO are the foremost problems of using it as substitute for diesel in CI engine [39]. The processed vegetable oil (biodiesel) obtained through transesterification solves the problems associated with SVO and can be used in CI engine [40]. Numerous researchers have showed the effective use of plant oil derived biodiesel as fuel in CI engines [39, 41–43].

Of the plant seeds, karanja is considered the most attractive source for biodiesel production due to its renewable, safe, and nonpollutant nature. Furthermore, it is cost-effective and diverse feedstock for biodiesel due to its higher recovery and quality than other seeds, no direct competition with edible food crops and with current farmland. The properties of biodiesel prepared from karanja oil are presented in Table 2.

Some researchers have tested the suitability of *Pongamia pinnata* oil as SVO in CI engine [44, 45]. Other studies have shown that the potentiality of *Pongamia pinnata* oil as a source of raw material for the production of biodiesel is well established [46–51]. However, some of them have mentioned the suitability of CKO compared to jatropha due to its less toxicity and economy [48–50]. In Bangladesh, a study has indicated the biodiesel production from CKO and its effective use in CI engine as diesel substitute [52]. Karanja biodiesel has no corrosion on piston metal whereas jatropha biodiesel has slight corrosive effect.

It has been estimated that approximately 550 karanja trees are planted in one hectare which yield about 7.7 tons of seeds and 1.8095 tons of oil [53]. It is considered that about 90% biodiesel is obtained through transesterification of karanja oil [54]. Hence, one hectare yields about 1.62855 tons of biodiesel. Bangladesh has a total marginal length of 4246 km comprised of 4053 km with India and 193 km with Myanmar [55]. Moreover, the country has about 0.32 million hectares of unused land [56]. Considering plant to plant spacing as 4 m, only marginal land of the country has a potentiality of planting about 1061501 plants which yield 3.49 kilotons of oil. Accordingly, expected biodiesel production in Bangladesh from karanja oil is estimated to be about 0.52 million tons per year utilizing the unused land. Hence, the country can reduce the import of diesel fuel by 21.67% (($0.52 \times 100)/2.4 \approx 21.67\%$) which will save approximately 508.53 million USD. Therefore, karanja can play an emerging role in the sector of liquid fuel.

3.2.1. Need of Biodiesel in Bangladesh Context. Liquid fuel especially diesel is the key input for the development of each and every sector in developing countries like Bangladesh. Excessive requirement of diesel in transport and power sector in Bangladesh forces to find alternative option. Biodiesel is considered as the prominent one due to the following reasons:

(i) It is needed for rapid depletion of fossil fuels and hikes in oil prices.

(ii) Biodiesel is a renewable and less polluting source of energy.

(iii) It is considered as ecofriendly and nontoxic.

(iv) Biodiesel industry can strengthen the domestic, rural, and agricultural economy.

(v) Biodiesel has positive energy balance ratio and can be used in CI engines.

(vi) It reduces import of petroleum products.

TABLE 3: Energy potential of karanja in Bangladesh.

Energy items	Quantity (ton ha^{-1} year^{-1})	Reference	Calorific value (GJ/ton)	Reference	Energy content (GJ ha^{-1} year^{-1})	Energy potential in Bangladesh (PJ year^{-1})
Fuel wood	5	[57]	19.25	[57]	96.25	30.8
Biodiesel	1.62855	[53, 54]	38.00	[145]	61.8849	19.8
Glycerin	0.18095	[53, 54]	18.05	[146]	3.2661	1.05
Seed cake	5.8905	[53]	18.98	[145]	111.8017	35.78
Pod shell	8.65	[58]	15	[145]	129.75	41.52
Total	—	—	—	—	402.9527	128.95

4. Energy Assessment of *Pongamia pinnata*

Like other trees, karanja has an enormous energy potential which comprises the energy of wood, energy of pod shell, energy of biodiesel, energy of glycerin, and energy of seed cake. It has been estimated that about 5 tons of fuel wood can be obtained from one hectare per year which has a calorific value of 19.25 MJ/ton [57]. One hectare can provide about 7.7 tons of seeds which yield 1.8095 tons of oil and 5.8905 tons of seed cake [53]. Furthermore, the seed cake can be utilized for biogas production and biogas slurry can be used as organic fertilizer. Considering 90% conversion factor about 1.62855 tons of biodiesel is estimated from oil and the remaining byproduct is glycerin which accounts for 0.18095 tons as shown in Table 3. Besides, one hectare produces about 8.65 tons of pod shell [58]. Total energy content of karanja per hectare was calculated by considering the calorific value of each item and estimated to be about 402.95 GJ/year. Accordingly, Bangladesh can utilize 128.95 PJ energy per year from karanja which is equivalent to 4.4 million tons of coal.

5. Engine Performance and Emission Analysis

In recent days, global warming is the foremost concern endorsed due to the large-scale use of fossil fuels. The use of vegetable oil ester which is biodiesel in CI engine shows the promising engine performance comparable with diesel fuel [59]. Besides, biodiesel is basically sulfur-free and emits considerably fewer particulates, hydrocarbons, and less carbon monoxide than conventional diesel fuel [32]. However, emissions of NO_x from biodiesel are slightly higher than diesel in CI engines [60]. Table 4 presents the summery of various emission statuses from biodiesel.

Many researchers have investigated the performance and emission characteristics of biodiesel in CI engine as a substitute for diesel fuel. Reduction in the power of CI engine due to the loss of heating value of biodiesel was reported in [61–79]. Some other researcher found no substantial variation between diesel and biodiesel performance in CI engine [80–84]. However, astonishing power increase due to use of pure biodiesel was noticed by [85, 86]. Karanja methyl ester (KOME) B100 which reduces the brake thermal efficiency of CI engine was investigated by some researcher [87, 88]. Minor difference between the engine efficiency of using KOME and pure diesel was found in [89, 90]. Surprisingly, high brake

TABLE 4: Emission characteristics of biodiesel on engine performances.

Variable	Emission status	Reference
PM	Increase	[61, 100, 114–116, 124, 125]
	Decrease	[62–69, 80–83, 85, 94, 95, 101–107, 119–123]
HC	Increase	[96, 114, 115]
	Decrease	[63, 65, 67–69, 80–82, 84–86, 95, 100–105, 107–112, 117, 118, 121–123, 126, 127]
NO_x	Increase	[62, 63, 65, 67–73, 80, 83, 85, 86, 101, 107, 109–113, 117, 120, 126, 128–130]
	Decrease	[61, 64, 66, 81, 92, 97, 100, 103, 105, 114, 115, 118, 123, 126, 131–136]
CO	Increase	[96, 107, 114–118]
	Decrease	[61–64, 66–71, 73, 81, 83, 84, 93–95, 100–113, 128]
CO_2	Increase	[67, 94–98]
	Decrease	[63, 64, 73, 82, 83, 92, 93]

PM: particulate matter, NO_x: nitrogen oxides, CO: carbon monoxide, HC: hydrocarbon, and CO_2: carbon dioxide.

thermal efficiency due to use of B20 KOME was observed by [91]. However, a study in Bangladesh showed the most promising result in CI engine performance with B25 and B100 KOME competitive to pure diesel as presented in Figure 6 [52].

Emission from CI engine is a concerning matter which is reduced by using biodiesel as substitute for pure diesel. Researches [63, 64, 73, 82, 83, 92, 93] revealed the reduction in CO_2 emission from CI engine with use of biodiesel. However, some researchers found an increase of CO_2 emission [67, 94–98], while, in literatures [73, 82], it was reported that biodiesel resulted in about 50–80% reduction in CO_2 emissions compared to diesel fuel. Table 5 presents the comparative summary of CO_2 emission from various fuels.

KOME biodiesel which resulted in reduction of CO_2 emission was presented in [90, 91] as shown in Figure 7 [91]. Karanja oil reduces the overall effect of CO_2 emissions by about 75% as it absorbs about 30 tons of CO_2 per hectare per year [99].

Many literatures [61–64, 66–71, 73, 81, 83, 84, 93–95, 100–113] reported the decreasing nature of CO emission from

TABLE 5: Emission of CO_2 per unit item.

Fuels	CO_2 emission (kg/kg fuel)	Reference
Biomass	1.19	[147]
Bituminous coal	2.46	[147]
Natural gas	1.93	[147]
Diesel	3.35	[148]
Biodiesel	0.67–1.675	[73, 82]

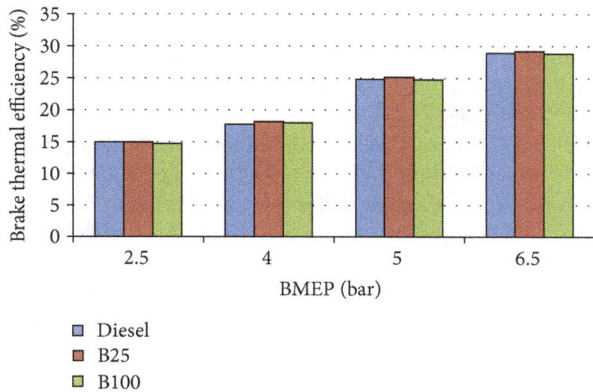

FIGURE 6: Brake thermal efficiency of CI engine with diesel fuel, B25, and B100 KOME [52].

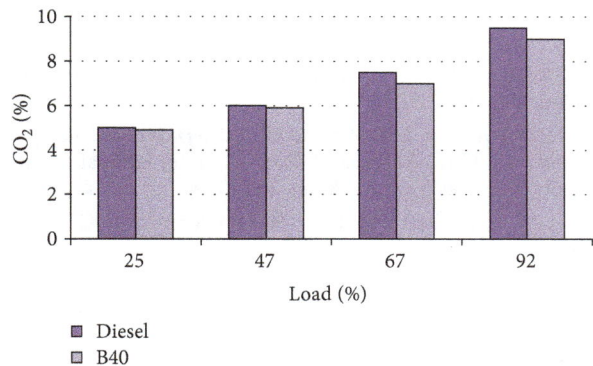

FIGURE 7: CO_2 emission of CI engine with diesel fuel and B40 KOME [91].

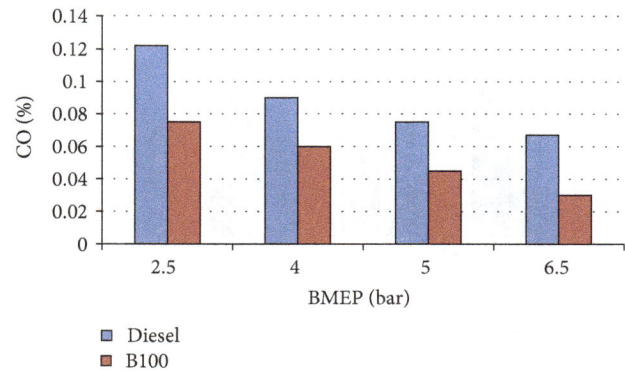

FIGURE 8: CO emission of CI engine with diesel fuel and B100 KOME [52].

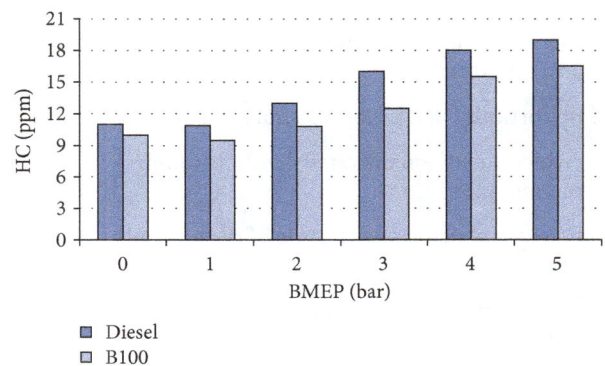

FIGURE 9: HC emission of CI engine with diesel fuel and B100 KOME [87].

biodiesel as CI engine fuel, whereas some researchers [96, 107, 114–118] observed the rise of CO emission. It was found that B100 KOME biodiesel reduces about 73–94% CO emission [66]. Besides, [52, 87, 89] agreed that KOME biodiesel reduces the CO emission as illustrated in Figure 8 [52]. Numerous studies also showed that biodiesel reduces PM emission in CI engine [62–69, 80–83, 85, 94, 95, 101–107, 119–123]. On the contrary, few reported the increase of PM emission [61, 100, 114–116, 124, 125].

It is predominantly investigated that use of biodiesel instead of pure diesel resulted in reduction of HC emission [63, 65, 67–69, 80–82, 84–86, 95, 100–105, 107–112, 117, 118, 121–123, 126, 127]. However, a very few number of

researchers noticed the increase of HC emission [96, 114, 115]. Approximately 63% of HC emission was reduced by using biodiesel compared to pure diesel fuel [97]. KOME biodiesel reduces the HC emission in CI engine [87, 89, 91] as shown in Figure 9 [87].

In many studies [62, 63, 65, 67–73, 80, 83, 85, 86, 101, 107, 109–113, 117, 120, 126, 128–130], it was reported that biodiesel causes the increase in NO_x emission. On the other hand, reduction of NO_x emission was presented in [61, 64, 66, 81, 92, 97, 100, 103, 105, 114, 115, 118, 123, 126, 131–136]. However, [52] observed a maximum of 15% increase in NO_x emissions for B100 KOME as presented in Figure 10 [52]. Besides, [87, 91] noticed the increasing trend wherein [89] noticed the decreasing trend of NO_x emission.

6. Rural Development in Bangladesh through *Pongamia pinnata*

The *Pongamia pinnata* system can play a pivotal role for the socioeconomic development of rural areas in Bangladesh. Cultivation of *Pongamia pinnata* covers the following main aspects of rural developments in the country:

(i) creation of job sector for rural unemployed people basically for women,

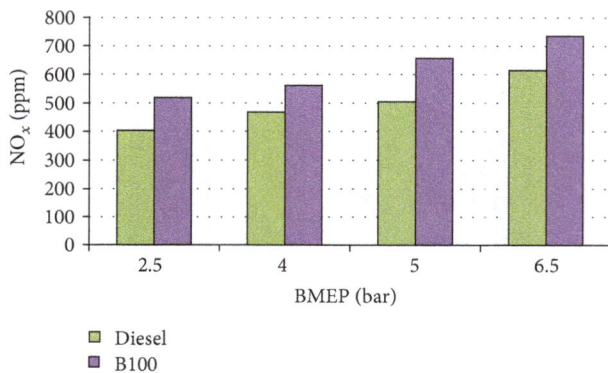

FIGURE 10: NO_x emission of CI engine with diesel fuel and B100 KOME [52].

(ii) increase of agricultural works like planting, weeding, and oil extraction,

(iii) opportunity of developing small size rural industry,

(iv) reduction of uncultivated lands through *Pongamia pinnata* plantation,

(v) increase of per capita income of rural people through biodiesel production,

(vi) energy supply for rural illumination and fuel for small stationery engines,

(vii) reduction in fossil fuel importing bill through biodiesel production.

7. Concluding Remarks

Pongamia pinnata, a versatile resource, shows the promising properties for the medical and biodiesel production industry. Many countries in the world are producing biodiesel in order to replace the fossil diesel fuel. Bangladesh has a considerable potential of biodiesel production from karanja as it has high growth rate in Bangladesh. The country has about 0.32 million hectares of unused lands which yield about 0.52 million tons of biodiesel per year and can reduce import of diesel fuel approximately by 21.67%. Besides, on energy basis country's unused lands provide about 128.95 PJ from karanja equivalent to 4.4 million tons of coal. However, the country has not started commercial cultivation of Pongamia yet. Furthermore, Pongamia biodiesel is environmentally friendly and causes fewer CO_2, CO, HC, and NO_x emission in CI engine as an alternative fuel to diesel. Karanja biodiesel blends of 20% with fossil diesel fuel produce approximately 70% less pollution. In conclusion, Bangladesh should take initiative to cultivate Pongamia and other nonedible seed plants for biodiesel production.

Conflict of Interests

The authors declare that there is no conflict of interests regarding the publication of this paper.

References

[1] Council of Scientific and Industrial Research, *Wealth of India-Raw Material*, vol. 8, Council of Scientific and Industrial Research, New Delhi, India, 1965.

[2] N. Mukta and Y. Sreevalli, "Propagation techniques, evaluation and improvement of the biodiesel plant, *Pongamia pinnata* (L.) Pierre-A review," *Industrial Crops and Products*, vol. 31, no. 1, pp. 1–12, 2010.

[3] World Agroforestry Centre (WAC), 2013, http://www.world-agroforestrycentre.org/sea/products/afdbases/af/asp/Species-Info.asp?SpID=1332#Addinfo.

[4] Satish Lele, http://www.svlele.com/karanj.htm.

[5] S. Sangwan, D. V. Rao, and R. A. Sharma, "A review on Ponga-mia Pinnata (L.) Pierre: a great versatile leguminous plant," *Nature and Science*, vol. 8, no. 11, pp. 130–139, 2010.

[6] Petroleum Conservation Research Association, "National Bio-fuel Centre," 2013, http://www.pcra-biofuels.org/Karanj.htm.

[7] The ARC Centre of Excellence for Integrative Legume Research (CILR), 2013, http://www.cilr.uq.edu.au/UserImages/File/Post-ers/Pongamia%20Biodiesel%20Poster.pdf.

[8] V. Lakshmikanthan, "Tree Borne Oilseeds. Directorate of Nonedible Oils & Soap Industry, Khadi & Village Industries Commission," Mumbai, India, 1978.

[9] N. V. Bringi, *Non-Traditional Oilseeds and Oils in India*, Oxford & IBH, New Delhi, India, 1987.

[10] The National Oilseed and Vegetable Oils Development Board (NOVOD), "Commercial exploitation of *Simarouba glauca*," in *Proceedings of the Workshop on Strategies for Development of Tree-Borne Oilseeds and Niger in Tribal Areas*, Gurgaon, India, 1995.

[11] E. H. Roberts, Ed., *In Viability of Seeds*, Chapman and Hall, London, Uk, 1972.

[12] T. D. Hong and R. H. Ellis, *A Protocol to Determine Seed Storage Behaviour*, IPGRI Technical Bulletin 1, International Plant Genetic Resources Institute, Rome, Italy, 1996.

[13] National Oilseeds and Vegetable Oils Development (NOVOD) Board, 2013.

[14] C. M. Misra and S. L. Singh, "Coppice regeneration of *Cassia siamea* and *Pongamia pinnata*," *Nitrogen Fixing Tree Research Reports*, vol. 7, no. 7, p. 4, 1989.

[15] A. K. Handa and D. Nandini, "An alternative source of biofuel, seed germination trials of *Pongamia pinnata*," *International Journal of Forest Usufructs Management*, vol. 6, no. 2, pp. 75–80, 2005.

[16] K. P. Singh, G. Dhakre, and S. V. S. Chauhan, "Effect of mechanical and chemical treatments on seed germination in *Pongamia glabra* L," *Seed Research*, vol. 33, no. 2, pp. 169–171, 2005.

[17] V. Manonmani, K. Vanangamudi, and R. S. Vinaya Rai, "Effect of seed size on seed germination and vigour in *Pongamia pinnata*," *Journal of Tropical Forest Science*, vol. 9, no. 1, pp. 1–5, 1996.

[18] D. K. Das and M. K. Alam, *Trees of Bangladesh*, Forest Research Institute, Chittagong, Bangladesh, 2001.

[19] NAS, *Irewood Crops: Shrub and Tree Species for Energy Production*, vol. 1, National Academy of Sciences, Washington, DC, USA, 1980.

[20] Government of India (GOI), *Troup's The Silviculture of Indian Trees*, vol. 4, Leguminosae. Government of India Press, Nasik, India, 1983.

[21] O. N. Allen and E. K. Allen, *The Leguminosae*, The University of Wisconsin Press, 1981.

[22] J. H. Burkill, "Dictionary of economic products of the Malay penimsula," *The Indian Journal of Hospital Pharmacy*, vol. 15, no. 6, pp. 166–168, 1996.

[23] V. Kesari, A. Das, and L. Rangan, "Physico-chemical characterization and antimicrobial activity from seed oil of *Pongamia pinnata*, a potential biofuel crop," *Biomass and Bioenergy*, vol. 34, no. 1, pp. 108–115, 2010.

[24] Shodhganga@INFLIBNET Centre, http://shodhganga.inflibnet.ac.in/bitstream/10603/3207/5/05_chapter%201.pdf.

[25] *Pongamia pinnata*, 2013.

[26] P. K. Warrier, V. P. K. Nambiar, and C. Ramakutty, *Indian Medicinal Plants*, Orient Longman, Madras, India, 1995.

[27] G. Prasad and M. V. Reshmi, *A Manual of Medicinal Trees*, Propagation Methods. Foundation for Revitalization for Local Health Tradition, Agrobios India, 2003.

[28] "*Pongamia pinnata*," http://bioweb.uwlax.edu/bio203/2011/bedard_emil/history_adaptation.htm.

[29] Bangladesh Petroleum Corporation (BPC), http://www.bpc.gov.bd/contactus.php?id=22.

[30] C. Collins, "Implementing phytoremediation of petroleum hydrocarbons," in *Methods in Biotechnology*, vol. 23, pp. 99–108, Humana Press, New York, NY, USA, 2007.

[31] P. K. Halder, M. U. H. Joardder, M. R. A. Beg, N. Paul, and I. Ullah, "Utilization of Bio-Oil for cooking and lighting," *Advances in Mechanical Engineering*, vol. 2012, Article ID 190518, 5 pages, 2012.

[32] P. Mahanta and A. Shrivastava, *Technology Development of Bio-Diesel as an Energy Alternative*, Department of Mechanical Engineering Indian Institute of Technology, 2004.

[33] M. N. Nabi, M. S. Akhter, and K. M. F. Islam, "Prospect of biodiesel production from *Jatropha curcas*, a promising non edible oil seed in Bangladesh," in *Proceedings of the International Conference on Mechanical Engineering (ICME '07)*, Paper no. ICME07-TH-06, Dhaka, Bangladesh, 2007.

[34] B. K. Barnwal and M. P. Sharma, "Prospects of biodiesel production from vegetable oils in India," *Renewable and Sustainable Energy Reviews*, vol. 9, no. 4, pp. 363–378, 2005.

[35] "Renewables 2012," Global Status Report REN21, 2012.

[36] A. Zhou and E. Thomson, "The development of biofuels in Asia," *Applied Energy*, vol. 86, pp. S11–S20, 2009.

[37] N. Sazdanoff, *Modeling and simulation of the algae to biodiesel fuel cycle [Undergraduate thesis]*, Department of Mechanical Engineering, College of Engineering, The Ohio State University, 2006.

[38] S. N. Bobade and V. B. Khyade, "Detail study on the properties of *Pongamia pinnata* (Karanja) for the production of biofuel," *Research Journal of Chemical Sciences*, vol. 2, no. 7, pp. 16–20, 2012.

[39] C. E. Goering, A. W. Schwab, M. J. Daugherty, E. H. Pryde, and A. J. Heakin, "Fuel properties of eleven vegetable oils," ASAE Paper Number 813579, St. Joseph, Mich, USA, ASAE, 1981.

[40] M. Senthil Kumar, A. Ramesh, and B. Nagalingam, "Investigations on the use of Jatropha oil and its methyl ester as a fuel in a compression ignition engine," *Journal of the Institute of Energy*, vol. 74, no. 498, pp. 24–28, 2001.

[41] A. K. Agarwal, "Vegetable oils versus diesel fuel: development and use of biodiesel in a compression ignition engine," *TIDE*, vol. 83, pp. 191–204, 1998.

[42] S. Sinha and N. C. Misra, "Diesel fuel alternative from vegetable oils," *Chemical Engineering World*, vol. 32, no. 10, pp. 77–80, 1997.

[43] A. Shaheed and E. Swain, "Combustion analysis of coconut oil and its methyl esters in a diesel engine," *Proceedings of the Institute of Mechanical Engineers*, vol. 213, no. 5, pp. 417–425, 1999.

[44] U. Shrinivasa, "A viable substitute for diesel in rural India," *Current Science Magazine*, vol. 80, no. 12, pp. 1483–1484, 2001.

[45] S. Dhinagar and B. Nagalingam, "Experimental investigation on non-edible vegetable oil operation in a LHR diesel engine for improved performance," SAE 932846, 1993.

[46] M. M. Azam, A. Waris, and N. M. Nahar, "Prospects and potential of fatty acid methyl esters of some non-traditional seed oils for use as biodiesel in India," *Biomass and Bioenergy*, vol. 29, no. 4, pp. 293–302, 2005.

[47] B. K. De and D. K. Bhattacharyya, "Biodiesel from minor vegetable oils like karanja oil and nahor oil," *Lipid/Fett*, vol. 101, no. 10, pp. 404–406, 1999.

[48] S. K. Karmee and A. Chadha, "Preparation of biodiesel from crude oil of *Pongamia pinnata*," *Bioresource Technology*, vol. 96, no. 13, pp. 1425–1429, 2005.

[49] Y. C. Sharma and B. Singh, "Development of biodiesel from karanja, a tree found in rural India," *Fuel*, vol. 87, no. 8-9, pp. 1740–1742, 2008.

[50] M. Naik, L. C. Meher, S. N. Naik, and L. M. Das, "Production of biodiesel from high free fatty acid Karanja (*Pongamia pinnata*) oil," *Biomass and Bioenergy*, vol. 32, no. 4, pp. 354–357, 2008.

[51] P. T. Scott, L. Pregelj, N. Chen, J. S. Hadler, M. A. Djordjevic, and P. M. Gresshoff, "*Pongamia pinnata*: an untapped resource for the biofuels industry of the future," *BioEnergy Research*, vol. 1, no. 1, pp. 2–11, 2008.

[52] M. N. Nabi, S. M. N. Hoque, and M. S. Akhter, "Karanja (*Pongamia Pinnata*) biodiesel production in Bangladesh, characterization of karanja biodiesel and its effect on diesel emissions," *Fuel Processing Technology*, vol. 90, no. 9, pp. 1080–1086, 2009.

[53] S. Khandelwal and R. Y. Chauhan, "Life cycle assessment of Neem and Karanja biodiesel: an overview," *International Journal of ChemTech Research*, vol. 5, no. 2, pp. 659–665, 2013.

[54] S. R. Kalbande, S. N. Pawar, and S. B. Jadhav, "Production of Karanja biodiesel and its utilization in diesel engine generator set for power generation," *Karnataka Journal of Agricultural Sciences*, vol. 20, no. 3, pp. 680–683, 2007.

[55] http://en.wikipedia.org/wiki/Geography_of_Bangladesh.

[56] M. N. Nabi, S. M. N. Hoque, and M. S. Uddin, "Prospect of *Jatropha curcas* and pithraj cultivation in Bangladesh," *Journal of Engineering and Technology*, vol. 7, no. 1, pp. 41–54, 2009.

[57] J. A. Duke, "Pongamia pinnata (L) Pierre," 2013, http://www.hort.purdue.edu/newcrop/duke_energy/Pongamia_pinnata.html.

[58] Heather Richman, "Asia Pacific Clean Energy Summit," 2012, http://www.ct-si.org/events/APCE2012/sld/pdf/30.pdf.

[59] S. S. Karhale, R. G. Nadre, D. K. Das, and S. K. Dash, "Studies on comparative performance of a compression ignition engine with different blends of biodiesel and diesel under varying operating conditions," *Karnataka Journal of Agricultural Science*, vol. 21, pp. 246–249, 2008.

[60] L. G. Schumacher, S. C. Borgelt, D. Fosseen, W. Goetz, and W. G. Hires, "Heavy-duty engine exhaust emission tests using methyl ester soybean oil/diesel fuel blends," *Bioresource Technology*, vol. 57, no. 1, pp. 31–36, 1996.

[61] H. Aydin and H. Bayindir, "Performance and emission analysis of cottonseed oil methyl ester in a diesel engine," *Renewable Energy*, vol. 35, no. 3, pp. 588–592, 2010.

[62] H. Hazar, "Effects of biodiesel on a low heat loss diesel engine," *Renewable Energy*, vol. 34, no. 6, pp. 1533–1540, 2009.

[63] A. N. Ozsezen, M. Canakci, A. Turkcan, and C. Sayin, "Performance and combustion characteristics of a DI diesel engine fueled with waste palm oil and canola oil methyl esters," *Fuel*, vol. 88, no. 4, pp. 629–636, 2009.

[64] Z. Utlu and M. S. Koçak, "The effect of biodiesel fuel obtained from waste frying oil on direct injection diesel engine performance and exhaust emissions," *Renewable Energy*, vol. 33, no. 8, pp. 1936–1941, 2008.

[65] H. Özgünay, S. Çolak, G. Zengin, Ö. Sari, H. Sarikahya, and L. Yüceer, "Performance and emission study of biodiesel from leather industry pre-fleshings," *Waste Management*, vol. 27, no. 12, pp. 1897–1901, 2007.

[66] H. Raheman and A. G. Phadatare, "Diesel engine emissions and performance from blends of karanja methyl ester and diesel," *Biomass & Bioenergy*, vol. 27, no. 4, pp. 393–397, 2004.

[67] Y. Ulusoy, Y. Tekin, M. Četinkaya, and F. Karaosmanoğlu, "The engine tests of biodiesel from used frying oil," *Energy Sources, Part A: Recovery, Utilization, and Environmental Effects*, vol. 26, no. 10, pp. 927–932, 2004.

[68] E. Buyukkaya, "Effects of biodiesel on a DI diesel engine performance, emission and combustion characteristics," *Fuel*, vol. 89, no. 10, pp. 3099–3105, 2010.

[69] S. H. Choi and Y. Oh, "The emission effects by the use of biodiesel fuel," *International Journal of Modern Physics B*, vol. 20, no. 25-27, pp. 4481–4487, 2006.

[70] M. Karabektas, "The effects of turbocharger on the performance and exhaust emissions of a diesel engine fuelled with biodiesel," *Renewable Energy*, vol. 34, no. 4, pp. 989–993, 2009.

[71] S. Murillo, J. L. Míguez, J. Porteiro, E. Granada, and J. C. Morán, "Performance and exhaust emissions in the use of biodiesel in outboard diesel engines," *Fuel*, vol. 86, no. 12-13, pp. 1765–1771, 2007.

[72] A. C. Hansen, M. R. Gratton, and W. Yuan, "Diesel engine performance and NOx emissions from oxygenated biofuels and blends with diesel fuel," *Transactions of the ASABE*, vol. 49, no. 3, pp. 589–595, 2006.

[73] C. Carraretto, A. Macor, A. Mirandola, A. Stoppato, and S. Tonon, "Biodiesel as alternative fuel: experimental analysis and energetic evaluations," *Energy*, vol. 29, no. 12–15, pp. 2195–2211, 2004.

[74] C. Kaplan, R. Arslan, and A. Sürmen, "Performance characteristics of sunflower methyl esters as biodiesel," *Energy Sources A: Recovery, Utilization and Environmental Effects*, vol. 28, no. 8, pp. 751–755, 2006.

[75] J. F. Reyes and M. A. Sepulveda, "PM-10 emissions and power of a diesel engine fueled with crude and refined Biodiesel from salmon oil," *Fuel*, vol. 85, no. 12-13, pp. 1714–1723, 2006.

[76] M. Çetinkaya, Y. Ulusoy, Y. Tekin, and F. Karaosmanoğlu, "Engine and winter road test performances of used cooking oil originated biodiesel," *Energy Conversion and Management*, vol. 46, no. 7-8, pp. 1279–1291, 2005.

[77] Y.-C. Lin, W.-J. Lee, T.-S. Wu, and C.-T. Wang, "Comparison of PAH and regulated harmful matter emissions from biodiesel blends and paraffinic fuel blends on engine accumulated mileage test," *Fuel*, vol. 85, no. 17-18, pp. 2516–2523, 2006.

[78] F. Neto da Silva, A. S. Prata, and J. R. Teixeira, "Technical feasibility assessment of oleic sunflower methyl ester utilisation in Diesel bus engines," *Energy Conversion and Management*, vol. 44, no. 18, pp. 2857–2878, 2003.

[79] H. S. Yücesu and C. Ilkiliç, "Effect of cotton seed oil methyl ester on the performance and exhaust emission of a diesel engine," *Energy Sources A: Recovery, Utilization and Environmental Effects*, vol. 28, no. 4, pp. 389–398, 2006.

[80] B.-F. Lin, J.-H. Huang, and D.-Y. Huang, "Experimental study of the effects of vegetable oil methyl ester on DI diesel engine performance characteristics and pollutant emissions," *Fuel*, vol. 88, no. 9, pp. 1779–1785, 2009.

[81] D. H. Qi, L. M. Geng, H. Chen, Y. Z. Bian, J. Liu, and X. C. Ren, "Combustion and performance evaluation of a diesel engine fueled with biodiesel produced from soybean crude oil," *Renewable Energy*, vol. 34, no. 12, pp. 2706–2713, 2009.

[82] M. Lapuerta, J. M. Herreros, L. L. Lyons, R. García-Contreras, and Y. Briceño, "Effect of the alcohol type used in the production of waste cooking oil biodiesel on diesel performance and emissions," *Fuel*, vol. 87, no. 15-16, pp. 3161–3169, 2008.

[83] A. Keskin, M. Gürü, and D. Altiparmak, "Influence of tall oil biodiesel with Mg and Mo based fuel additives on diesel engine performance and emission," *Bioresource Technology*, vol. 99, no. 14, pp. 6434–6442, 2008.

[84] B. Ghobadian, H. Rahimi, A. M. Nikbakht, G. Najafi, and T. F. Yusaf, "Diesel engine performance and exhaust emission analysis using waste cooking biodiesel fuel with an artificial neural network," *Renewable Energy*, vol. 34, no. 4, pp. 976–982, 2009.

[85] J.-T. Song and C.-H. Zhang, "An experimental study on the performance and exhaust emissions of a diesel engine fuelled with soybean oil methyl ester," *Proceedings of the Institution of Mechanical Engineers, Part D: Journal of Automobile Engineering*, vol. 222, no. 12, pp. 2487–2496, 2008.

[86] M. I. Al-Widyan, G. Tashtoush, and M. Abu-Qudais, "Utilization of ethyl ester of waste vegetable oils as fuel in diesel engines," *Fuel Processing Technology*, vol. 76, no. 2, pp. 91–103, 2002.

[87] N. Shrivastava, S. N. Varma, and M. Pandey, "Experimental study on the production of Karanja oil methyl ester and its effect on diesel engine," *International Journal of Renewable Energy Development*, vol. 1, no. 3, pp. 115–122, 2012.

[88] H. Dharmadhikari, P. R. Kumar, and S. S. Rao, "Performance and emissions of CI engine using blends of biodiesel and diesel at different injection pressures," *International Journal of Applied Research in Mechanical Engineering*, vol. 2, no. 2, pp. 1–6, 2012.

[89] Y. Singh, A. Singla, and N. Lohani, "Production of biodiesel from oils of Jatropha, Karanja and performance analysis on CI engine," *International Journal of Innovative Research and Development*, vol. 2, no. 3, pp. 286–294, 2013.

[90] M. V. Nagarhalli, V. M. Nandedkar, and K. C. Mohite, "Emission and performance characteristics of karanja biodiesel and its blends in a CI engine and its economics," *ARPN Journal of Engineering and Applied Sciences*, vol. 5, no. 2, pp. 52–56, 2010.

[91] N. Panigrahi, M. K. Mohanty, and A. K. Pradhan, "Non-edible Karanja biodiesel—a sustainable fuel for C.I. engine," *International Journal of Engineering Research and Applications*, vol. 2, no. 6, pp. 853–860, 2012.

[92] P. K. Sahoo, L. M. Das, M. K. G. Babu, and S. N. Naik, "Biodiesel development from high acid value polanga seed oil and performance evaluation in a CI engine," *Fuel*, vol. 86, no. 3, pp. 448–454, 2007.

[93] C.-Y. Lin and H.-A. Lin, "Engine performance and emission characteristics of a three-phase emulsion of biodiesel produced by peroxidation," *Fuel Processing Technology*, vol. 88, no. 1, pp. 35–41, 2007.

[94] A. S. Ramadhas, C. Muraleedharan, and S. Jayaraj, "Performance and emission evaluation of a diesel engine fueled with methyl esters of rubber seed oil," *Renewable Energy*, vol. 30, no. 12, pp. 1789–1800, 2005.

[95] M. Canakci, "Performance and emissions characteristics of biodiesel from soybean oil," *Proceedings of the Institution of Mechanical Engineers, Part D: Journal of Automobile Engineering*, vol. 219, no. 7, pp. 915–922, 2005.

[96] G. Fontaras, G. Karavalakis, M. Kousoulidou et al., "Effects of biodiesel on passenger car fuel consumption, regulated and non-regulated pollutant emissions over legislated and real-world driving cycles," *Fuel*, vol. 88, no. 9, pp. 1608–1617, 2009.

[97] S. Puhan, N. Vedaraman, G. Sankaranarayanan, and B. V. B. Ram, "Performance and emission study of Mahua oil (madhuca indica oil) ethyl ester in a 4-stroke natural aspirated direct injection diesel engine," *Renewable Energy*, vol. 30, no. 8, pp. 1269–1278, 2005.

[98] G. Labeckas and S. Slavinskas, "The effect of rapeseed oil methyl ester on direct injection Diesel engine performance and exhaust emissions," *Energy Conversion and Management*, vol. 47, no. 13-14, pp. 1954–1967, 2006.

[99] 2013, http://www.himalayaninstitute.org/humanitarian/tibetan-settlements/about-pongamia/.

[100] O. Armas, K. Yehliu, and A. L. Boehman, "Effect of alternative fuels on exhaust emissions during diesel engine operation with matched combustion phasing," *Fuel*, vol. 89, no. 2, pp. 438–456, 2010.

[101] H. Kim and B. Choi, "The effect of biodiesel and bioethanol blended diesel fuel on nanoparticles and exhaust emissions from CRDI diesel engine," *Renewable Energy*, vol. 35, no. 1, pp. 157–163, 2010.

[102] X. Meng, G. Chen, and Y. Wang, "Biodiesel production from waste cooking oil via alkali catalyst and its engine test," *Fuel Processing Technology*, vol. 89, no. 9, pp. 851–857, 2008.

[103] A. Hull, I. Golubkov, B. Kronberg, and J. Van Stam, "Alternative fuel for a standard diesel engine," *International Journal of Engine Research*, vol. 7, no. 1, pp. 51–63, 2006.

[104] M. Gumus and S. Kasifoglu, "Performance and emission evaluation of a compression ignition engine using a biodiesel (apricot seed kernel oil methyl ester) and its blends with diesel fuel," *Biomass and Bioenergy*, vol. 34, no. 1, pp. 134–139, 2010.

[105] D. Sharma, S. L. Soni, and J. Mathur, "Emission reduction in a direct injection diesel engine fueled by neem-diesel blend," *Energy Sources A: Recovery, Utilization and Environmental Effects*, vol. 31, no. 6, pp. 500–508, 2009.

[106] M. Gürü, A. Koca, Ö. Can, C. Çinar, and F. Şahin, "Biodiesel production from waste chicken fat based sources and evaluation with Mg based additive in a diesel engine," *Renewable Energy*, vol. 35, no. 3, pp. 637–643, 2010.

[107] J. M. Luján, V. Bermúdez, B. Tormos, and B. Pla, "Comparative analysis of a DI diesel engine fuelled with biodiesel blends during the European MVEG-A cycle: performance and emissions (II)," *Biomass and Bioenergy*, vol. 33, no. 6-7, pp. 948–956, 2009.

[108] M. A. Kalam and H. H. Masjuki, "Testing palm biodiesel and NPAA additives to control NOx and CO while improving efficiency in diesel engines," *Biomass and Bioenergy*, vol. 32, no. 12, pp. 1116–1122, 2008.

[109] S. Godiganur, C. Suryanarayana Murthy, and R. P. Reddy, "Performance and emission characteristics of a Kirloskar HA394 diesel engine operated on fish oil methyl esters," *Renewable Energy*, vol. 35, no. 2, pp. 355–359, 2010.

[110] S. Godiganur, C. H. S. Murthy, and R. P. Reddy, "6BTA 5.9 G2-1 cummins engine performance and emission tests using methyl ester mahua (*Madhuca indica*) oil/diesel blends," *Renewable Energy*, vol. 34, no. 10, pp. 2172–2177, 2009.

[111] S. J. Deshmukh and L. B. Bhuyar, "Transesterified Hingan (Balanites) oil as a fuel for compression ignition engines," *Biomass and Bioenergy*, vol. 33, no. 1, pp. 108–112, 2009.

[112] D. M. Korres, D. Karonis, E. Lois, M. B. Linck, and A. K. Gupta, "Aviation fuel JP-5 and biodiesel on a diesel engine," *Fuel*, vol. 87, no. 1, pp. 70–78, 2008.

[113] N. Usta, "Use of tobacco seed oil methyl ester in a turbocharged indirect injection diesel engine," *Biomass and Bioenergy*, vol. 28, no. 1, pp. 77–86, 2005.

[114] M. S. Kumar, A. Ramesh, and B. Nagalingam, "A comparison of the different methods of using jatropha oil as fuel in a compression ignition engine," *Journal of Engineering for Gas Turbines and Power*, vol. 132, no. 3, Article ID 032801, pp. 32801–32811, 2010.

[115] N. R. Banapurmath, P. G. Tewari, and R. S. Hosmath, "Performance and emission characteristics of a DI compression ignition engine operated on Honge, Jatropha and sesame oil methyl esters," *Renewable Energy*, vol. 33, no. 9, pp. 1982–1988, 2008.

[116] N. R. Banapurmath and P. G. Tewari, "Performance of a low heat rejection engine fuelled with low volatile Honge oil and its methyl ester (HOME)," *Proceedings of the Institution of Mechanical Engineers, Part A: Journal of Power and Energy*, vol. 222, no. 3, pp. 323–330, 2008.

[117] P. K. Sahoo, L. M. Das, M. K. G. Babu et al., "Comparative evaluation of performance and emission characteristics of jatropha, karanja and polanga based biodiesel as fuel in a tractor engine," *Fuel*, vol. 88, no. 9, pp. 1698–1707, 2009.

[118] M. A. R. Nascimento, E. S. Lora, P. S. P. Correa et al., "Biodiesel fuel in diesel micro-turbine engines: modelling and experimental evaluation," *Energy*, vol. 33, no. 2, pp. 233–240, 2008.

[119] A. Pal, A. Verma, S. S. Kachhwaha, and S. Maji, "Biodiesel production through hydrodynamic cavitation and performance testing," *Renewable Energy*, vol. 35, no. 3, pp. 619–624, 2010.

[120] L. Zhu, W. Zhang, W. Liu, and Z. Huang, "Experimental study on particulate and NOx emissions of a diesel engine fueled with ultra low sulfur diesel, RME-diesel blends and PME-diesel blends," *Science of the Total Environment*, vol. 408, no. 5, pp. 1050–1058, 2010.

[121] K. Ryu, "The characteristics of performance and exhaust emissions of a diesel engine using a biodiesel with antioxidants," *Bioresource Technology*, vol. 101, no. 1, pp. S78–S82, 2010.

[122] D. Agarwal, S. Sinha, and A. K. Agarwal, "Experimental investigation of control of NO_x emissions in biodiesel-fueled compression ignition engine," *Renewable Energy*, vol. 31, no. 14, pp. 2356–2369, 2006.

[123] S. Puhan, N. Vedaraman, B. V. B. Ram, G. Sankarnarayanan, and K. Jeychandran, "Mahua oil (Madhuca Indica seed oil) methyl ester as biodiesel-preparation and emission characterstics," *Biomass and Bioenergy*, vol. 28, no. 1, pp. 87–93, 2005.

[124] L. Turrio-Baldassarri, C. L. Battistelli, L. Conti et al., "Emission comparison of urban bus engine fueled with diesel oil and "biodiesel" blend," *Science of the Total Environment*, vol. 327, no. 1–3, pp. 147–162, 2004.

[125] D. H. Qi, H. Chen, L. M. Geng, and Y. Z. Bian, "Experimental studies on the combustion characteristics and performance of a direct injection engine fueled with biodiesel/diesel blends,"

Energy Conversion and Management, vol. 51, no. 12, pp. 2985–2992, 2010.

[126] C. S. Cheung, L. Zhu, and Z. Huang, "Regulated and unregulated emissions from a diesel engine fueled with biodiesel and biodiesel blended with methanol," *Atmospheric Environment*, vol. 43, no. 32, pp. 4865–4872, 2009.

[127] A. Tsolakis, A. Megaritis, M. L. Wyszynski, and K. Theinnoi, "Engine performance and emissions of a diesel engine operating on diesel-RME (rapeseed methyl ester) blends with EGR (exhaust gas recirculation)," *Energy*, vol. 32, no. 11, pp. 2072–2080, 2007.

[128] N. Usta, "An experimental study on performance and exhaust emissions of a diesel engine fuelled with tobacco seed oil methyl ester," *Energy Conversion and Management*, vol. 46, no. 15-16, pp. 2373–2386, 2005.

[129] N. Usta, E. Öztürk, Ö. Can et al., "Combustion of bioDiesel fuel produced from hazelnut soapstock/waste sunflower oil mixture in a Diesel engine," *Energy Conversion and Management*, vol. 46, no. 5, pp. 741–755, 2005.

[130] N. R. Banapurmath, P. G. Tewari, and R. S. Hosmath, "Effect of biodiesel derived from Honge oil and its blends with diesel when directly injected at different injection pressures and injection timings in single-cylinder water-cooled compression ignition engine," *Proceedings of the Institution of Mechanical Engineers Part A: Journal of Power and Energy*, vol. 223, pp. 31–40, 2009.

[131] M. P. Dorado, E. Ballesteros, J. M. Arnal, J. Gómez, and F. J. López, "Exhaust emissions from a diesel engine fueled with transesterified waste olive oil," *Fuel*, vol. 82, no. 11, pp. 1311–1316, 2003.

[132] B. Baiju, M. K. Naik, and L. M. Das, "A comparative evaluation of compression ignition engine characteristics using methyl and ethyl esters of Karanja oil," *Renewable Energy*, vol. 34, no. 6, pp. 1616–1621, 2009.

[133] S. Kalligeros, F. Zannikos, S. Stournas et al., "An investigation of using biodiesel/marine diesel blends on the performance of a stationary diesel engine," *Biomass and Bioenergy*, vol. 24, no. 2, pp. 141–149, 2003.

[134] M. Lapuerta, O. Armas, R. Ballesteros, and J. Fernández, "Diesel emissions from biofuels derived from Spanish potential vegetable oils," *Fuel*, vol. 84, no. 6, pp. 773–780, 2005.

[135] K. Dincer, "Lower emissions from biodiesel combustion," *Energy Sources Part A: Recovery, Utilization and Environmental Effects*, vol. 30, no. 10, pp. 963–968, 2008.

[136] R. Ballesteros, J. J. Hernández, L. L. Lyons, B. Cabañas, and A. Tapia, "Speciation of the semivolatile hydrocarbon engine emissions from sunflower biodiesel," *Fuel*, vol. 87, no. 10-11, pp. 1835–1843, 2008.

[137] P. Mahanta, S. Mishra, and Y. Kushwah, "A comparative study of pongamia pinnata and jatropha curcus oil as diesel substitute," *International Energy Journal*, vol. 7, no. 1, pp. 1–12, 2006.

[138] M. S. Uddin, M. U. H. Joardder, and M. N. Islam, "Design and construction of fixed bed pyrolysis system and plum seed pyrolysis for bio-oil production," *International Journal of Advanced Renewable Energy Research*, vol. 7, no. 1, pp. 405–409, 2012.

[139] M. Kader, M. U. H. Joardder, M. R. Islam, B. K. Das, and M. M. Hasan, "Production of liquid fuel and activated carbon from mahogany seed by using pyrolysis technology," in *Proceedings of the International Conference on Mechanical Engineering (ICME '11)*, Bangladesh University of Engineering and Technology, Dhaka, Bangladesh, December 2011.

[140] M. U. H. Joardder, M. R. Islam, and M. R. A. Beg, "Pyrolysis of coconut shell for bio-oil," in *Proceedings of the International Conference on Mechanical Engineering (ICME '11)*, Bangladesh University of Engineering and Technology, Dhaka, Bangladesh, December 2011.

[141] K. Annamalai and I. K. Puri, *Combustion Science and Engineering*, CRC Press, New York, NY, USA, 2006.

[142] AGF Seeds, 2013, http://www.agfseeds.com/pongamia/.

[143] R. G. Andrews and P. C. Patniak, "Feasibility of utilizing a biomass derived fuel for industrial gas turbine applications," in *Bio-Oil Production & Utilization*, A. V. Bridgwater and E. N. Hogan, Eds., pp. 236–245, CPL Press, Berkshire, UK, 1996.

[144] F. Rick and U. Vix, "Product standards for pyrolysis products for use as fuel in industrial firing plants," in *Biomass Pyrolysis Liquids Upgrading and Utilization*, A. V. Bridgwater and G. Grassi, Eds., Elsevier, London, UK, 1991.

[145] Subbarao, *Gasification of De-Oiled Cakes of Jatropha and Pongamia Seeds*, 2010.

[146] J. A. Neto, R. S. Cruz, J. M. Alves, M. Pires, S. Robra, and E. Parente Jr., "Energy balance of ester methyl and ethyl of castor oil plant oil," in *Proceedings of the Brazilian Congress of Castor Oil Plant*, Campina Grande, Brazil, 2004.

[147] "Carbon emissions from burning biomass for energy," http://www.pfpi.net/wp-content/uploads/2011/04/PFPI-biomass-carbon-accounting-overview_April.pdf.

[148] T. Thamsiriroj and J. D. Murphy, "Is it better to import palm oil from Thailand to produce biodiesel in Ireland than to produce biodiesel from indigenous Irish rape seed?" *Applied Energy*, vol. 86, no. 5, pp. 595–604, 2009.

15

Sustainable Design of a Nearly Zero Energy Building Facilitated by a Smart Microgrid

Gandhi Habash,[1] **Daniel Chapotchkine,**[1] **Peter Fisher,**[2] **Alec Rancourt,**[2] **Riadh Habash,**[2] **and Will Norris**[3]

[1]*Azrieli School of Architecture and Urbanism, Carleton University, Ottawa, ON, Canada K1S 5B6*
[2]*School of Electrical Engineering and Computer Science, University of Ottawa, Ottawa, ON, Canada K1N 6N5*
[3]*DEI & Associates Inc., Waterloo, ON, Canada N2L 4E4*

Correspondence should be addressed to Gandhi Habash; gandhihabash@cmail.carleton.ca

Academic Editor: Joydeep Mitra

One of the emerging milestones in building construction is the development of nearly zero energy buildings (NZEBs). This complex concept is defined as buildings that on a yearly average consume as much energy as they generate using renewable energy sources. Realization of NZEBs requires a wide range of technologies, systems, and solutions with varying degrees of complexity and sophistication, depending upon the location and surrounding environmental conditions. This paper will address the role of the above technologies and solutions and discusses the challenges being faced. The objective is to maximize energy efficiency, optimize occupant comfort, and reduce dependency on both the grid and the municipal potable water supply by implementing sustainable strategies in designing a research and sports facility. Creative solutions by the architectural and engineering team capitalize on the design of a unique glazing system; energy efficient technologies; water use reduction techniques; and a combined cooling, heating, and power (CCHP) microgrid (MG) with integrated control aspects and renewable energy sources.

1. Introduction

The building sector currently accounts for about one-third of the total worldwide energy use and much of this consumption is directly attributed to building design and construction [1]. A wide array of measures have been adopted and implemented to actively promote a better energy performance of buildings, including the nearly zero energy building (NZEB) concept, which is a realistic solution for the mitigation of CO_2 emissions. The NZED concept is also a viable way of reducing energy use in buildings, in order to alleviate the current worldwide energy challenges of rising prices, climate change, and security of supply [2]. The NZEB implies that the energy demand for electrical power is reduced, and this reduced demand is met on an annual basis from renewable energy supply which can be either integrated into the building design or provided, for example, as part of a community renewable energy supply system. It also implies that the grid is used to supply electrical power when there is no renewable power available and that the building will export power back

to the grid when it has excess power generation, in many cases, selling this exported power to the local utility company through a Feed-In Tariff program. The objective of NZEBs is not only to minimize the energy consumption of the building with passive design methods but also to design a building that balances energy requirements with active techniques and renewable technologies.

The increasing number of NZEB demonstration and research projects [3–12] highlights the growing attention given to NZEBs. Goals for the implementation of NZEBs are discussed and proposed at the international level, for example, in the USA within the Energy Independence and Security Act of 2007 (EISA 2007) and at the European level within the recast of the Directive on Energy Performance of Buildings (EPBD) adopted in May 2010. The EISA 2007 authorizes the net zero energy commercial building initiative for all new commercial buildings by 2030. It further specifies a zero energy target for 50% of US commercial buildings by 2040 and net zero for all US commercial buildings by 2050 [13]. There are also several advanced sustainable building design standards such as Ecohomes (BRE, UK), PassivHaus (Germany), and the US Green Building Council's Leadership in Energy and Environmental Design (LEED). These standards provide different ranking criteria to evaluate energy efficiency and/or NZEBs. However, there are no specific strategies or design guidelines provided for achieving NZEB designs. Specific design guidelines and strategies are extremely important for architects or engineers to popularize NZEBs.

Characterizing the energy demand of a building involves the initial identification of specific energy end uses involved including lighting, space heating and cooling, ventilation, water heating, refrigeration, and others such as mechanical and computing systems. Energy optimization (efficiency) of each of the above elements is the first step to address the overall efficiency as the ratio of the energy demand handled by the building to the energy consumed by the building. European energy performance indicators (EPI) [14] and American energy intensities (EI) [15] are valid and synonymous efficiency indicators, since both are ratios of energy use input to energy service output. However, this does not always result in the most optimal overall building performance because several of these functions interact with one another and result in energy wastage due to competing processes.

This work discusses the whole design process of a defined NZEB. First, passive design methods which reduce heating and cooling loads have been investigated. Various energy-efficient techniques including lighting; heating, ventilation, and air conditioning (HVAC); and water performance improvement have been detailed. A combined cooling, heating, and power (CCHP) microgrid (MG) combined with photovoltaic (PV) and wind energy components is integrated in the design of the building to provide additional energy and enable system design optimizations.

The paper is laid out as follows. Section 2 addresses the architectural design aspects of the building. Section 3 discusses the engineering performance aspects including efficient electricity and water usage. Section 4 describes a grid-connected MG for the building with various loads and on-site renewable components. Finally, Section 5 concludes the work presented in the paper.

2. Architectural Design

The building chosen to serve as a case study is a sports research facility for the city of Toronto. It is classified by the Ontario Building Code as a Group A, Division 2 occupancy type commercial building under the "education" activity category, which is regarded as one of most energy intensive type after the "office" category.

The main objectives of the design team are to provide interior spaces with as much illumination as possible through natural daylighting, reduce system and plug loads, maximize energy efficiency, optimize occupant comfort, and decrease dependency on both the grid and the municipal potable water supply through integrating leading-edge technologies and NZEB design strategies. The analysis of the case study

was performed for one climate scenario: Toronto (Canada), representing a climate with a cold winter. By integrating the above technologies and strategies, the design offsets its dependency on the grid significantly.

2.1. Building Envelope. There are several parameters in a building design that could be controlled to achieve low building energy consumption, including the building orientation and structure (size, layout, partition, etc.), constructions and their materials, whether or not to allow natural ventilation, and natural ventilation control mode [16]. Appropriate siting of the facility has much to do with its energy savings. The footprint is wide and shallow, capitalizing on the unobstructed land mass near Downsview airport. Such a footprint allows natural light to illuminate key areas during daytime hours, thereby reducing the lighting load. Attention has also been paid to the design of the building envelope to maximize thermal performance as a best practice. This is carried out by taking local energy policy, urban planning, and industry standards into account. The building envelope was organized primarily to maintain privacy between different zones. The office and gathering spaces are located at the southern face of the building, capitalizing on daylighting since they will most likely be used during business hours. The research and sports facility components are located along the northern side of the building, since they require more controlled environments and benefit less from natural daylighting and ventilation. The design approach starts by reducing energy demands within the building envelope by minimizing losses and optimizing solar gains. The architectural rendering of the building is shown in Figure 1, while the building major zones are shown in Figure 2.

The target of net zero energy use is a requirement of the living building challenge. Our approach to achieve it is through a "conserve, capture, and create" concept. The heavily articulated building form increases the area of the external envelope with a related increase in heat loss. However, by allowing the building to capture winter solar gain, daylight, and natural ventilation, the final result is a net benefit to the overall energy equation. The south orientation of the glazed entrance wing and some of the classrooms allow low-level winter solar gain to be captured, reducing the heating loads on the building. The capture of daylight and natural ventilation reduces lighting and ventilation requirements, further reducing the energy demands of the building.

2.2. Aspects of Passive Design. When designing NZEB buildings, architects control some of the most important aspects of passive design: the geometric expression of massing and program, glazing type, size, assemblies, orientation, and location. These affect nearly every load and energy performance of a building such as heat gain and shading, daylighting, operability, and envelope conduction. Creative solutions by the architectural team capitalized on the large building mass, namely, in the design of a unique glazing system along the south side of the building. The glazing system maximizes daylighting while maintaining occupant comfort. Through the use of three different panels on a pyramid structure,

FIGURE 1: Architectural rendering of the Downsview Plain research and sports building.

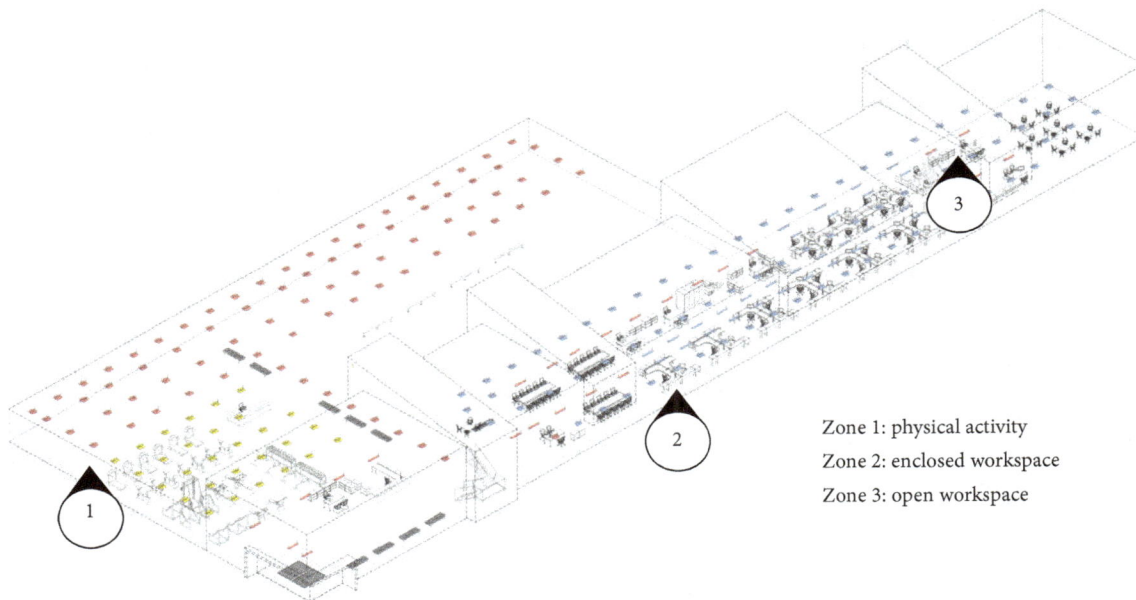

Zone 1: physical activity
Zone 2: enclosed workspace
Zone 3: open workspace

FIGURE 2: The building major zones.

the glazing system minimizes glare and internal heat gains while maximizing light trespass where appropriate. Additionally, the glazing system incorporates PV cells which produce a substantial amount of energy due to their southern orientation.

In NZEB, the roof often becomes the project's densest design challenge in terms of daylight, energy generation, and equipment for photovoltaic system. For the proposed facility, the roof design ingeniously balances the desire for optimally oriented PV panels, skylights, and sun tunnels for daylighting and ventilation.

3. Engineering Performance Improvement

There are various energy consuming systems required in a building in order to maintain ideal environmental conditions, including lighting, HVAC loads, environmental loads, and plug loads. The proposed design targets the two largest loads in the building, namely, lighting and HVAC loads, in order to maximize energy savings.

3.1. Metric of Energy Use. Putting metrics on evaluating building energy performance is a necessary step to make any progress on NZEBs. In such evaluation, it is important to consider building use and climate. One useful metric of energy use in buildings is the energy use intensity (EUI). The EUI is typically defined as the ratio of annual site energy use to building floor area. It is calculated by dividing the total energy consumed by the building in one year (measured in kBtu or GJ) by the total gross floor area of the building. It can be used to refer just to the electricity use in a building in which case it is usually defined in units of $kBTU/ft^2$ or kWh/m^2 [17].

3.2. Efficient Lighting System. Lighting is the most pervasive element and is of essential need in modern buildings. It is a significant component of the total energy consumption in a building, often comprising 20–30% of total building energy consumption. In addition to the advancements in light source technologies, from incandescent and high intensity discharge (HID) to fluorescent and compact fluorescent (CFL) to light emitting diodes (LEDs), controlling the artificial light sources to provide illumination of the right kinds (adequate light levels, colour temperature, and colour rendering), to the right places (offices, recreation areas, etc.) and at the right times, provides significant opportunity for energy savings. The Illuminating Engineering Society (IES) provides a collection of recommended lighting levels for various tasks and occupancy types in the IES Lighting Handbook, but provided that minimum lighting levels and energy targets are met in accordance with the OBC lighting levels are at the final discretion of the building owner and the designers. For this reason, establishing the required lighting levels for each space type then becomes an important design decision, as many of these opportunities for energy savings can only be realized provided that the end users' visual comfort is not negatively affected during standard operating conditions. By completing a photometric study using computer aided modelling software such as AGI32 or Visual 3D, it is possible to design a lighting layout that achieves the desired light levels, while maximizing the energy savings by providing an accurate comparison of fixtures' performances in specific applications and determining optimal fixture placements/aiming within a space. Modelling the lighting also allows the average solar light trespass throughout the days of the year to be taken into account, providing a basis for determining the average contributions from artificial light required to maintain consistent illumination in each space, making it possible to calculate the typical dimming levels for luminaires. Additionally, integrating the lighting control system with the HVAC controls through the Building Automation System (BAS) and with individual plug load controls can provide substantial energy savings and leads the way to the commercial realization of NZEBs.

To estimate power consumption, let d_m, where $0 \leq d_m \leq 1$, be the dimming level of the mth LED luminaire. The value $d_m = 0$ means that the LED is dimmed off completely, whereas $d_m = 1$ represents that the LED is at its maximum luminance. Dimming of LEDs is typically done using pulse width modulation (PWM) [18], where the dimming level corresponds to the duty cycle of the PWM waveform. Let d be the $M \times 1$ dimming vector of the lighting system given by

$$d = [d_1, d_2, \ldots, d_M]^T \qquad (1)$$

indicating that the mth LED is at dimming level d_m. The power consumption of a LED is directly proportional to the dimming level. Denote the average power consumption of the mth LED at dimming level d_m by $P_m(d_m)$ [19]. Then

$$P_m(d_m) = P_{on} d_m, \qquad (2)$$

where P_{on} is the power consumption of the LED in the on-state.

The occupancy sensing, daylight (photocell) sensing, and communication elements are integral to the lighting control system. In this project, an intelligent lighting system using Zigbee protocol and control capability has been proposed. In addition, luminaires are used to provide the basic function of illumination rendering and for adaptation to sensing information inputs and user preferences. A lighting control system with multitechnology sensors, both standing alone and integral to specific LED luminaires, has been proposed. LEDs are currently the primary illumination source due to their longer lifetimes and better design flexibility. In particular, LED luminaires offer easy and accurate dimming capability. In order to capitalize on the large amount of daylighting afforded by vast glazed areas, an automated control system is employed. The automated lighting control system not only saves energy but also improves indoor environmental quality by reducing unnecessary artificial lighting. The control system uses a network of sensors which automatically control artificial lighting levels depending on the amount of natural sunlight within designated areas.

For the building, the digital addressable lighting interface (DALI) reference design has been adopted. DALI is a concept that stands for an intelligent lighting management system that provides increased energy savings, easier installation and maintenance, and maximum control and retrofit flexibility in an entirely open standard [20]. It is defined in IEC 60929 and has been updated in IEC 62386. One of the main reasons for this update is the inclusion of the LED device type. DALI is an affordable "open systems architecture" that allows any manufacturer's devices to interface with any dimming, control, sensor, or fixture to create a room or area lighting system.

The use of DALI devices with wireless sensor network provides many parameters about the efficient lighting; this is very useful for saving energy and maintenance purposes, as it can detect any single lamp fault allowing a predictive maintenance and group replacement or schedule power consumption rules enabling the integration of the lighting system in the building into smart MG approach, due to monitoring and acting capability.

For simpler projects or for cost savings measures, 0–10 V dimming control through a Digital Lighting Management (DLM) system is an effective alternative to a DALI control system. Commercially available LED fixtures are increasingly being supplied with 0–10 V dimming control compatible drivers as a standard feature. 0–10 V control is another "open systems architecture" and utilizes a DC analogue control signal ranging from 0 V, lowest lumen output, to 10 V, highest lumen output, with the protocol defined by the IEC-60929 Annex E standard. This control system was originally developed for dimmable fluorescent fixtures approximately 30 years ago, being adopted into the IEC standards in 1992, but due to the nature of the circuitry within a DC LED driver it is proving to be cost-effective controls solution for LED fixtures as well.

DLM systems consist of local relays packs (or small-scale relay panels), controllers (wall switches/dimmers), and sensors that can either be networked together or act as stand-alone systems. DLM systems providing 0–10 V control

TABLE 1: Energy-efficient HVAC system and projected energy savings.

Implementation	Specifications	Saving (%)
Waste heat recovery system for outside air	For colder climate applications, outside air ventilation must be preheated before entering the vents. The waste heat from exhausted air may be used for such purpose. Accomplished with the integration of a thermal wheel or rotary heat exchanger with the HVAC system.	21
Cooling heat recovery system	Heat exhausted from a cooling system could be recovered for ventilation and space heating. Accomplished using heat recovery condensers.	28
Air-source heat pump	Extracts heat from the air and then transfers heat to either the inside or the outside of your home depending on the season.	15
	Total projected energy saving	**64**

are readily available, allowing fixtures to be switched and dimmed in control groupings while minimizing wiring by locating the relay packs centrally to the associated fixtures.

The main separation between DLM and DALI systems is that DALI communicates directly with each individual fixture, whereas DLM controls fixtures in groupings. As DLM systems do not communicate directly with each individual fixture there are a number of limitations when compared to a DALI system; namely, there is no mechanism to directly detect faults or maintenance issues, and DLM systems cannot inherently meter each individual fixture's energy usage; however, power monitoring of individual relay packs is achievable as is implementing scheduled power consumption rules. The result of this is overall less flexibility but for most practical applications a very effective solution.

The lighting energy consumption based on DALI/DLM LED industry standard protocol is estimated as 16,418 kWh while the same facility with nonefficient lighting system requires 109,744 kWh. Based on the above, the projected energy saving is 84%. Lighting is selected based on energy efficiency, fixture lifespan, and the degree to which the fixture's composition could potentially harm the environment.

A Central Emergency Lighting Inverter (CELI) is proposed to provide the building with an easily maintained and tested emergency lighting system in accordance with the OBC requirements, providing designated lighting fixtures with battery backed-up AC power produced by a pulse width modulation inverter (creating sinusoidal output at designated voltage from a battery source). The CELI will maintain its charge utilizing locally generated power whenever available. By employing a CELI it is possible to nearly entirely eliminate the need for emergency battery units and remote heads, reducing overall quantity of materials by employing only the fixtures part of the general lighting design to provide emergency lighting as well.

3.3. HVAC Energy Efficiency. Nearly every energy model requires the user to select a HVAC system. The heating and cooling demand is generally driven by the goal of maintaining the occupied space within a comfortable temperature and humidity range. The most widely promoted guidance for thermal comfort comes from the American Society for Heating, Refrigeration and Air Conditioning Engineers (ASHRAE) in the form of the ASHRAE 55 standard for

thermal environmental conditions for human occupancy (ANSI/ASHRAE 2010) [17].

Among building energy services, HVAC system is the most energy consuming segment, accounting for about 10–20% of final energy use in developed countries [21]. The HVAC system generally contains a source of heating and cooling, a distribution system, and a technique for supplying fresh air. These include boilers, furnaces, and electric resistance heaters which are typically used to add heat to buildings while cooling is typically accomplished via air conditioning units, heat pumps, or cooling towers.

While the architect's goal with early design simulation and energy modeling is generally to reduce loads passively, the HVAC system selection and design can have a large impact on resulting energy use. The heating and cooling demand is generally driven by the goal of maintaining the occupied space within a comfortable temperature and humidity range. Ventilation service uses thermal energy for outdoor air treatment and electrical energy for filtration and distribution to conditioned spaces.

The number of occupants needs to be defined so as to determine the hot water needs as well as the thermal internal gains. The cooling oversize recommended for the building is estimated as 581,625 Btuh [22]. This includes base cooling and additional cooling required for an assumed occupancy of 75 people and 4 kitchens, as proposed in the case study. Accordingly, 5 rooftop units are recommended to share the load with supplemental electric heating. To minimize the building's heating and cooling energy consumption, two systems were integrated into the HVAC design. First, an air-source heat pump is used to extract heat from the outdoor air, resulting in 15% of additional energy saving. Second, waste heat and cooling recovery systems are used to preheat fresh outdoor air. The largest percentage of energy savings comes from the specification of energy-efficient HVAC equipment. Implementing these systems results in an overall savings of 64% when compared to traditionally designed systems as shown in Table 1.

3.4. Water Performance Improvement. Externally, considerations for irrigation have been completely eliminated through lower-technology, low-cost, and high-impact strategies such as xeriscaping. Native plant species and natural habitat are maintained and promoted throughout the site, thus saving

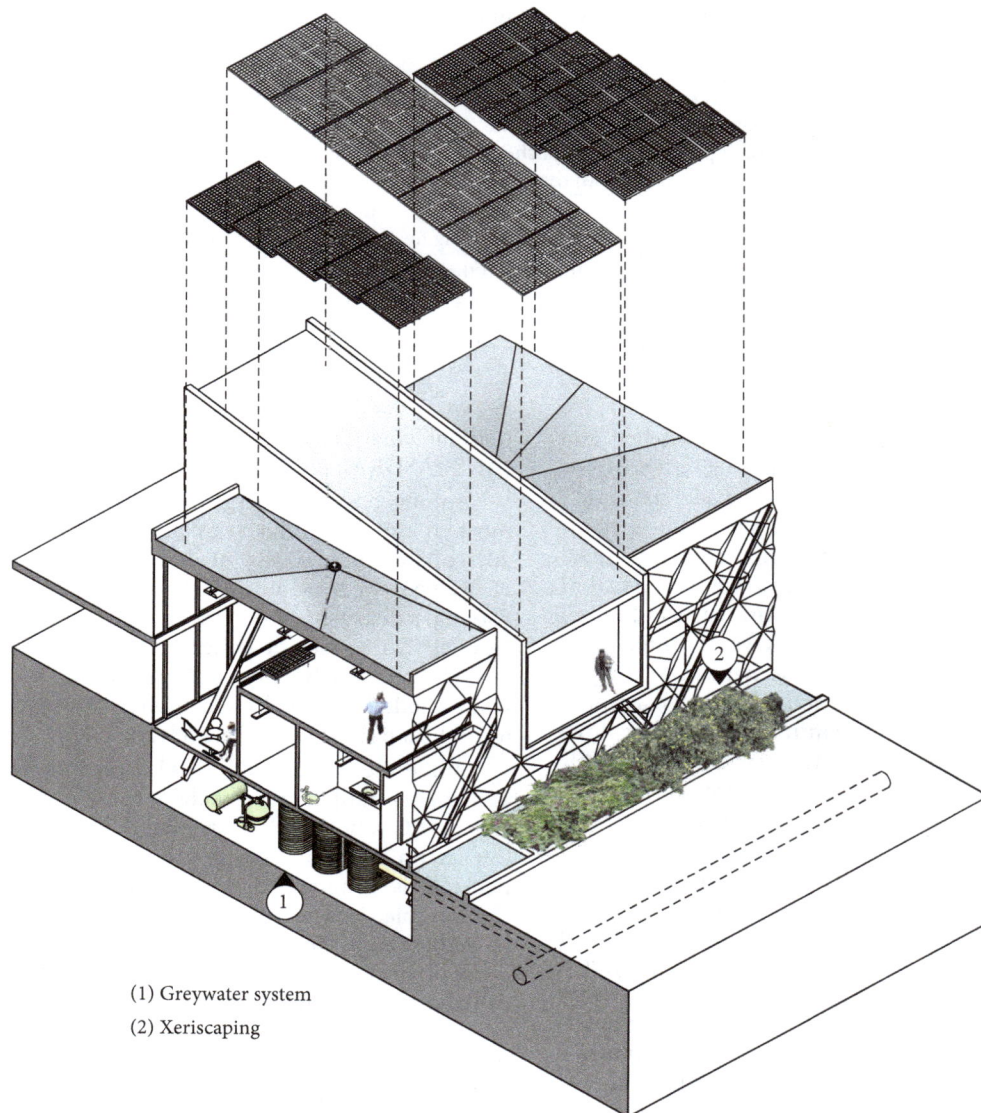

FIGURE 3: Water use reduction proposal for the building.

(1) Greywater system
(2) Xeriscaping

tens of thousands of gallons of water annually when compared to traditional landscaping techniques.

Internally, the full-time equivalent (FTE) for the building was calculated at 38 FTE including closed offices, workstations, and hoteling stations based on 250 days worked per year. Using default EPAct figures for flush and flow fixtures, the baseline value for daily water use was calculated as 55,425 gals/year. While substituting the EPAct flush and flow rates with flush and flow rates for water-efficient fixtures (waterless urinals, dual-flush toilets, and low-flow lavatories) the baseline water use was reduced to 29,687 gals/year. This reduces the need for municipally provided potable water by 54% annually.

As a result of the above considerations, the proposal is more healthful, energy-efficient, and sustainable alternative to a traditionally built complex. Water saving strategies such as the incorporation of water-efficient fixtures typically have little added cost. Figure 3 shows the water use reduction proposal for the building.

4. CCHP Microgrid

MG is the cornerstone and indispensable infrastructure of smart grid [23]. It is defined as a cluster of loads and energy generation sources operating as a single controllable system that provides both power and heat to its local area [24, 25]. Development of the CCHP MG by using various loads and renewable energy sources has drawn considerable research attention recently. Compared with conventional CCHP systems, the CCHP MG has novel and greater functionality, because the CCHP MG not only satisfies the cooling, heating, and power demands of buildings but also interacts with

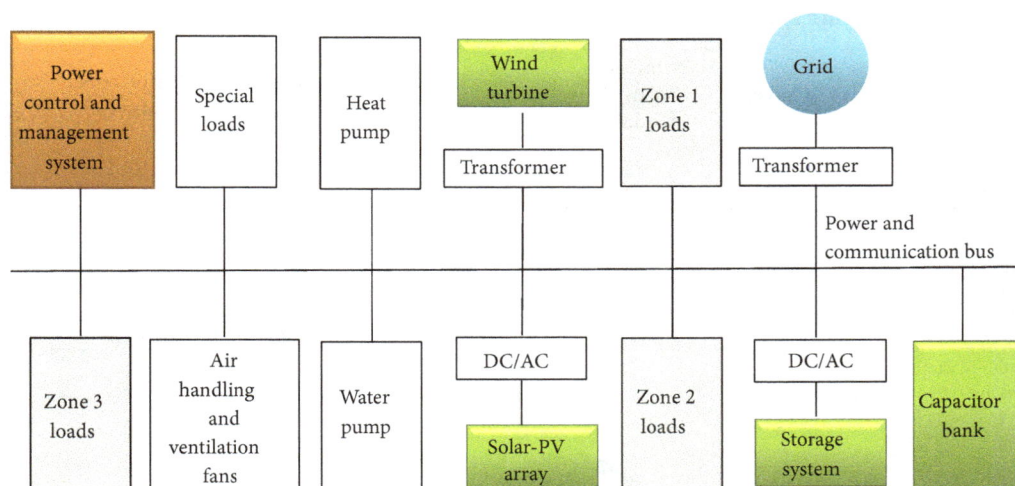

FIGURE 4: Schematic of the proposed building CCHP MG.

the main grid to provide reserve, peak-saving, and demand response services and provides improved capability for integration of renewable energy sources [26].

MGs, if designed properly, are capable of operating in either grid-connected mode or islanded mode [27]. MG technology provides an opportunity and a desirable infrastructure for improving the efficiency of energy consumption in buildings [28]. Recent research work shows that 20%–30% of building energy consumption can be saved through optimized operation and management without changing the structure and hardware configuration of the building energy supply system [29].

The majority of electrical loads within a building can be grouped into lighting subsystems, computing equipment, individual plug-loads, and HVAC related equipment. All these subsystems must be optimized to improve the energy efficiency of the building. A schematic of the MG model proposed for the building is shown in Figure 4. Through optimal operation control of the MG, this system will enable the building to maximize the gain of the renewable sources, to improve energy efficiency, to decrease the energy bill, and to reduce greenhouse gas emission.

4.1. Annual Energy Demand. As opposed to residential buildings, commercial buildings have energy consumption profiles that are driven primarily by the defined work week. That means having very low occupancies on weekends, holidays, early morning hours, and weekday evenings. Characterizing the energy demand of a building involves firstly identifying the specific energy end uses involved. Each major end use that forms the energy demand can be influenced by a number of design variables, and each building design variable has a wide range of likely values or alternatives. The load profile of the facility is proposed to closely match that of similar typical buildings for which a measurement campaign was undertaken. Figure 5 illustrates the projected power demand of the building. The annual energy demand for the building was estimated by making seasonal adjustments to the weekly

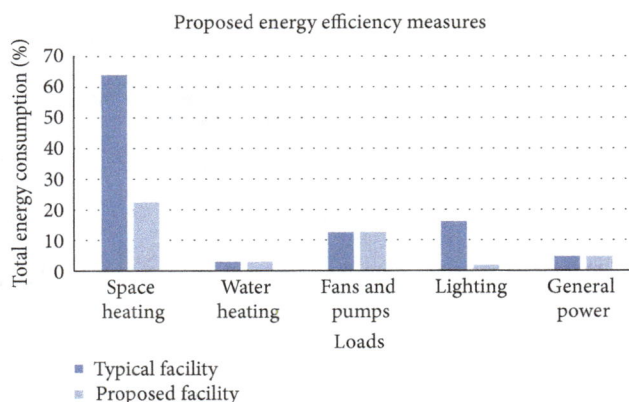

FIGURE 5: Load distribution of the building: typical versus proposed.

demand and typical day profiles and summating it over a year. The total power consumption in various components of the building (sport performance laboratory, strength and conditioning facility, ancillary rooms, reception, server room, showers/locker rooms/washrooms, offices, boardrooms, hoteling stations, staff and demonstration kitchens, athlete career lounge, hallways, track, and field floodlights) has been estimated based on size, lumens/luminary, type of fixtures, utilization factor, maintenance factor, and lighting on 18 hours/day. Based on the above estimation, an expected total approximated energy demand of 336.79 MWh per year was established.

4.2. On-Site Renewable Energy. A number of factors are considered to estimate the amount of solar energy available on a building. First is the area of roof that is available to install solar-PV array. Due to the placement of mechanical systems on the roof, the available area is often less than the actual roof area. One way to increase it is by installing solar-PV array on the wall. Besides the amount of space available, other

TABLE 2: Technical details of the three locations of solar-PV array.

Solar-PV location	Racking type	Tilt (degree)	Number of solar-PV modules[1]	Module area (m^2)	Nominal power (kW)	Performance ratio (%)	Yearly production (MWh/year)[2]
Solar wall	Building integrated	36.86	85	137	21.68	79.4	25.45
Sloped roof	Extruded aluminum	10	234	376	59.67	81.5	68.30
Flat roof	Ballasted KB EKONORACK	10	740	1190	188.70	77.6	206.00
	Total		**1059**	**1703**	**270.05**	**78.6**	**299.75**

[1] Actual number of modules is based on the area.
[2] Yearly production values are based on simulation.

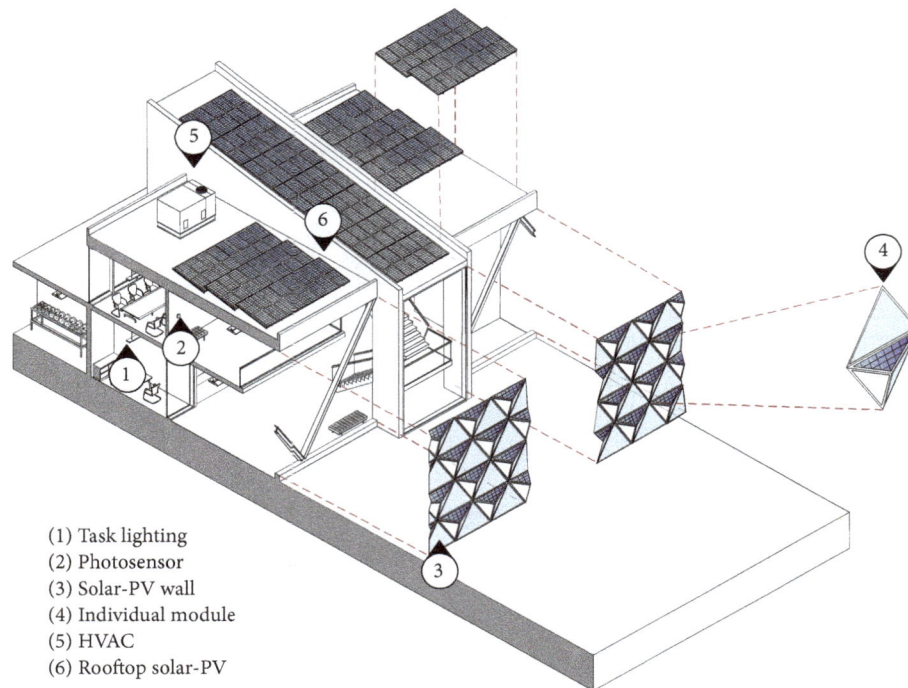

(1) Task lighting
(2) Photosensor
(3) Solar-PV wall
(4) Individual module
(5) HVAC
(6) Rooftop solar-PV

FIGURE 6: Solar-PV array in the building.

important factors include the amount of energy production available from the solar panels themselves. This depends on the efficiency of the panels and the amount of solar radiation exposure. For this project, solar-PV array is proposed on three parts of the building including the triangular sections on the wall and sloped and flat roofs as shown in Figure 6. Based on photovoltaic software calculations, the total estimated capacity for the three solar-PV arrays is 270 kW as shown in Table 2. This includes margins for factors such as panel availability, load estimation, equipment efficiency, and solar-PV array and inverter losses. The energy produced by the solar-PV array allows the building to achieve a total value of 89% energy offset.

The generating resources in the proposed MG include a solar-PV array and a small-scale wind generator. The wind turbine associated with the MG is treated as an experimental

generation unit and thus was not included in the energy calculations. A small-scale contrarotating induction generator-based wind turbine is proposed with a nominal rating of 75 kW [30]. Its presence provides some additional margin for meeting the energy requirements and also assists in eliminating required energy storage. It is anticipated that the proposed wind turbine will provide approximately 340 MWh per year at a wind speed of 7.5 m/s [31]. The proposed MG is expected to operate in either grid-connected mode or islanded mode in order to maximize the energy harvesting from the solar-PV array and ensure economically feasible operation.

4.3. Energy Storage System. Energy storage is designed not only to conserve energy but also to allow demand shifting where energy can be stored during lower demand times and used during peak demand periods. Several strategies can be considered when sizing on-site energy storage including peak demand reduction, energy harvesting to reduce export, mitigating intermittent grid outages, and standalone operation. Each strategy has differing demands on energy storage sizing and economic constraints. For the proposed facility, sizing energy storage for prolonged intentional islanding is not required, and in order to minimize the energy storage requirement (due to financial constraints), intentional islanded operation is limited to a single day demonstration.

There are several forms of energy storage available that can be used in MGs. These include batteries, supercapacitors, and flywheels. Conventional lithium-ion batteries are the selected energy storage technology for the building due to their high density (90–150 Wh/kg), high efficiency (80–85% DC to DC), and long cycle life (3000 cycles at 80% DOD). The main disadvantages are the high cost (about $600/kWh), complex battery management circuitry, and safety matters related to thermal management [25]. For preliminary design, 150 kWh is selected as the capacity, with intentional transition to islanded mode dependent on state of charge (SOC), available generation (weather forecast), and load shedding. Extending islanded mode operation in the future may be made possible by increasing energy storage or localized generation.

4.4. Capacitor Bank.
Shunt capacitor banks are used to improve the quality of the electrical supply and the efficient operation of the power system. To meet utility connection requirements, the reactive power demand of a building is compensated using dynamically switched banks to provide capacitive reactive compensation/power factor correction (PFC). Eight segmented PFC stages are proposed for the building. The peak reactive power demand of the sample building is able to be delivered by the first six segments of PFC capacitors, for example, 60 kVAR. An additional 20 kVAR segment is provided to deliver margin for motor start requirements of additional equipment for the building.

4.5. Surge Protective Devices (SPDs).
The sophistication of the electrical equipment to be installed on this and other similar projects is achieved mostly through the introduction of microprocessors/microcontrollers as control units for the variety of electrical devices such as LED drivers, lighting control systems, and variable-frequency drives. By implementing these microelectronic devices, this equipment has become as sensitive to transient voltages and voltage surges as any computer load. In order to protect the significant investment this equipment represents as much as possible, surge protective devices (SPDs) in a tiered approach have been proposed. The tiered approach refers to the implementation of SPDs on each level of distribution, that is, the facility-wide protection on the service entrance switchboard and the branch level protection on the individual panel boards. This tiered approach allows for externally and internally generated surges and transients to be reduced to harmless levels by knocking the impulse waveforms down in stages based on amplitudes, greatly reducing their impact on the desired

waveforms and leaving no single point of exposure upon failure of an individual SPD.

4.6. DC Bus.
In the future, it is worthwhile to consider a DC bus to supply the required power to the appliances where MGs employing renewable energy sources and LED lighting with a large fraction of appliances are internally DC powered. This would eliminate the need for DC to AC and back from AC to DC conversions, thereby saving energy on the whole. The literature shows that a large segment of the building's electric load could be fed directly with DC power including devices based on microprocessors, computer system power supply, switched-mode power supply, electronic ballasts for the fluorescent lighting, variable-frequency drives for the speed variation of the motors which equip the systems of HVAC, and lighting based on LEDs [32–34].

4.7. Control and Management.
The MG control and management system synchronizes multiple energy sources with the grid and with building loads. The MG is based on key components including a local monitoring and control system, communication network, and central controller integrated with an intelligent energy management system (IEMS). The controller consists of algorithms for real-time monitoring and control of the network. Sources and loads need to communicate quickly with each other. This is an opportunity for short messaging protocols such as extensible messaging and presence protocol (XMPP) which is widely used by Twitter, Google Talk, Facebook, and other large scale applications. XMPP may use a connection through a cloud server to link an external energy source management network with an internal building load management network. The integrated IEMS regulates energy consumption and storage according to the availability of its resources.

Electricity and cold and heating demands fluctuate both daily and seasonally. Therefore, taking operational strategies into account during the design stage should prove beneficial for the candidate CCHP system. Consequently, a robust technique for sizing and identification of the optimum operational strategies for such systems is needed [25].

Requirements for control systems to achieve efficient performance and regulation of various facilities can be organized in several categories including zone controls (thermal zoning, zone isolation, temperature, and humidity controls); air handling unit controls (variable air volume and temperature); water loop controls (pump isolation and load management); primary equipment controls (cooling towers and heat pumps); and automatic meter reading and data collection facilities. Therefore, intelligent control of the above elements offers additional energy savings otherwise not possible with independent operation.

5. Concluding Remarks

The rising problems of energy shortages and environmental concerns have boosted the development of the NZEBs and CCHP MGs. As a consequence of the expected widespread development of NZEBs and CCHP MGs, there is an urgent need to improve the available interdisciplinary skills for design and operation of the systems to realize energy savings, environmental protection, and economical operation. In this context, this paper presented a methodology with the goal of assisting the choice of economically efficient NZEB and CCHP MG solutions, right from the early design stage. This can be used for any residential or commercial building in any part of the world considering the local climate, the endogenous energy resources, and the local economic conditions that lead to a nearly zero annual energy balance. Achieving nearly zero is not only a matter of design; it requires careful attention to operations and maintenance and to occupancy patterns and loads. While such buildings are possible with existing technologies, this research uncovered the challenges associated with achieving nearly zero energy in this building with today's onsite renewable energy and MG technologies. However, the encouragement of nearly zero at the building scale sets the stage for future technology solutions and the removal of barriers to energy and water systems to move buildings to energy and water independence.

The objective of this work is to maximize energy efficiency, optimize occupant comfort, and reduce dependency on both the grid and the municipal potable water supply by implementing sustainable strategies in designing a research and sports building. Creative solutions by the architectural and engineering team capitalize on the design of a unique glazing system; energy-efficient technologies; water use reduction technique; and a CCHP MG with integrated control aspects and renewable energy component. The CCHP MG provides an effective solution to energy-related problems, including increasing energy demand, higher energy costs, energy supply security, and environmental concerns. The key challenge to improve building energy efficiency in operation is to coordinate and optimize the operation of various energy sources and loads. As a result of the above considerations, the proposed building is a more healthful, energy-efficient, and sustainable alternative to a traditionally built complex.

It must nevertheless be noted that the design tool presented in this paper intends to provide guidance in the early design stage more than to replace a detailed dynamic assessment of the solution that ends up being elected by the architecture and engineering design teams.

Conflict of Interests

The authors declare that there is no conflict of interests regarding the publication of this paper.

Acknowledgments

The authors are grateful to Royal Bank of Canada (RBC) Foundation and B+H Architects for the opportunity to participate in the annual Evolve Sustainable Design Competition and for their financial support.

References

[1] IEA (International Energy Agency), "Energy Balances of OECD Countries—2010 edition," Paris, France, 2010.

[2] A. J. Marszal, P. Heiselberg, J. S. Bourrelle et al., "Zero Energy Building—a review of definitions and calculation methodologies," *Energy and Buildings*, vol. 43, no. 4, pp. 971–979, 2011.

[3] P. Torcellini, S. Pless, and M. Deru, "Zero energy buildings: a critical look at the definition," in *Proceedings of the ACEEE Summer Study on Energy Efficiency in Buildings*, Pacific Grove, Calif, USA, August 2006.

[4] M. Noguchi, A. Athienitis, V. Delisle, J. Ayoub, and B. Berneche, "Net zero energy homes of the future: a case study of the EcoTerraTM house in Canada," in *Renewable Energy Congress*, Glasgow, Scotland, July 2008.

[5] "The Active House project," http://www.activehouse.info.

[6] M. Heinze and K. Voss, "Goal: Zero energy building—exemplary experience based on the solar estate solarsiedlung freiburg am schlierberg, Germany," *Journal of Green Building*, vol. 4, no. 4, pp. 93–100, 2009.

[7] E. Musall, T. Weiss, A. Lenoir, K. Voss, F. Garde, and M. Donn, "Net zero energy solar buildings: an overview and analysis on worldwide building projects," in *Proceedings of the EuroSun Conference*, Graz, Austria, 2010.

[8] E. Doub, *Solar Harvest: City of Boulder's First Zero Energy Home*, Ecofutures Building, Boulder, Colo, USA, 2009.

[9] R. Hawkes, *Crossway Eco-House*, Hawkes Architecture, Kent, UK, 2009.

[10] K. B. Wittchen, J. S. Ostergaard, S. Kamper, and L. Kvist, *BOLIG+ an Energy Neutral Multifamily Building*, EuroSun, Graz, Austria, 2010.

[11] M. Kapsalaki and V. Leal, "Recent progress on net zero energy buildings," *Advances in Building Energy Research*, vol. 5, no. 1, pp. 129–162, 2011.

[12] A. Ferrante and M. T. Cascella, "Zero energy balance and zero on-site CO_2 emission housing development in the Mediterranean climate," *Energy and Buildings*, vol. 43, no. 8, pp. 2002–2010, 2011.

[13] D. Crawley, S. Pless, and P. Torcellini, "Getting to net zero," *ASHRAE Journal*, vol. 51, no. 9, pp. 18–25, 2009.

[14] EN 15217, "Energy performance of buildings—methods for expressing energy performance and for energy certification of buildings," in *European Committee for Standardization*, 2007.

[15] Energy Information Administration (EIA), *Measuring Energy Efficiency in the United States Economy: A Beginning*, Department of Energy, 1995.

[16] H. Liu, Q. Zhao, N. Huang, and X. Zhao, "A simulation-based tool for energy efficient building design for a class of manufacturing plants," *IEEE Transactions on Automation Science and Engineering*, vol. 10, no. 1, pp. 117–123, 2013.

[17] D. J. Sailor, "Energy buildings and urban environment," in *Vulnerability of Energy to Climate*, vol. 3, pp. 167–182, 2013.

[18] Y. Gu, N. Narendran, T. Dong, and H. Wu, "Spectral and luminous efficacy change of high-power LEDs under different dimming methods," in *Sixth International Conference on Solid State Lighting*, vol. 6337 of *Proceedings of SPIE*, pp. 63370J-1–63370J-7, 2006.

[19] A. Pandharipande and D. Caicedo, "Adaptive illumination rendering in LED lighting systems," *IEEE Transactions on Systems, Man, and Cybernetics*, vol. 43, no. 5, pp. 1052–1062, 2013.

[20] M. Moeck, "Developments in digital addressable lighting control," *Journal of Light and Visual Environment*, vol. 28, no. 2, pp. 104–106, 2004.

[21] L. Pérez-Lombard, J. Ortiz, and C. Pout, "A review on buildings energy consumption information," *Energy and Buildings*, vol. 40, no. 3, pp. 394–398, 2008.

[22] Natural Resources Canada, Air-Source Heat Pumps.

[23] X. Tan, Q. Li, and H. Wang, "Advances and trends of energy storage technology in Microgrid," *International Journal of Electrical Power and Energy Systems*, vol. 44, no. 1, pp. 179–191, 2013.

[24] W. Y. Habash, V. Groza, T. McNeill, and I. Roberts, "Lightning risk analysis of a power microgrid," *British Journal of Advanced Science and Technology*, vol. 3, no. 1, pp. 107–122, 2013.

[25] S. X. Chen, H. B. Gooi, and M. Q. Wang, "Sizing of energy storage for microgrids," *IEEE Transactions on Smart Grid*, vol. 3, no. 1, pp. 142–151, 2012.

[26] W. Gu, Z. Wu, R. Bo, W. Liu, G. Zhou, and W. Chen, "Modeling, planning and optimal energy management of combined cooling, heating and power microgrid: a review," *International Journal of Electrical Power and Energy Systems*, vol. 54, pp. 26–37, 2014.

[27] L. G. Meegahapola, D. Robinson, A. P. Agalgaonkar, S. Perera, and P. Ciufo, "Microgrids of commercial buildings: strategies to manage mode transfer from grid connected to islanded mode," *IEEE Transactions on Sustainable Energy*, vol. 5, no. 4, pp. 1337–1347, 2014.

[28] R. O'Neill, "Smart grids: sound transmission investments," *IEEE Power and Energy Magazine*, vol. 5, no. 5, pp. 101–104, 2007.

[29] X. Guan, Z. Xu, and Q.-S. Jia, "Energy-efficient buildings facilitated by microgrid," *IEEE Transactions on Smart Grid*, vol. 1, no. 3, pp. 243–252, 2010.

[30] R. W. Y. Habash and P. Guillemette, "Harnessing the winds: trends and advances," *IEEE Potentials*, vol. 31, no. 1, pp. 16–21, 2012.

[31] http://www.greenengineers.ca/REMLab/ZECWP%20-%20customer%20Package%20v5.3.pdf.

[32] K. Engelen, E. L. Shun, P. Vermeyen et al., "The feasibility of small-scale residential DC distribution systems," in *Proceedings of the 32nd Annual Conference on IEEE Industrial Electronics (IECON '06)*, pp. 2618–2623, November 2006.

[33] D. Salomonsson and A. Sannino, "Low-voltage DC distribution system for commercial power systems with sensitive electronic loads," *IEEE Transactions on Power Delivery*, vol. 22, no. 3, pp. 1620–1627, 2007.

[34] M. Sechilariu, B. Wang, and F. Locment, "Building-integrated microgrid : advanced local energy management for forthcoming smart power grid communication," *Energy and Buildings*, vol. 59, pp. 236–243, 2013.

16

Design of an Energy System Based on Photovoltaic Thermal Collectors in the South of Algeria

K. Touafek, A. Khelifa, M. Adouane, and H. Haloui

Unité de Recherche Appliquée en Energies Renouvelables (URAER), Centre de Développement des Energies Renouvelables (CDER), 47133 Ghardaïa, Algeria

Correspondence should be addressed to K. Touafek; khaledtouafek@uraer.dz

Academic Editor: Zuhal Oktay

The objective of this work is the design of a new energy system where the energy source will be provided by solar photovoltaic thermal (PV/T) hybrid collectors. This system will be applied to a habitation in the region of Ghardaïa in the south of Algeria. The cold water reaches the thermal storage tank and then will be heated by the hybrid collector. The hot water will be used directly as sanitary water. The electric power produced by the hybrid collector will be used to charge the battery and will be delivered to the load (electrical appliances, lamps, etc.). Two types of loads are considered: a DC load and the other alternating current. The fans located adjacent to the radiators supplied with hot water will provide warm air to the house in winter.

1. Introduction

Research on hybrid solar collectors began in the 70s and was intensified in the 80s. Thus, the work of Wolf [1] in 1976 performs the analysis of a solar thermal collector with PV modules based on silicon and coupled to a heat storage system. Subsequently, the study of Kern and Russel in 1978 provides the basics of using solar water or air as coolant in PV/T systems. Hendrie, in 1982 [2], develops a theoretical model of PV/T hybrid based on correlations related to solar standards. In 1981, Raghuraman [3] presents numerical methods for predicting the performance of flat solar PV/T water or air. In 1986, Lalovic et al. [4] proposed a new type of amorphous a-Si cells as transparent economic solution for the construction of PV modules. Various experimental and theoretical studies have been conducted, then, for the development of PV/T hybrid [5]. In 2005, Zondag [6, 7] proposes a state of the art on the solar PV/T hybrid based on the report of the European Project PV-Catapult. Among the first studies reviewed by Zondag [6], some focus on the evolution of the geometry and other components of the modeling methods are studied. In 2007, Tiwari and Sodha [8] proposed a parametric comparative study of four types of solar air close to the system presented above. Tripanagnostopoulos [9] conducted, at the University of Patras, the study of solar PV/T hybrid of which the coolant is either air or water and can be integrated to the frame. The objective of this work was to reduce the operating temperature of PV modules. In Algeria, the work of Touafek et al. [10–12] is the important research done on PV/T systems. They have studied various configurations in many conditions. This work is the application of the hybrid collectors studied in detail in previous papers [11, 12].

2. General Outline and Constitution of the System

We begin by studying a single system composed of a single new hybrid collector configuration discussed in [11, 12]. Figure 1 shows an overview of the system.

Figure 1 shows an energy system that can be applied to a habitation. The cold water reaches the thermal storage tank and will be heated by heat convection transfer fluid that transports heat from the hybrid collector. The hot water will be used directly as sanitary water. The electric power produced by the hybrid collector will be used to charge the battery and will be delivered to the load (electrical appliances, lamps, etc.). Two types of loads are considered: a DC load and the other alternating current. The fans located adjacent to the

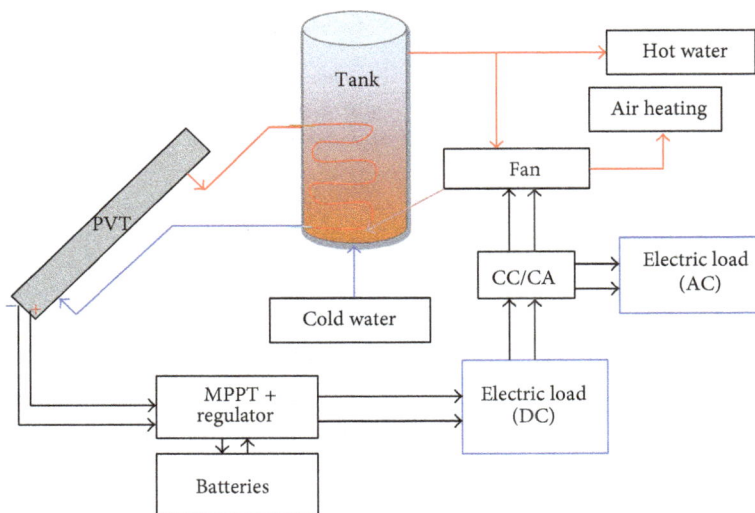

FIGURE 1: Diagram of the energy system based on the new collector PV/T.

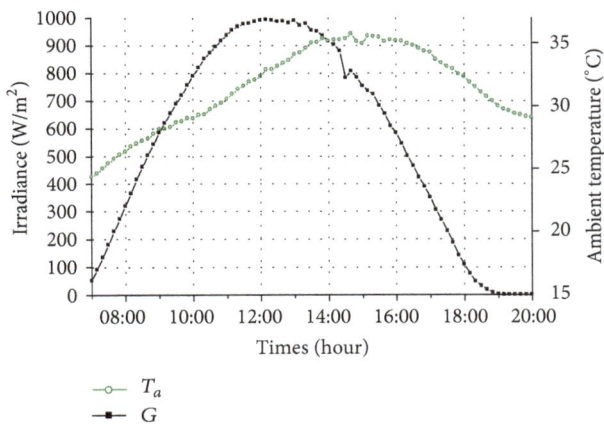

FIGURE 2: Global irradiance and temperature tests.

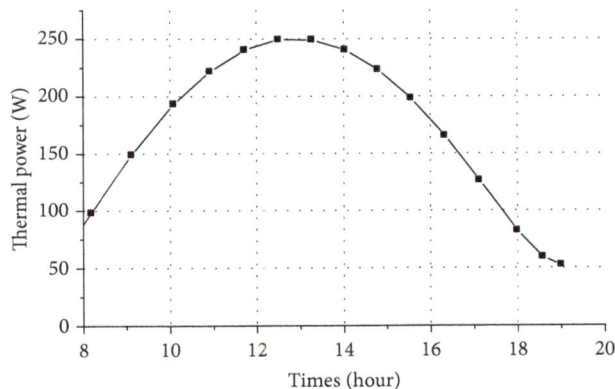

FIGURE 3: Thermal power produced daily by the hybrid collector.

radiators supplied with hot water will provide a warm air to the house in winter.

3. Daily Production of Hybrid Collector

The daily production of hybrid collector includes a heat energy and electrical energy. The change in global irradiance and ambient temperature of that day is shown in Figure 2. It is a sunny day.

Under these conditions, the hybrid collector produces heat energy in addition to electric power.

3.1. Daily Heat Production. The daily thermal energy is the amount issued by a hybrid collector for a given day. The hybrid collector used is studied in [11, 12].

Figure 3 shows the variation of the instantaneous thermal power of a single hybrid collector for the day of September 14, 2008. For an input temperature of water of 25°C, the power output reaches 250 W. It is the instantaneous power generated

by the hybrid collector. The day chosen is a typical day. The experimental thermal efficiency based on the reduced temperature is shown in Figure 4.

The instantaneous thermal efficiency of the collector used is equal to 68% when the input temperature of water is equal to ambient temperature.

The energy production is determined by multiplying the average power by the hours of sunshine. Note in Figure 3 that, during the period from 10 h to 17 h, the instantaneous power exceeds 150 W. We can take this value as a reference and we can say that the hybrid collector can produce at least (150 W) × (7 hours) = 1050 Wh. This value is for the collector surface of 0.42 m^2. So the daily production is 1050 Wh/0.42 m^2 = 2500 Wh/m = 2.5 KWh/m^2. For one month, the collector can produce (1050 Wh) × (30 days) = 31.5 KWh.

3.2. Daily Electric Production. The hybrid collector mainly produces electric power. This energy is determined by multiplying the instantaneous power for the duration of sunshine daily. The instantaneous power can be calculated by two

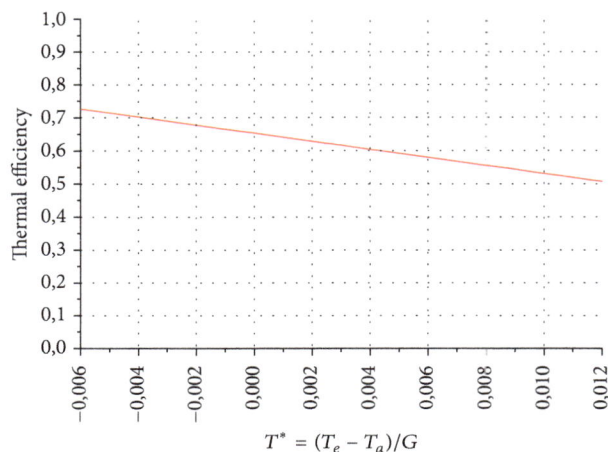

FIGURE 4: Thermal efficiency as a function of reduced temperature for the day of testing.

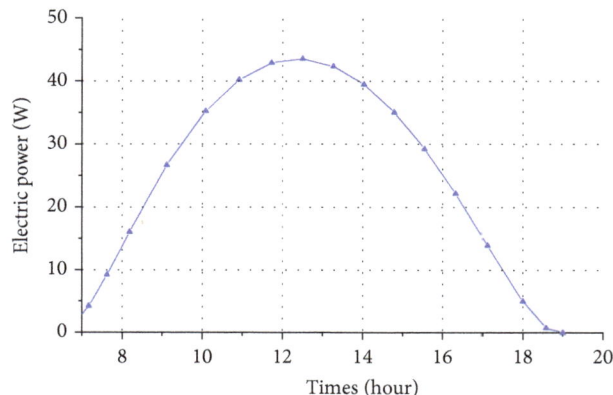

FIGURE 5: Electric power produced by the collector hybrid.

ways: the first is by calculating the voltage time's current product and the second using mathematical method and will be based on the temperature of solar cells. The second method is used in this study. Figure 5 shows the variation of the instantaneous electrical power during the day on September 14, 2008.

Note that, for at least seven hours in succession, the electrical power exceeds 25 W. The hybrid collector produces at least $(25\,W) \times (7\,hours) = 175\,Wh$. The daily production of the collector is at least equal to 175 Wh. For one month, the collector can produce 5250 KWh. For an area of $1\,m^2$, we have $(175\,Wh)/(0.42\,m) = 416.67\,Wh/m^2$ (daily). The energy produced during a month is $12.5\,kWh/m^2$. The energy produced during a year is $150\,kWh/m^2$.

The hybrid collector thus produces a daily electric energy of 175 Wh and thermal energy of 1050 Wh.

The daily production of electric heat collector is the basis for sizing the energy system.

4. System Modeling

The proposed energy system consists mainly of hybrid collectors for heating water and water for electrical power generation and air collectors for space heating.

4.1. Water Heating. The tank of hot water contain a serpentine and an additional electrical resistor. The latter provides heat when the tank temperature is not high enough to heat the hot water temperature desired output. It is not easy to know the needs of hot water (DHW) of a family in a house. In general, it is estimated as 50 liters to 50°C per day per person but this can vary for about 20%. The volume of hot water tank mixed with electric backup should be able to cover 1.5 times the daily needs. The volume which must have the hot water tank is calculated by the following formula:

$$V = \left(\frac{\left(B_p \cdot N_p \cdot (T_{es} - T_{ef}) \right)}{(T_{st} - T_{ef})} \right) \cdot 1.5, \tag{1}$$

where B_p = volume required per person per day, N_p = number of persons occupying habitation, T_{es} = temperature of water extraction (water for direct use), T_{ef} = temperature of cold water entering the reservoir, and T_{st} = average temperature of the water stored.

For example for 4 people with the needs of 50 liters/person/day

$$T_{es} = 40°C,$$
$$T_{ef} = 15°C,$$
$$T_{st} = 45°C,$$
$$V = ((50 \times 4 \times (40 - 15))/(45 - 15)) \times 1.5 = 250 \text{ liters.}$$

In general, the storage tanks have too much capacity and thus the maximum temperature reached with solar energy for the period from October to April is not large enough and necessarily requires the operation of the booster which could be avoided with a lower capacity.

The system studied in this paper consists of hybrid photovoltaic thermal collectors that produce a flow of heat for heating the water container in the thermal storage tank. The thermal power transferred between the system useful PV/T and thermal tank is given by

$$Q_u = \dot{m} \cdot C_p \left(T_{eres} - T_{res} \right), \tag{2}$$

where T_{eres} is the input temperature of the storage tank. The tank temperature decreases linearly with T_{res}, Q_u. The mass flow can be increased to greater heat production system.

T_{resmax} is the maximum stored temperature; it is determined by the thermal energy transferred by the PV/T collectors and can be given as follows:

$$T_{resmax} = T_{eres} - \left(\frac{Q_{PVTmax}}{\dot{m}_{max} \cdot C_p} \right), \tag{3}$$

where Q_{PVTmax} is the total thermal energy available at the output of the hybrid PV/T collector per hour and m_{max} is the maximum mass flow rate max.

A heat balance around the thermal storage tank provides

$$Q_u - Q_{es} - Q_p = m_{res} \cdot C_p \cdot \left(\frac{dT_{res}}{dt} \right), \qquad (4)$$

where Q_p is the heat flux which represents the heat loss of tank to the surroundings; Q_{es} is the heat flux transferred from the hot water tank; What is the heat flux transferred by the collectors PV/T; M_{RES} is the mass of water in the tank. C_p is the specific heat of water. The flow of heat transfer from the hot water tank is as follows:

$$\dot{Q}_{ES} = \dot{m}_C C_P (T_C - T_F), \qquad (5)$$

where T_c is the temperature of hot water and T_F is the temperature of cold water.

The temperature at the input of the tank increases during the day as there is sunlight, and because the cell temperature increases. As for the temperature inside the tank T_{res}, it depends on the load, that is to say, the use of hot water. The temperature inside the tank during the day may be less compared to that of the night because there is a prolonged effect of thermal storage during the night.

4.2. Space Heating. The space is heated by the hot air produced by the collectors PV/T air. The fan AC supplied ensures the distribution of air in the house.

5. Electric Power Generated by the PV/T System

The total energy generated by the collectors PV/T to power various loads of the house is as follows:

$$\dot{E}_{PVT} = \dot{E}_E + \dot{E}_V + \dot{E}_{EA} + \dot{E}_{ES}, \qquad (6)$$

where $\dot{E}_E$ is the electrical energy consumed by lighting lamps (DC current). $\dot{E}_V$ is the electrical energy consumed by the fans (for space heating). $\dot{E}_{EA}$ is the electrical energy consumed by the load (A current). $\dot{E}_{ES}$ is the electrical energy consumed for heating domestic water (the electric boost).

6. Generalisation of the Model

The hybrid PV/T collectors are systems cogeneration of electricity and heat. They generate electricity that is used for lighting the home and the fans needed to heat the interior in winter or cooling in summer. For general application, the energy system studied previously is applied. The space heating is provided by hot water produced by the collectors PV/T through radiators combined with low power fans fed from converters continuous alternative.

The system consists of photovoltaic cells to heat water and air collector's hybrid-type single crystal and storage tank water is heated by the collectors and a PV/T extra electric power storage system (batteries) and DC to AC fans and extra strength.

Energy needs of the house are the electrical (DC lamps and fans) and thermal loads (heating water and space).

7. Habitation Energy Needs

To calculate the energy needs of a habitation, we must have two types of information, the first on climate data from the site of habitation's implantation (room temperature, solar irradiance, humidity, etc.) and data of profile of electrical and thermal load, that is to say, the type of load to power and level of comfort you choose. It should be understood that the reference temperatures allow the calculation of heating power for the worst case (winter in general). We will take the average temperatures recorded to ensure that our facility is not oversized and therefore not profitable.

7.1. Dimensioning of the Electrical System. The design follows a phased approach that can be summarized as follows.

  (i) Step 1: determination of user requirements: voltage, power equipment, and service life.

  (ii) Step 2: encryption of solar energy recoverable by location.

  (iii) Step 3: setting the battery capacity.

  (iv) Step 4: design of hybrid solar photovoltaic thermal collectors: operating voltage, total power to be installed, and number of branches.

  (v) Step 5: selecting the controller and the inverter.

7.2. Needs Assessment. Since the system provides energy during the day, it is natural to take the 24-hour period as time unit. The load profile for an habitation in Ghardaïa, witch is composed by: 2 litles rooms, 1 big room, 1 kitchen and 1bathroom (see Table 1). The power requirements are those necessary for the internal lighting and feeding of some appliances (TV, refrigerator). Electrical energy is required to supply the extra electrical hot water tank and the fans that are attached to the wall heat exchangers (radiators) for heat removal in the space of home. Hybrid air collectors for the preheating of the space can be used to reduce dependence on heating water radiators (Figure 6).

The position of the hybrid collectors relative to the sun affects their production aggressively. It is very important to place them for maximum use.

  (i) Orientation to the south in the northern hemisphere.

  (ii) Orientation to the north in the southern hemisphere.

Electrical energy needs of the home are estimated at 5262 Wh/day.

These needs are determined for the winter season for the heating of domestic water and space. For the summer season, the power of the electric boost will be added to meet fans for ventilation that creates the cold housing. So the calculated energy is constant all along the year.

7.3. Battery Capacity. Electrical energy needs of the habitation are estimated at 5262 Wh/day.

In terms of Ah, the consumption becomes

$$C = 5262/48 = 109\,625 \text{ (if it works under 48 V)},$$

TABLE 1: Electrical load profile for a habitation in Ghardaïa.

Device	Number	Tension (V)	Power (W)	Usage time/day (hour)	Daily consumption (Wh/j)
Lamps	05	12	18	05	450
Lamp	02	12	18	02	72
Television	01	12	75	04	300
Refrigerator	01	12	60	24	1440
Extra power (for heating water)	01	220 (AC)	1000	02	2000
Fan (for space heating)	02	220 (AC)	100	05	1000

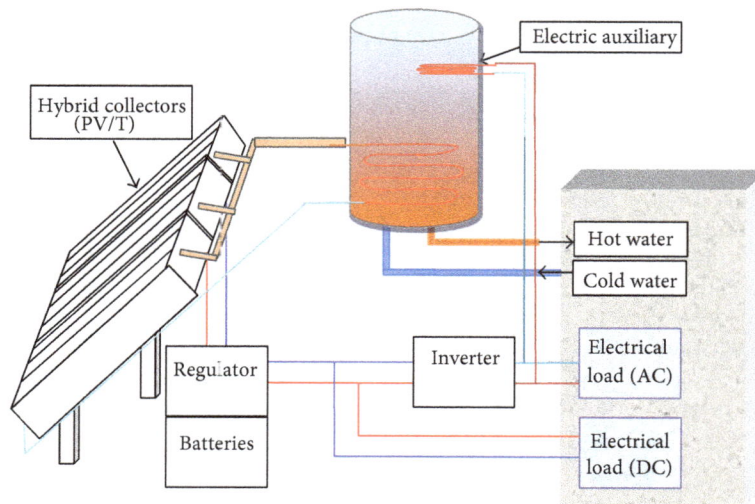

FIGURE 6: Diagram of the energy system based on collectors PV/T applied to habitation.

$$C = 5262/24 = 219.25 \text{ (if it works at 24 V)},$$

$$C = 5262/12 = 438.5 \text{ Ah (if it works at 12 V)}.$$

Knowing that the hybrid collector is used that has been studied previously [12], it delivers 175 Wh/day. In terms of power, it must be less than 30 collectors to meet the needs of the habitation application studied. Depending on the voltage we want to use, one calculates the capacity of the battery and according to the autonomy of the system determines the exact number of batteries to be installed. Once the number of batteries is determined, one determines the number of branches of hybrid PV/T collectors installed and the number of PV/T collectors in series in each branch.

For the purpose of hot water, we found that a family of 4 people needs a thermal storage tank of 250 L.

8. Conclusion

We have proposed a cogeneration system based on hybrid solar photovoltaic thermal collectors for supplying electric power and heat, particularly the water heating and space and the electric charge needed for a comfortable habitation located in Ghardaïa. The results suggest that the cogeneration system based on the hybrid collectors PV/T is a complete energy system to supply electricity and heat a home.

Conflict of Interests

The authors declare that there is no conflict of interests regarding the publication of this paper.

References

[1] M. Wolf, "Performance analyses of combined heating and photovoltaic power systems for residences," *Energy Conversion*, vol. 16, no. 1-2, pp. 79–90, 1976.

[2] S. D. Hendrie, "Photovoltaic/thermal collector development program," Rapport Final, Massachusetts Institute of Technology, Cambridge, Mass, USA, 1982.

[3] P. Raghuraman, "Analytical predictions of liquid and air photovoltaic/ thermal, flat-platz collector performance," *Journal of Solar Energy Engineering*, vol. 103, no. 2, pp. 291–298, 1981.

[4] B. Lalovic, Z. Kiss, and H. Weakliem, "A hybrid amorphous silicon photovoltaic and thermal solar collector," *Solar Cells*, vol. 19, no. 2, pp. 131–138, 1986.

[5] Y. Tripanagnostopoulos, D. Tzavellas et al., "Hybrid PV/T systems with dual heat extraction operation," in *Proceedings of the 17th European PV Solar Energy Conference*, pp. 2515–2518, Munich, Germany, 2001.

[6] H. A. Zondag, "Flat-plate PV-thermal collectors and systems: a review," *Renewable and Sustainable Energy Reviews*, vol. 12, no. 4, pp. 891–959, 2008.

[7] "PV/T roadmap-a European guide for the development and market introduction of PV-thermal technology," Rapport Eu-Project PV-Catapult, 2005.

[8] A. Tiwari and M. S. Sodha, "Parametric study of various configurations of hybrid PV/thermal air collector: experimental validation of theoretical model," *Solar Energy Materials and Solar Cells*, vol. 91, no. 1, pp. 17–28, 2007.

[9] Y. Tripanagnostopoulos, "Aspects and improvements of hybrid photovoltaic/thermal solar energy systems," *Solar Energy*, vol. 81, no. 9, pp. 1117–1131, 2007.

[10] K. Touafek, A. Malek, and M. Haddadi, "Etude expérimentale du capteur hybride photovoltaïque thermique," *Revue des Energies Renouvelables*, vol. 9, no. 3, pp. 143–154, 2006.

[11] K. Touafek, M. Haddadi, and A. Malek, "Modeling and experimental validation of a new hybrid photovoltaic thermal collector," *IEEE Transactions on Energy Conversion*, vol. 26, no. 1, pp. 176–183, 2011.

[12] K. Touafek, M. Haddadi, and A. Malek, "Experimental study on a new hybrid photovoltaic thermal collector," *Applied Solar Energy*, vol. 45, no. 3, pp. 181–186, 2009.

17

Decentralized Autonomous Hybrid Renewable Power Generation

Prakash Kumar and Dheeraj Kumar Palwalia

Department of Electrical Engineering, Rajasthan Technical University Kota, Rajasthan 324010, India

Correspondence should be addressed to Prakash Kumar; prakash.ucertu@gmail.com

Academic Editor: Jing Shi

Power extension of grid to isolated regions is associated with technical and economical issues. It has encouraged exploration and exploitation of decentralized power generation using renewable energy sources (RES). RES based power generation involves uncertain availability of power source round the clock. This problem has been overcome to certain extent by installing appropriate integrated energy storage unit (ESU). This paper presents technical review of hybrid wind and photovoltaic (PV) generation in standalone mode. Associated components like converters, storage unit, controllers, and optimization techniques affect overall generation. Wind and PV energy are readily available, omnipresent, and expected to contribute major future energy market. It can serve to overcome global warming problem arising due to emissions in fossil fuel based thermal generation units. This paper includes the study of progressive development of standalone renewable generation units based on wind and PV microgrids.

1. Introduction

Development of innovative power solution, capable of minimizing environmental concerns, has been key point of interest for power system researchers. These sources have gained attention since the oil crisis faced in early 1970. Depleting fossil fuel reserves need to be replaced by alternate economic and environment friendly power generation sources. Wind energy conversion system (WECS) and PV generation have proven to be potential power generation sources but its nonpromising nature has been of major concern. As these sources are climate and environment dependent, it may not efficiently meet load demand for specified time duration. Overall delivery cost of centralized electricity generation and grid distribution counts up to four times the cost of generation by stand-alone and minigrid options for "minimum threshold" demand scenario.

Most common sources of energy currently utilized worldwide for generating electricity include coal (39.3%), petroleum (0.7%), natural gas (27.6%), nuclear power (19.5%), hydro power (6.7%), wind (4.2%), and other renewable power (2.1%) that covers mainly geothermal, biomass, and PV energy [1]. PV generation has been found more promising than wind generation for small scale generation [2] but it is sure to remain unavailable during night time. So, in wind rich regions, standalone minigrid WECS is preferred more than PV generation. WECS-PV integrated storage unit causes accountable power loss in conversion equipment during conversion process [3–7]. Complementary characteristics of PV and wind reduce overall requirement of storage unit. In specific locations, hybrid wind-PV generation with storage unit can provide highly reliable power [8] to isolated loads like satellite earth station, broadcasting stations, hill top load stations, and so forth. Overall power reliability depends on optimal sizing of conversion equipment, optimization techniques, meteorological data, and load forecast. RES involves efficiency and economical issues. Among reported solutions, ensuring spinning reserve [9–12] and suitable storage unit facilities [13–16] have been considered as the most effective ones.

Due to variable nature of PV and wind round the clock and throughout the year, it becomes a herculean task to obtain regulated power supply [17]. Generation gets affected by weather and climatic condition. Tsoutsos et al. discussed impact of environment on PV generation and proposed necessary measures for proper project designing so as to ensure public acceptance [18]. Haruni et al. proposed a novel operation and control strategy for hybrid wind-fuel cell-electrolyzer-battery and a set of loads as standalone unit. Overall control strategy is based on a two-level structure, namely, energy management and power regulation system, to avoid system blackout. Depending on reference dynamic operating points of individual subsystems, local controllers control wind turbine, fuel cell, electrolyzer, and battery storage units [19]. Zhou et al. proposed an autonomous unified var controller to address system voltage issues and unintentional islanding problems associated with distributed PV generation systems. The controller consisted of features of both voltage regulation (VR) and islanding detection (ID) functions in a PV inverter based on reactive power control to ensure (1) fast VR due to autonomous control, (2) enhanced system reliability because of capability to distinguish between temporary grid disturbances and islanding events, (3) negligible nondetection zone (NDZ) and no adverse impact on system power quality for ID, and (4) no interferences among multiple PV systems during ID [20]. Hong et al. proposed a novel multiobjective nonlinear programming to determine shed loads of UFLS 81L relays in a hybrid wind-PV-gas turbine microgrid. Method incorporates GA to ensure decrease in load shed [21]. Koutroulis and Kolokotsa presented a methodology for optimal sizing of hybrid wind-PV power generation as standalone unit. He suggested a list of commercially available system devices, optimal number, and type of units ensuring that total system cost for about 20 years is minimized subject to constraint that load energy requirements are completely covered, resulting in zero load rejection implementing cost (objective) function minimization using gas [22]. Eftekharnejad et al. investigated impact of increased penetration of PV systems on static performance and transient stability of large power system. Advantages and problems associated with utility scale and residential rooftop PVs have also been identified for steady state stability and transient stability performance [23]. Kadda et al. discussed optimal sizing issue of a hybrid wind-PV-battery as standalone unit along with diesel generator, located at Oujda/Angad, in order to minimize overall generation cost [24]. Lund evaluated problems, associated with hybrid wind-PV-tidal power generation as standalone unit, in terms of excess or scarce electricity production due to fluctuating RES [25]. Kim et al. presented hybrid wind-PV-SMES (superconducting magnetic energy storage) system to operate under abnormal conditions, such as reactive power or current fluctuations. SMES can significantly enhance dynamic security of such distributed power systems due to its high energy density and quick response characteristics during fault or surge conditions [26].

PV power industry has gained attention due to its easy installation at domestic and commercial level. Associated problems have been minimized to great extent in consequent researches, yet a lot more problems need to be shorted out [27–30].

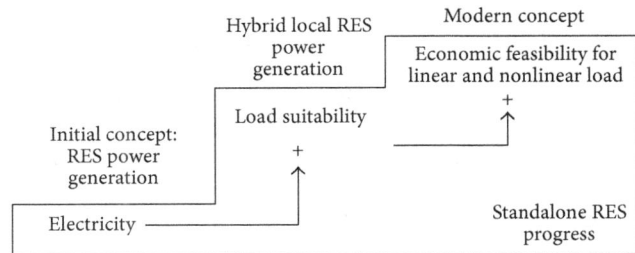

FIGURE 1: Progressive load demand dependent step by step RES standalone system development.

2. Decentralized Power Generation

Decentralized mode of power generation or distributed generation that depends on locally available resources, mostly RES, is either in standalone mode or connected to utility grid. This paper investigates standalone mode of power generation. The concept of standalone or grid isolated system has been revised from time to time, as shown in Figure 1. Initially RES based power generation only aimed to obtain an alternate source of electrical energy; but with increasing load demand, hybrid RES power generation gained attention to satisfy load suitability and ensure regulated electrical power. Presently, researchers have been trying to obtain regulated power from hybrid RES-ESU and meet liner as well as nonlinear load demand economically.

2.1. Decentralized WECS. Wind energy is supposed to be a major contributor in future world energy scenario and continues to be one of the fastest growing energy resources round the globe. Major challenge associated with WECS is uneven distribution of wind energy. Fluctuating wind brings voltage and frequency regulation problem. These problems can be dealt by dividing wind study into three time frames, namely, regulation, load following, and unit commitment [31]. The regulation time frame includes the period during which generation automatically compensates minute-by-minute deviations in load. Load following time frame is generally longer than regulation and refers to time required to obtain different set points of capacity to cope up with the load. Dedicated peak load generating units either have ready to use power or can be started quickly. Load following time frame generally ranges from 10 minutes to a few hours, depending upon time required to move generating unit to different set points of capacity and involved cost constraints. Unit commitment ranges from several hours to several days depending on scheduling dedicated generation to meet required electric demand.

At present, standalone wind system is more economical than standalone PV system for off-grid regions due to continuous ongoing research. Main components required for wind power generation is turbine, gearbox, generator, step-up transformer, nacelle, and tower. Santoso et al. described design and construction of wind power in terms of steady-state and dynamic operation of induction machines (IMs), speed of alternator, and modeling of aerodynamic, mechanical, and electrical components [32].

For WECS commonly employed alternators include self-exited induction generators, doubly fed induction generator (DFIG), permanent magnet (PM) brushless generators [33], PM synchronous generator (PMSG), switched reluctance generators, and doubly salient PM generators [34]. These alternators are not concerned with maximum power generation. Liu et al. proposed doubly excited PM brushless generator to tap maximum wind energy using online flux control [35].

Abu-Elhaija and Muetze discussed effects of fluctuating wind speed on minimum capacitance requirement to self-excite single phase self-excited reluctance generators by analyzing overall system damping and amplifying the components of Eigen values with lower and upper natural frequencies [36]. Singh and Sharma presented design, development, and analysis of voltage and frequency controllers (VFCs) for standalone WECS. An isolated asynchronous generator, a synchronous generator (SG), and a PMSG are used with these WECS [37]. These VFCs are developed with three-phase generators driven through a wind turbine to feed three-phase and single-phase loads. A battery energy storage system is used invariably with each system configuration to facilitate load leveling during change in wind speeds and/or consumer loads [38]. Performance of VFCs has been demonstrated to validate their operation as a load leveler, load balancer, phase balancer, neutral current compensator and an active filter along with a VFC.

2.2. Decentralized PV Generation.

In 1839, French physicist named Edmund Bacquerel discovered PV effect [39]. Modern PV module consists of PV cells, mounting structure, MPPT mechanism, converters, storage unit, and electrical and mechanical connections to regulate and utilize electrical output [40]. Electrical power output of PV module depends on electrical, thermal, PV spectral, and optical property of PV cell array, PV angle, and irradiance [41]. Technical aspects and environmental factors affect optimal PV generation [42]. Zhao et al. discussed PV energy conversion standards and processes involved [4]. Main role player countries like Germany, Italy, Japan, Spain, USA, and South Korea contribute to about 30 to 50 percent of PV annual growth rate [43, 44]. Technical problems such as islanding detection, harmonic distortion, electromagnetic interference, and low efficiency of PV cells are major bottleneck for widespread application of PV systems [45].

Conventionally, PV cell designs are based on band gap energy (eV). Low band gap energy has high current ($I = eNA$) but low voltage ($V = E_g/e$), and vice versa. Here, e is electron charge, N is number of photons, E_g is energy gap, and A is surface area of solar cell. It is preferred to use materials with energy gap between 1 and 1.8 eV like crystal silicon (1.12 eV), amorphous silicon (1.75 eV), copper indium diselenide (1.05 eV), cadmium telluride (1.45 eV), gallium arsenide (1.42 eV), and indium phosphate (1.34 eV) [46–48]. These modules are rated in terms of peak kilowatts (kWp), that is, amount of expected electrical power output when sun is directly overhead on a clear day. Kosten et al. improved efficiency of silicon PV cell by restricting light escape angle. Restricting light escape angle to 2.767° in silicon PV cell of 3 μm thickness improved light trapping and efficiency by 3% [49].

There has been progressive growth in PV cell material [50–54]. Adamian et al. investigated possibility of porous silicon layers application as antireflection coating in common silicon PV cells (ZnS) [50]. Hanoka discussed a silicon ribbon growth method by comparing string Ribbon with two other vertical ribbon technologies and discussed characteristics of this ribbon, specially dislocation distribution, and explained growth progress of 100 m ribbon [51]. Yang et al. explored amorphous-Si PV technology and achieved an AM 1.5, 13% stable cell efficiency for splitting triple-junction spectrum made with roll-to-roll continuous deposition process [52]. Fave et al. compared epitaxial growth of silicon thin film on double porous sacrificial layers obtained by liquid or vapor phase epitaxy and found that mobility and diffusion length are slightly higher with vapor phase epitaxy compared to liquid phase epitaxy. Fabricating PV cells using a detached film obtained with vapor phase epitaxy and without any surface passivation treatment or antireflective coating gives an efficiency of 4.2% with a fill factor of 0.69 [53]. Dobrzański and Drygała explored laser texturization for PV cells made of multicrystalline silicon to improve interaction between laser light and test PV cell [54].

2.3. Decentralized Hybrid WECS-PV Generation.

Wind and PV are complement to a certain extent. Due to individual merits and demerits of PV and WECS, hybrid PV-wind generation system with storage backup unit has proved to be reliable power source [55, 56] to feed electrical loads that need high reliability [57] and uninterrupted power supply [58]. Hybrid generation has been considered preferred choice for remote systems like radio telecommunication, satellite earth stations, or sites isolated from conventional power system [59, 60].

Hybrid system has also been an optimal choice for locations where grid connection has been farfetched idea due to economical and technical reasons [61–63]. Daniel and Gounden presented an isolated hybrid system consisting of a three-phase square wave inverter integrated with PV array and a wind-driven induction generator. They developed mathematical model for hybrid scheme consisting of variables in terms of synchronous reference frame [64]. Chen et al. proposed a multi-input inverter for hybrid PV-wind power system to obtain regulated supply and reduce overall power cost. Multi-input inverter consisted of a buck-boost fused multi-input dc–dc converter and a full bridge dc–ac inverter [65]. Kim et al. discussed power-control strategies for hybrid PV-wind generation with versatile power transfer. Hybrid system consisted of PV array, wind turbine, and battery storage connected to a common dc bus. Versatile power transfer has been defined as multimode operation, including normal operation without use of battery, power dispatching, and power averaging, which enables grid- or user-friendly operation [66]. Further Chiang et al. presented a hybrid regenerative power system consisting of PV-WECS hybrid generation with grid-tie system and uninterruptible power supply (UPS) for critical load applications. System included six-arm converter topology with three arms for rectifier-inverter, one arm each for battery charging/discharging and third arm for power conversion of the PV module and WECS alternator [58]. Liu et al. discussed PV-wind hybrid generation in standalone mode employing doubly excited PM brushless machine used for maximum electrical power extraction by using online flux control [35].

3. Maximum Power Point Tracking (MPPT) Strategies

Maximum power extraction, commonly known as maximum power point tracking (MPPT), includes maximum mechanical and electrical power extraction from wind. Mechanical power extraction is obtained by regulating tip speed ratio of the wind turbine, whereas electrical power extraction is associated with voltage and frequency regulation. In order to track maximum power, suitable control strategy, depending on site, needs to be implanted. MPPT's estimated usable efficiency (EUE) should be as high as possible. Table 1 shows some commonly used control strategies for WECS-MPPT and EUE obtained.

In order to track maximum power, useful MPPT techniques for PV applications have been developed and applied [67]. Commonly used methods for MPPT includes perturb and observe (PO), power matching, incremental conductance, fractional open-circuit voltage/short-circuit current, power differential feedback control, curve fitting, dc-link capacitor droop control, intelligent control, and some other special control methods [68, 69]. Due to simplicity and robust nature, PO and curve fitting techniques are widely used. MPPT techniques and algorithms have gained attention of power system researchers [70–87] due to its dominant advantages over conventional techniques. Table 2 shows some commonly used control strategies for PV-MPPT and EUE obtained.

4. Modeling of Hybrid Generation

Optimal sizing of conversion equipment is necessary to meet load demand. A number of useful simulations and optimization techniques have been used to evaluate performance of hybrid PV-wind systems [88, 89]. Commonly employed software tools include hybrid optimization model for electric renewables (HOMER), HYBRID2, hybrid optimization by GA (HOGA), HYBRIDS, hydrogen energy models (HYDROGEMS), transient systems simulation program (TRNSYS), village power optimization for renewable (ViPOR), Dymola, and matrix laboratory (MATLAB) simulink tool, which are employed for cost and performance analysis [90, 91]. These optimization tools have helped in optimizing simulation configuration in terms of production cost and reliability. Table 3 shows year of development, developer, merits, and demerits of some commonly used optimization tools used to simulate hybrid RES generation.

HOMER is public domain software used for hourly simulations to obtain optimum target. It is a time-step simulator using hourly load and environmental data inputs for RES assessment; it facilitates optimization of RES based on net present cost for a given set of constraints and sensitivity variables. HOMER has been used extensively in previous renewable energy system case studies [92, 93] and in renewable energy system validation tests [94]. Although simulations can take a long time, depending on number of variables used, their operation is simple and straightforward. Program's limitation is that it does not enable user to intuitively select appropriate components for a system, as algorithms and calculations are not visible or accessible. HYBRID2 is hybrid system simulation software with precise simulation, as it can define time intervals from 10 min to 1 h. NREL recommends optimizing system with HOMER and then once optimum system is obtained, designing is improved by using HYBRID2. HOGA is a hybrid system optimization program. Optimization is carried out by means of GA and can be single objective or multiobjective. Simulation is carried out using 1 h intervals, during which all of the parameters remained constant. Control strategies can also be optimized using GA. HYBRIDS assesses technical potential of RES for a given configuration to determine potential renewable fraction and evaluate economic viability based on net present cost. HYBRIDS is a Microsoft Excel spreadsheet based RES assessment application and design tool, requiring daily average load and environmental data estimated for each month of the year. Unlike HOMER, HYBRIDS can only simulate one configuration at a time and is not designed to provide an optimized configuration. HYBRIDS is comprehensive in terms of RES variables, level of detail required, and need of higher level of knowledge of RES configurations as compared to HOMER. It is designed so as to improve renewable energy system design skills through its application.

5. Energy Storage Technology

Renewable hybrid generation is incorporated with ESU to ensure better reliability and meet energy gap between generation and load demand.

Depending on application, classification of ESU technology has been shown in Figure 2. It can be classified as electrochemical [dry batteries: lithium ion (LI), nickel metal hydride (NMH), metal air (MA), nickel cadmium (NiCd), polysulphide bromide (PSB), and electrochemical capacitor (EC); wet batteries: lead acid (LA), valve regulated lead acid (VRLA), sodium sulphur (NaS), all-vanadium redox (AVR), zinc bromine (ZnBr), vanadium bromide redox (VBR), and zero emission battery research activity (ZEBRA); flow batteries (FB)], chemical [fuel cell (FC), electrolyzer (EZ), and synthetic natural gas (SNG)], electromagnetic [capacitors, super capacitor (SC), superconducting magnetic energy storage (SMES), and super conducting coil (SCC)], mechanical [flywheel energy storage (FES), pumped storage arrangement (PSA), and compressed air storage (CAS)], thermal [cryogenic energy storage (CES), electric thermal heaters (ETH), ice based technology (IBT), and pumped heat storage (PHS)] energy storage.

In large-scale nonregulated hybrid renewable electricity networks, energy storage (mostly bulk systems) is required to absorb the shock of energy overproduction and compensate

TABLE 1: Some MPPT technique literatures for standalone WECS.

Author	EUE (%)	Control strategy	Content	References
Pan and Juan	about 90.2	ACC	Presents reduced harmonics, reliable, and cost effective adaptive compensation control (ACC) based MPPT for a microscale WECS. ACC improves dynamic response and more wind energy can be extracted during variable wind speed.	[154]
Nishida et al.	>95	PWM converter	Discusses cost effective, reliable, and wide speed range variable-speed wind-turbine MPPT controller using PWM inverter cascaded in series with a series-type 12-pulse rectifier for interior PMSG based WECS. The system has reduced voltage and current ($V\&I$) harmonics and total losses in WECS is minimized.	[155]
Lo et al.	>90	OTC, PCM, CVM, DCM, and CCM	Investigates buck-type power converter based MPPT controller using pulsating-current battery charger for small-size PM wind turbine in standalone mode to obtain improved charging efficiency. Battery charger operates in discontinuous conduction mode (DCM) with constant on-time control (OTC) to achieve the desired pulsating current mode (PCM) operation. At the end of battery charging state, charger operates in the constant voltage mode (CVM) to prevent battery overcharging. Over speed protection of the wind turbine can be naturally obtained when the charger enters continuous conduction mode (CCM) operation.	[156]
Mendis et al.	>97	Power curve, vector control	Proposes tip speed ratio and pitch angle based MPPT for PMSG and DFIG based hybrid WECS-battery storage as standalone unit. Different control strategies have been developed and proposed for system module to achieve AC voltage and frequency regulation, DC-link voltage stability, and maximum power extraction in proposed standalone unit.	[157]
Zou et al.	>95	Characteristic power curve, DSP kit	Investigates power-curve based MPPT algorithm to obtain robust and cost effective control method for wind turbine systems. Conditions for stable MPPT operation have been determined based on the small-signal model. The transfer function for variation of wind speed to generator speed is determined to be of the first order. The simulation and experimental results confirm validity of proposed transfer function. Dynamic behavior of generator speed is independent of instantaneous wind speed but dependent on dynamics of the wind speed.	[158]
Cirrincione et al.	>90	GNG algorithm, FOC, VOC, DS1103, and DSP TMS320F240	Presents growing natural gas (GNG) based MPPT for variable-pitch WECS with IMs to meet need of maximum power range and constant power range. To cope up with constant power region, the blade pitch angle has been controlled on the basis of closed-loop control of mechanical power absorbed by the IM. MPPT technique included field-oriented control (FOC) for induction generator and voltage oriented control (VOC) for grid-connected inverter.	[159]
Dalala et al.	>90	PO algorithm	Discusses perturb and observe (PO) based MPPT algorithm for small scale WECS, using DC-side current as perturbing variable. Algorithm is best suited for both slow and high wind speed fluctuation, attaining enhanced stability as well as fast tracking capability.	[160]
Urtasun et al.	About 99.7	$V\&I$ control during PCM, CVM, DCM, and CCM modes	Evaluates robustness and power loss in a sensorless MPPT, for PMSG based small WECS incorporating a diode bridge, as compared to conventional curve based MPPT techniques. Due to fluctuating source, it is difficult to obtain optimum power curve and precise relation between dc current and the dc voltage, thus causing power loss.	[161]

energy gap during low generation or blackouts [95]. This is also applicable renewable energy based distributed generation system (either isolated or connected to any distribution network), to deal with power quality issues and efficient power flow management [96, 97]. Optimal sizing of storage unit is important to obtain reliable power generation [98, 99].

In domestic and microgrid renewable generation, energy storage units are employed for satisfaction of electricity needs [100–104]. Maclay et al. presented a PV–hydrogen powered model for both standalone and grid-connected operation employing Matlab/Simulink tool to access computability of a regenerative fuel cell (RFC) as energy storage device with PV electrical generation and discussed issues like battery sizing, charge/discharge rates, and state of charge limitations [101]. Xu et al. proposed an improved optimal sizing method for wind-PV-battery hybrid power system for standalone

TABLE 2: Some MPPT technique literatures for standalone PV.

Author	EUE (%)	Control strategy	Content	References
Rub et al.	94–100	flyback dc-dc converter	Discusses parallel dc-dc converter based current compensated DMPPT for partially shaded series connection of PV modules. Current compensation schemes are either too complex or inaccurate. In this DMPPT scheme, current compensation has been simplified with accurate compensation to assure MPPT by special arrangement of shunt-connected flyback dc–dc converter.	[70]
Gules et al.	>98	neurofuzzy inference	Proposes neurofuzzy inference based artificial intelligent (AI) MPPT for PV generation in standalone operation. It incorporated quasi-Z-source (qZS) inverter to regulate duty ratio and the modulation index to ensure required voltage, current, and frequency.	[71]
Elgendy et al.	>95	PO, voltage perturbation, and direct duty ratio perturbation	Presents incremental conductance MPPT algorithm for standalone PV pumping system using 1080 Wp PV array connected to a 1 kW PM dc motor-centrifugal pump set. System has been investigated for fluctuating weather conditions using comparative study with PO algorithm. Results exhibited better stability for slow transient response and worse performance at rapidly changing irradiance, using direct duty ratio control.	[72]
Elgendy et al.	97–99	Reference voltage and direct duty ratio perturbation based PO MPPT algorithms	Evaluates reference voltage and direct duty ratio perturbation based PO MPPT algorithms. Reference voltage perturbation provides better response to rapidly changing irradiance and temperature transients but exhibits poor stability. Direct duty ratio perturbation provides better stability and energy utilization at a slower transient response but poor performance for rapidly changing irradiance. Algorithms have been justified on the basis of system stability, performance characteristics, and energy utilization for standalone PV pumping systems (1080 Wp PV array connected to a 1 kW PM dc motor-centrifugal pump set) in variable weather conditions.	[73]
Cristaldi et al.	98–100	PO, CVM, and MPPT	Proposes model based (MB) MPPT for single-series-diode model of PV module. This MPPT method has been found suitable and better alternative to traditional module integrated converter (MICs) topologies in terms of cost, robustness, and accuracy. Traditional PO or incremental conductance MPPT algorithms have low efficiency for rapidly changing weather conditions, whereas MB-MPPT offers better dynamic performance. This model can estimate solar radiation with adequate accuracy and does not need radiometer or dedicated cell, as required in conventional MB-MPPT techniques.	[74]
Lian et al.	Approx. 99	PO, PSO	Presents hybrid PO-PSO based MPPT for standalone PV generation. PO is cheap, robust, and good at exploration but not at exploitation; that is, it only tracks first local maximum point and stops progressing to next maximum. PSO works good to obtain global maximum point (GMP) but needs long time for convergence. Thus hybrid PO-PSO works as complement and provides optimized MPPT.	[75]
Konstantopoulos and Koutroulis	Approx. 99	HCPSO	Investigates hybrid chaotic-PSO (HCPSO) algorithm based global MPPT technique for flexible PV modules using effect of geometrical installation parameters like bending angle, tilt angle, orientation, and power-voltage characteristics. Application of proposed HCPSO algorithm minimizes power loss and maximizes energy production of the flexible PV module during global MPPT process.	[76]
Al Nabulsi and Dhaouadi	97–100	Fuzzy logic, PO	Evaluates fuzzy logic and a dual MPPT controller based digital MPPT control scheme for standalone PV system. Duel MPPT controller consisted of an astronomical two-axis sun tracker to track maximum solar radiation, power converter to control power flow between the PV panel and the load. This proposed technique reduces steady state oscillations and enhances the operating point convergence speed.	[77]
Alajmi et al.	98–100	Fuzzified hill climbing algorithm	Investigates fuzzy-logic based MPPT controller for PV systems under fluctuating weather conditions. Hill climbing MPPT has been improved by fuzzifying its rules. This provides less oscillations and fast convergence.	[78]

TABLE 2: Continued.

Author	EUE (%)	Control strategy	Content	References
Sundareswaran et al.	99.2–99.8	ABC	Proposes ABC algorithm based MPPT for partially shaded PV generation. ABC algorithm has been compared with other genetic algorithms (GA) and found ABC as superior solution.	[79]
Zhang et al.	>95	Duty cycle control	Investigates adaptive PO MPPT based on duty cycle modulation, to balance the tracking speed and oscillation requirements of resonant converters. Resonant converters, especially the LLC converter with soft switching, have high gain range and wide load and input voltage range for microinverter applications.	[80]
Badawy et al.	93.6–100	Duty ratio, RBB converter	Presents converter topology based MPPT technique for standalone battery charging PV module. Battery charging system included reversed buck-boost (RBB) converter enabled parallel power processing topology.	[81]
Balasubramanian et al.	>95.8	Boost converter	Addresses boost converter based MPPT for partially shaded PV generation. Boost converter has been designed to operate with high efficiency at MPPT voltage of the array by assuming a single peak power point on the PV characteristics.	[82]
Ghaffari et al.	98–100	Newton-based ES algorithm, dc-dc converter	Evaluates extremum seeking (ES) in dc-dc microconverter based MPPT for partially shaded standalone PV generation, where each PV module is coupled with its own dc/dc converter. PV generation dependents on variable parameter like irradiance and temperature, thus obtaining nonuniform transients in convergence to MPPT. This method uses Newton-based ES algorithm to estimate instantaneous irradiance and temperature variation for MPPT, thereby improving overall performance and reducing cost of power extraction.	[83]
Olalla et al.	90–98	Sub-MIC	Discusses distributed MPPT (DMPPT) architecture for partially shaded PV module in terms of conversion efficiencies and power constraints. DMPPT solutions based on submodule integrated converters (sub-MIC) offer 6.9–11.1% improvement in annual energy yield compared to baseline centralized MPPT scenario. Sub-MIC architecture eliminates insertion loss and provides higher granularity to DMPPT to track more power.	[84]
Singh et al.	>97	ILST, VSC	Investigates voltage source converter (VSC) based MPPT for PV distributed generation. A linear sinusoidal tracer (ILST) based control algorithm has been used for control of VSC and a variable dc link voltage is used for MPPT. This improved overall power quality and VSC utilization.	[85]
Raj and Jeyakumar	99.76–100	Power triangle	Evaluates power triangle based low cost MPPT in standalone operation of PV generation. A background online sweeping technique has been used in power region of I-V characteristic without disturbing actual PV module. This method offers robust control, with almost no divergence, upon change in irradiation and has no oscillations at steady state.	[86]
Boztepe et al.	97–99.33	GVS, POT, and VW	Presents global voltage step (GVS), power operating triangle (POT), and voltage window (VW) based global MPPT (GMPPT) algorithm for string PV system with shaded cells. Such GMPPT algorithms need to scan wide voltage ranges of PV array (nearly zero to open circuit voltage), which needs more scanning time and, in turn, more energy loss. Proposed GMPPT algorithm needs narrow VW search space and thus lower scanning time.	[87]

and grid-connected operation to ensure (a) high power supply reliability; (b) full utilization of complementary characteristics of wind and PV; (c) small fluctuation of power injected into the grid; (d) optimization of battery's charge and discharge state; (e) minimization of total cost of system [102]. Whittingham discussed evaluation of energy storage systems for gigawatt pumped hydro to smallest watt-hour battery and future predictions. Energy storage system can reduce peak power demands and intermittent nature of PV and wind power [103]. Trifkovic et al. presented system integration and controller design for power management of a standalone renewable energy hybrid system consisting of five main

TABLE 3: Optimization tools for hybrid RES study.

Tool	Year	Developer	Merits	Demerits	References
HOMER	2000	National Renewable Energy Laboratory (NREL)	Allows comparison between DC and AC coupled systems	Cannot enable user to intuitively select appropriate system components	[162]
HYBRID2	1996	Renewable Energy Research Laboratory (RERL)	Allows very detailed analysis for energy sources, system architectures, and dispatch strategies	Does not consider short term system fluctuation due to system dynamics or component transients, not suitable for economic and multiobjective optimization	[163]
HOGA	2005	Electrical Engineering Department (University of Zaragoza, Spain)	Carried out by gas; can be single or multi objective; can evaluate all possible combinations for components and control variable strategies	Not suitable for economic optimization	[164]
HYDROGEMS	1995	Institute for Energy Technology, (Norwegian University of Science and Technology, Trondheim)	Used to analyze performance of hydrogen energy systems; simulate hydrogen mass flows; and estimate electrical power flow in standalone and hybrid generation system	Now merged with TRANSYS	[165]
HYBRIDS	—	Solaris Homes	Comprehensive in terms of optimization variables; require higher level of knowledge of system configurations	Only simulates one configuration at a time	[166]
RETScreen	1997	Canadian Government (Ministry of Natural Resources)	Supports basic dimensioning calculations for PV-diesel off-grid systems preliminary feasibility study and general dimensioning	Limits available options for energy sources, system architectures, and dispatch strategies	[167]
PV-SPS	2001	Australian Business Council for Sustainable Energy (BCSE)	Consumption and power generation give a good general impression of system performance over course of the year	Layout of generator input form is not always optimal in terms of clarity, even after feeding PV irradiation, temperature, generator size, values of shiny and cloudy months, annual mean value, and so forth	[167]
PV*SOL	1998	Energy Software, Berlin, Germany	Time step simulation program for off-grid and grid coupled solar generation systems; capable of performing energy calculations, analysis of economic efficiency, and analysis of influences of shadowing	Limits available options for energy sources, system architectures, and dispatch strategies	[167]
TRNSYS	1975	University of Wisconsin-Solar Energy Laboratory	Provides customized performance simulation by splitting entire energy system into individual components	Does not include nuclear, wave, tidal, and hydro power. It includes BES as only electrical energy storage.	[168]
MATLAB/Simulink	1970	MathWorks	Allows much more flexibility in defining energy sources, system architectures, and dispatch strategies	More effort to learn software and develop models	[169]
Dymola	1978	Lund Institute of Technology (Lund University)	Allows much more flexibility in defining energy sources, system architectures, and dispatch strategies	More effort to learn software and develop models	[170]
PVsyst	1994	University of Geneva, Switzerland	Provides dimensioning proposals for standalone installations (PV and battery size) and warns user if chosen component combinations are not technically feasible	No inverter models for off-grid system simulation; only DC modeling is possible	[171]

TABLE 3: Continued.

Tool	Year	Developer	Merits	Demerits	References
ViPOR		NREL	Can decide power supply distribution layout for loads such as houses and minigrid	Able to access local loads and needs skilled labors.	[167]
PV-DesignPro	1998	Maui Solar Energy Software Corporation in Hawaii, USA	Designed to simulate both grid coupled and off-grid systems with PV and wind generators	Additional generator (e.g., diesel generator) is used to match the shortfalls; that is, realistic additional generator cannot be modeled	[167]

Other reported tools are Jpelect, PV-DesignPro, PowerSim, Off Grid Pro, Power Factory, Off Grid Sizer, Sunny Island Design, TALCO, INSEL, ARES, RAPSIM, SOMES, SOLSIM, Simplorer, Solar Pro, and HSWSO (Hybrid Solar-Wind System Optimization Sizing).

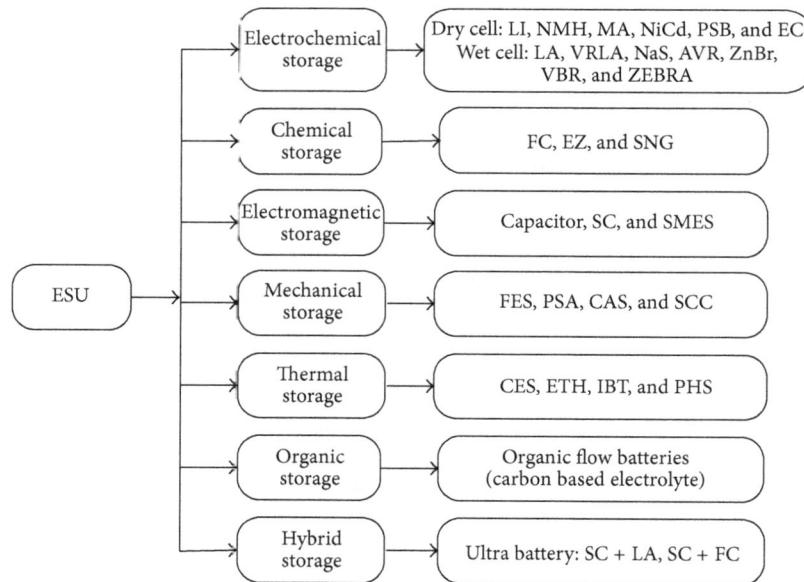

FIGURE 2: Electrical storage unit classifications.

components, namely, PV arrays, wind turbine, electrolyzer, hydrogen storage tanks, and fuel cell. They considered a two-level control system consisting of a supervisory controller to ensure power balance between intermittent renewable generation, energy storage, and dynamic load demand, as well as local controllers for PV, wind, electrolyzer, and fuel cell unit [104].

Therefore depending upon the storage need of respective RES, suitable storage technology is put into application. While considering suitable storage technology, optimal size, economical, and technical specifications are the dominant factors. The specifications and data enlisted in Tables 4 and 5 are among the key specifications which can be considered for different RES integration with suitable storage unit.

6. Barriers and Market Challenges

Due to high investment and low efficiency of renewable energy resources, generation unit should generate sufficiently high enough power to ensure economic power generation. This can be assured by using optimization techniques in

hybrid generation system [105]. Optimization techniques include optimization scenario based on different meteorological data [106], graphic construction method [107], probabilistic approach [108], iterative technique [109], AI methods [110, 111], genetic algorithm [112], system control for energy flow and management [113–115], and multiobjective design [116–118] to obtain optimum size of generation unit to guarantee lowest investment with full use of system component. In order to obtain reliable optimum system configurations quickly and accurately, feasible optimization technique should be incorporated [119].

Nishioka et al. discussed variation in electrical characteristics due to temperature in InGaP/InGaAs/Ge triple junction PV cells under concentration and found that conversion efficiency decreases with increase in temperature and increased with increase in concentration ratio resulting in an increase in open-circuit voltage for these PV cells [106]. Hernández et al. presented a systematic algorithm to determine optimal allocation and sizing of PV grid-connected systems in feeders. It could efficiently compromise technical and economical aspect of multiobjective optimization approach and is robust

TABLE 4: Storage technology capital cost, advantages, disadvantages, and applications [172, 173].

Technology	Capital cost $/kW	$/kWh	kWh/cycle	Advantages	Disadvantages	Application Energy	Application Power
PSA	600–2000	5–200	0.1–2	High capacity, low cost	Special site requirement	Fully capable and reasonable	Not feasible or economical
MA	2800–5000	500–950	90–100	Very high energy density	Electric charging is difficult	Fully capable and reasonable	Not feasible or economical
FB	400–2900	110–2000	6–90	High capacity, independent power and energy ratings	Low energy density	Fully capable and reasonable	Reasonable for this application
NaS	1000–3000	300–950	8–50	High power and energy densities, high efficiency	High production costs and safety concerns	Fully capable and reasonable	Fully capable and reasonable
LA	300–900	200–1500	20–100	Low capital cost	Limited cycle life when deeply discharged	Feasible but not quite practical or economical	Fully capable and reasonable
NiCd	500–1500	800–3000	20–100	High power and energy densities, high efficiency	High production cost requires special charging circuit	Reasonable for this application	Fully capable and reasonable
LI	1200–4000	600–5000	15–100	High power and energy densities, high efficiency	High production cost requires special charging circuit	Feasible but not quite practical or economical	Fully capable and reasonable
FES	250–800	1000–7000	3–40	High power	Low energy density	Feasible but not quite practical or economical	Fully capable and reasonable
FC	10000+		6000–20000	High efficiency, fuel flexibility, and solid electrolyte reduce corrosion and management problems, quick start-up	High temperature enhances corrosion and breakdown of cell components	Fully capable and reasonable	Fully capable and reasonable
ZnBr	700–2500	150–1000	5–80	High capacity, independent power and energy ratings	Low energy density	Fully capable and reasonable	Reasonable for this application
AVR	600–1500	150–1000	5–80	High capacity, independent power and energy ratings	Low energy density	Fully capable and reasonable	Reasonable for this application
SC	100–700	100–2000	2–40	Long cycle life, high efficiency	Low energy density	Reasonable for this application	Fully capable and reasonable
SMES	200–300	1000–850000	350–489	Useful for power regulation on smaller, highly critical equipment such as computer systems	Very short timescales (<10 s), not for bulk power storage	Reasonable for this application	Fully capable and reasonable
CAS	400–1000	2–110	2–6	High capacity, low cost	Special site requirement needs gas fuel	Fully capable and reasonable	Not feasible or economical

TABLE 5: Storage technology ratings.

Technology	Power		Volume energy density (kWh/m³)	Efficiency (%)	Time		Self-discharge/day (%)	Lifetime	
	Rating (MW)	Density (kW/m³)			Discharge	Response		80% DoD (cycles)	Years
PSA	100–1000	0.1–0.2	0.2–2	70–85	1–24 h+	min	Negligible	15000–50000	50+
CAS	5–1000	0.2–0.6	2–6	41–79	1–24 h+	min	Small	9000–30000	25+
MA	0.001–0.01		20–30	40–50				100–300	
FB	0.01–100		20–30	72–85	min–10 h+	<s	0.1–0.8	2000–14000	
NaS	0.05–10	120–160	15–300	70–90	s–h+	<s	0.5–20	2100–4500	10–15
LA	0.001–20	90–700	20–80	72–90	min–h+	<s	0.1–0.5	200–1500	3–15
NiCd	0.001–40	75–700	15–80	60–80	s–h+	<s	0.2–0.6	1000–4000	5–20
LI	0.001–1.1	1300–10000	200–450	65–98	min–h+	<s	0.1–0.3	600–7000	5–100
FES	0.001–0.25	5000	10–80	80–97	s–h+	<s	50–100	1000–60000	15–25
FC	0.001–50	0.2–20	600	30–60	s–24 h+	s–min+	Negligible	250–2000	10–30
ZnBr	0.05–2	1–25	65	65–75	s–10 h+	<s	Small	1000–4000	5–10
AVR	0.001–5	0.5–2	20–70	60–75	s–10 h+	s	Small	>8000	5–20
SC	0.005–0.5	40000–120000	10–30	85–99	ms–h+	<s	20–40	1000–100000	4–12
SMES	0.1–10	2600	6	75–90	ms–s+	<s	10–15	500–10000	

with moderate computer requirements [120]. Shatter et al. investigated a hybrid generation system consisting of PV, wind, and fuel cells incorporating fuzzy based controller to ensure maximum power tracking for both PV and wind energies in order to obtain maximum power at fixed dc voltage bus [121]. Koutroulis and Kolokotsa proposed methodology to obtain optimum number of commercially available system devices and units so that overall system cost is minimized using GA with constraint like load energy requirements causing zero load rejection [122].

7. Policy Development

Consumers prefer low cost and reliable electricity irrespective of environmental concerns. Policy should ensure basic needs of users and reduce burden of fossil fuel based energy sources. Therefore respective policy should foot higher subsidy on renewable energy generating equipment to encourage common users to use green energy. Furthermore, large subsidies need to be offered to rural users to meet their economy. Rural users use very low efficient fuel like kerosene (about 6%), causing loss of a large amount of kerosene. Renewable energy is not only cost saving but also reduces carbon emission. Chaurey and Kandpal estimated in rural household of India that 373 kg carbon dioxide emission per year can be avoided by installing PV panels of 20–53 W [123]. For promoting decentralized rural electrification projects, India has provided capital subsidy up to a 90% for installation of new plants in some regions.

Power system researchers have continuously been working to make electricity market user friendly. Jia et al. evaluated a joint schedule problem for PV power, wind power, combined cooling, heating, power generation, high temperature chiller, liquid desiccant fresh air unit, battery, and power grid in order to satisfy electricity load in buildings with minimal expected cost. They concluded with two important results; that is, (1) simulation-based policy improvement (SBPI) methods are developed to improve from given base policies and (2) performance of these methods is systematically analyzed through numerical experiments. For sufficient computing budget, SBPI methods improve given base policies [124]. Shaahid and Elhadidy assessed technoeconomic feasibility of hybrid PV-diesel-battery power systems for a typical residential building at Dhahran (East-Coast, KSA) and evaluated that hybrid PV-diesel model configuration with battery storage decreases overall cost of diesel with increase in PV capacity [125]. Sarkar and Ajjarapu studied a stochastic planning approach for assessing MW resource of three wind and PV hybrid models by fixing varying penetration ratio level at 10%, 20%, and 30%. Method found applicability for different parameters such as cut-in speed, rated speed, furling speed, power rating of wind turbines, efficiencies of heat exchanger, steam turbine and electric generator, and maximum load [126]. Sun et al. studied a joint schedule problem to schedule PV power, wind power, combined cooling, heating, power generation, high temperature chiller, liquid desiccant fresh air unit, battery, and power grid in order to satisfy electricity load, sensible heat load, and latent heat load in buildings with minimal expected cost. Two major contributions have been presented; that is, three simulation-based policy improvement methods are developed to improve from given base policies and performance of these methods is systematically analyzed through numerical experiments [127].

Algorithm based policy approach has helped in maximizing efficiency and reducing cost function [128–131]. Arabali et al. proposed strategy to meet controllable heating, ventilation, and air conditioning (HVAC) load with a hybrid-renewable generation and energy storage system. GA based optimization approach is incorporated with a two-point estimate method to minimize cost and increase efficiency. Minimized cost function ensured minimum PV and wind generation installation as well as storage capacity selection to supply HVAC load [128]. Lannoye et al. proposed insufficient ramping resource expectation (IRRE) metrics to measure power system flexibility for long-term planning and derive adequacy metrics from traditional generation. A flexibility metrics can identify time intervals over which a system is most likely to face a shortage of flexible resources and can measure relative impact of changing operational policies and addition of flexible resources [129]. Mei et al. proposed game approaches for hybrid power system planning to model planning of a grid-connected hybrid power system consisting of wind turbines, PV panels, and storage batteries [130]. El-Tamaly and Mohammed investigated a fuzzy logic technique to calculate and study reliability index of PV-wind hybrid power system to determine impact of interconnecting system with utility grid [131].

8. Financial Approaches

Several economic criteria exist in providing useful power to utility grid. The costing structure should be simple so as to make user understand it when levied upon them. Until overall cost is user friendly, users may not prefer to adopt the system. Marí and Nabona splitted wind-PV hybrid generation cost into five parts, namely, initial investment cost, operating and maintenance cost, replacement cost of equipment, cost of power exchange between hybrid power generation unit and grid, and regulation cost of utility grid [107]. Different approaches have been made to encourage dependence on renewable energy and reduce fossil fuel based energy dependence [132–134]. Tezuka et al. suggested method to reduce amount of CO_2 emission by imposing carbon-tax revenue and give subsidy on PV-system installations and concluded that amount of CO_2 emission reduces by advertising PV system with subsidy policy even under the same tax-rate and CO_2 payback time [135]. Nelson et al. discussed unit sizing and made an economical evaluation of hybrid wind-PV-fuel cell generation system. They obtained a clear economic advantage of hybrid wind-PV-fuel cell-electrolyzer system over traditional hybrid wind-PV-battery system for a typical home in US Pacific Northwest [136]. Bilal et al. proposed a methodology of optimal sizing of hybrid systems PV/wind/battery in order to minimize annual cost system (ACS) and loss of power supply probability (LPSP) using multiobjective GA. The obtained results show that cost of optimal configuration strongly depends on LPSP. For example, cost of optimal configuration decreases by 25% when LPSP grows to 1% from 0% [116].

9. Microgrid and Equipment

9.1. Converters. In order to obtain regulated power supply from fluctuating power supply, converters need to be designed to meet frequency and voltage standards. Commonly employed converters include AC/DC (rectifiers), DC/DC (choppers), and DC/AC (inverters). Inverters are commonly employed at point of common coupling and need to be designed optimally. Depending upon output waveform, inverters are classified as square wave, modified square wave (quasi square wave or modified sin wave), and multilevel

(multistep) and sin wave (high frequency PWM) inverters. Daniel and Gounden proposed three-phase square wave inverter for an isolated wind-PV hybrid scheme for the first time. They presented a dynamic mathematical model of the hybrid scheme in terms of synchronous reference frame and verified it for transient load conditions [64]. But this converter had high harmonic components and was not fully controlled. Park et al. presented five-level PWM inverter employing dead beat control for minimizing harmonic components of output voltage and load current [137]. Park et al. described an assembly of multilevel PWM inverter and cascaded transformer scheme for standalone generation to obtain high quality output voltage waveform. They validated the proposed system for 11-level and 29-level PWM output [137]. Nasrudin A. Rahim et al. presented five-level and seven-level single-phase multilevel PWM scheme for fluctuating reference input. They employed fluctuating output voltage as reference and compared it with triangular carrier signal to generate desired PWM signals for the switches of converter [138, 139]. Kumaravel and Ashok studied a diode-clamped multilevel inverter using bidirectional buck-boost choppers using single-pulse, multipulse, and hysteresis band current control schemes. Single-pulse scheme involves slow switching actions but needs high current rated chopper devices whereas multipulse scheme involves faster switching actions and low current rated chopper devices but has slower response. The hysteresis band current control scheme has faster switching action and lower current rating of the chopper devices and can nullify the initial voltage imbalance as well [16]. Further Gautam and Gupta discussed cascaded H-bridge multilevel PWM Inverter using multiband hysteresis modulation employing current control scheme [140]. Liu et al. proposed control scheme for ZS and qZS cascaded multilevel inverter (CMI). A multilevel space vector modulation integrated with shoot through states for single-phase qZS-CMI synthesizes staircase type multilevel voltage waveforms having low harmonic contents [141]. Fatu et al. discussed a variable-speed motion sensorless duel converter PI current controlled control scheme PM synchronous generator for WECS. They presented a voltage control scheme with selective harmonic compensation for standalone mode operation [142].

9.2. Power Flow Controllers (PFCs). PFCs are essential for promising reliable and economical power supply to connected load in microgrid for standalone operation [32]. Fluctuating variation in source causes stability and power quality problems in terms of voltage and frequency regulation. Situation becomes worse for reactive power demand due to limitation of reactive capability of wind generating system. Mendis et al. proposed a standalone hybrid system consisting of a PMSG, hybrid energy storage (battery storage and a supercapacitor), a dump load and a mains load for obtaining voltage and frequency regulation. Energy management algorithm has been used to improve performance of battery storage and active-reactive power flows [143].

Therefore, an efficient and intelligent PFC is necessary to ensure balance between load and source of generation. This can be assured by forecasting load demand and scheduling regulated power [144]. Depending on power flow, Chauhan and Saini divided PFCs into three categories: centralized control arrangement, distributed control arrangement, and hybrid control arrangement [145]. In centralized control arrangement, system consists of one master controller (centralized controller) and several slave controllers for various individual power sources and energy storage unit. Master controller operates in close coordination with all sources and slave controllers. In distributed control arrangement, each power source sends measurement signals to its local controller. Local controllers communicate with one another to take appropriate decision for global optimal solution [146, 147]. Hybrid control arrangement is combination centralized and distributed control schemes. In hybrid control scheme, RES are grouped within integrated system [148, 149]. Centralized control scheme is applied within each group and distributed control scheme is used to coordinate each group. In such hybrid control scheme, local optimization is achieved through centralized control within group and global optimization among different groups of energy sources is achieved by distributed control [145]. In this quest, Santoso et al. proposed a hierarchical control including a master and slave controllers for hybrid generation. Master controller selects power generation source and slave controller maintained constant DC bus voltage regulating duty cycle of DC/DC converters [32]. Zeng et al. presented a reduced switch count multiport dc–dc converter for standalone PV-wind hybrid generation. Converter has been applied for simultaneous MPPT control of a wind/PV hybrid generation system consisting of one wind turbine generator (WTG) and two different PV panels [150]. Botterón et al. proposed a high reliability DC/DC converter and a single-phase PWM inverter for standalone microgrids. It aims at designing two different controllers for two converters for reliable operation of microgrid and results have been validated for a 5 kVA PWM modulated single-phase inverter, fully controlled by DSP TMS320F2407 [151].

10. Conclusion

In order to meet pollution-free future power demands, dependence on renewable energy should be encouraged by providing subsidy on installation products. Energy policy-makers need to encourage RES based usage and research to facilitate latest technology for power extraction. Narula et al. investigated probability of achieving 100% electricity demand in South Asia region by year 2030 by increasing dependence on RES based distributed generation [152]. Timilsina et al. estimated to achieve 1845 GW, 1330 GW, and 2033 GW power from PV by year 2030, 2040, and 2050, respectively [153]. More research activity needs to be carried out to improve reduced generation cost, improve storage unit facility, and load forecasting and efficiency of conversion equipment. By increasing dependence on DC microgrids, conversion losses can be minimized and exploitation of power could be optimized.

Hybrid PV-wind integrated with storage unit has been a feasible solution to meet off-grid load demand. A comprehensive review of PV-WECS standalone generation, converter topologies, storage facility options, hybrid simulation tools,

and challenges associated with RES generation has been investigated in this paper. It has been found that hybrid generation is supposed to be major contributor in electrification of isolated regions to feed loads which need reliable power source. Suitable optimization techniques using GA and AI can optimize global optimal generation.

High cost of installation has been major issue in widespread RES based power generation. Hence power policy needs to be made liberal to encourage power dependence on RES. More research work needs to be carried out to improve overall durability and performance of storage facility and power conversion equipment.

Conflict of Interests

The authors declare that there is no conflict of interests regarding the publication of this paper.

References

[1] "'Total energy', U.S. Energy Informatio Administration," http://www.eia.gov/totalenergy/data/monthly/#electricity.

[2] Z. Jiang and H. Rahimi-Eichi, "Design, modeling and simulation of a green building energy system," in *Proceedings of the IEEE Power & Energy Society General Meeting (PES '09)*, pp. 1–7, IEEE, Calgary, Canada, July 2009.

[3] R. Sternberger and D. Jovcic, "Theoretical framework for minimizing converter losses and harmonics in a multilevel STATCOM," *IEEE Transactions on Power Delivery*, vol. 23, no. 4, pp. 2376–2384, 2008.

[4] G. G. Oggier, G. O. García, and A. R. Oliva, "Switching control strategy to minimize dual active bridge converter losses," *IEEE Transactions on Power Electronics*, vol. 24, no. 7, pp. 1826–1838, 2009.

[5] S. Ben-Yaakov, "On the influence of switch resistances on switched-capacitor converter losses," *IEEE Transactions on Industrial Electronics*, vol. 59, no. 1, pp. 638–640, 2012.

[6] L. Schwager, A. Tuysuz, C. Zwyssig, and J. W. Kolar, "Modeling and comparison of machine and converter losses for PWM and PAM in high-speed drives," *IEEE Transactions on Industry Applications*, vol. 50, no. 2, pp. 995–1006, 2014.

[7] A. Vidal, A. G. Yepes, F. D. Freijedo, J. Malvar, Ó. López, and J. Doval-Gandoy, "A technique to estimate the equivalent loss resistance of grid-tied converters for current control analysis and design," *IEEE Transactions on Power Electronics*, vol. 30, no. 3, pp. 1747–1761, 2014.

[8] H. X. Yang, W. Zhou, and C. Z. Lou, "Optimal design and techno-economic analysis of a hybrid solar-wind power generation system," *Applied Energy*, vol. 86, no. 2, pp. 163–169, 2009.

[9] S. S. Reddy, B. K. Panigrahi, R. Kundu, R. Mukherjee, and S. Debchoudhury, "Energy and spinning reserve scheduling for a wind-thermal power system using CMA-ES with mean learning technique," *International Journal of Electrical Power and Energy Systems*, vol. 53, no. 1, pp. 113–122, 2013.

[10] R. Ghaffari and B. Venkatesh, "Options based reserve procurement strategy for wind generators—using binomial trees," *IEEE Transactions on Power Systems*, vol. 28, no. 2, pp. 1063–1072, 2013.

[11] M. A. Delucchi and M. Z. Jacobson, "Providing all global energy with wind, water, and solar power. Part II. Reliability, system

and transmission costs, and policies," *Energy Policy*, vol. 39, no. 3, pp. 1170–1190, 2011.

[12] A. A. Khatir and R. Cherkaoui, "A probabilistic spinning reserve market model considering DisCo' different value of lost loads," *Electric Power Systems Research*, vol. 81, no. 4, pp. 862–872, 2011.

[13] B. W. Jones and R. Powell, "Evaluation of distributed building thermal energy storage in conjunction with wind and solar electric power generation," *Renewable Energy*, vol. 74, pp. 699–707, 2015.

[14] M. Parastegari, R.-A. Hooshmand, A. Khodabakhshian, and A.-H. Zare, "Joint operation of wind farm, photovoltaic, pump-storage and energy storage devices in energy and reserve markets," *International Journal of Electrical Power and Energy Systems*, vol. 64, pp. 275–284, 2015.

[15] K. Wu, H. Zhou, S. An, and T. Huang, "Optimal coordinate operation control for wind–photovoltaic–battery storage power-generation units," *Energy Conversion and Management*, vol. 90, pp. 466–475, 2015.

[16] S. Kumaravel and S. Ashok, "Optimal power management controller for a stand-alone solar PV/wind/battery hybrid energy system," *Energy Sources*, vol. 37, no. 4, pp. 407–415, 2015.

[17] G. Delille, B. François, and G. Malarange, "Dynamic frequency control support by energy storage to reduce the impact of wind and solar generation on isolated power system's inertia," *IEEE Transactions on Sustainable Energy*, vol. 3, no. 4, pp. 931–939, 2012.

[18] T. Tsoutsos, N. Frantzeskaki, and V. Gekas, "Environmental impacts from the solar energy technologies," *Energy Policy*, vol. 33, no. 3, pp. 289–296, 2005.

[19] A. M. O. Haruni, M. Negnevitsky, M. E. Haque, and A. Gargoom, "A novel operation and control strategy for a standalone hybrid renewable power system," *IEEE Transactions on Sustainable Energy*, vol. 4, no. 2, pp. 402–413, 2013.

[20] Y. Zhou, H. Li, and L. Liu, "Integrated autonomous voltage regulation and islanding detection for high penetration PV applications," *IEEE Transactions on Power Electronics*, vol. 28, no. 6, pp. 2826–2841, 2013.

[21] Y.-Y. Hong, M.-C. Hsiao, Y.-R. Chang, Y.-D. Lee, and H.-C. Huang, "Multiscenario underfrequency load shedding in m Microgrid consisting of intermittent renewables," *IEEE Transactions on Power Delivery*, vol. 28, no. 3, pp. 1610–1617, 2013.

[22] E. Koutroulis and D. Kolokotsa, "Design optimization of desalination systems power-supplied by PV and W/G energy sources," *Desalination*, vol. 258, no. 1, pp. 171–181, 2010.

[23] S. Eftekharnejad, V. Vittal, G. T. Heydt, B. Keel, and J. Loehr, "Impact of increased penetration of photovoltaic generation on power systems," *IEEE Transactions on Power Systems*, vol. 28, no. 2, pp. 893–901, 2013.

[24] F. Z. Kadda, S. Zouggar, and M. L. Elhafyani, "Optimal sizing of an autonomous hybrid system," in *Proceedings of the International Renewable and Sustainable Energy Conference (IRSEC '13)*, pp. 269–274, IEEE, Ouarzazate, Morocco, March 2013.

[25] H. Lund, "Large-scale integration of optimal combinations of PV, wind and wave power into the electricity supply," *Renewable Energy*, vol. 31, no. 4, pp. 503–515, 2006.

[26] S.-T. Kim, B.-K. Kang, S.-H. Bae, and J.-W. Park, "Application of SMES and grid code compliance to wind/photovoltaic generation system," *IEEE Transactions on Applied Superconductivity*, vol. 23, no. 3, Article ID 5000804, 2013.

[27] S. Pacca, D. Sivaraman, and G. A. Keoleian, "Parameters affecting the life cycle performance of PV technologies and systems," *Energy Policy*, vol. 35, no. 6, pp. 3316–3326, 2007.

[28] V. H. M. Quezada, J. R. Abbad, and T. G. S. Román, "Assessment of energy distribution losses for increasing penetration of distributed generation," *IEEE Transactions on Power Systems*, vol. 21, no. 2, pp. 533–540, 2006.

[29] L. H. Stember, "Reliability considerations in the design of solar photovoltaic power systems," *Solar Cells*, vol. 3, no. 3, pp. 269–285, 1981.

[30] J. Schmidt, B. Lim, D. Walter et al., "Impurity-related limitations of next-generation industrial silicon solar cells," *IEEE Journal of Photovoltaics*, vol. 3, no. 1, pp. 114–118, 2013.

[31] J. DeCesaro and K. Porter, "Wind energy and power system operations: a review of wind integration studies to date," Subcontract Report NREL/SR-550-47256, National Renewable Energy Laboratory, 2009.

[32] S. Santoso, M. Lwin, J. Ramos, M. Singh, E. Muljadi, and J. Jonkman, "Designing and integrating wind power laboratory experiments in power and energy systems courses," *IEEE Transactions on Power Systems*, vol. 29, no. 4, pp. 1944–1951, 2014.

[33] S. Niu, K. T. Chau, J. Z. Jiang, and C. Liu, "Design and control of a new double-stator cup-rotor permanent-magnet machine for wind power generation," *IEEE Transactions on Magnetics*, vol. 43, no. 6, pp. 2501–2503, 2007.

[34] Y. Fan, K. T. Chau, and M. Cheng, "A new three-phase doubly salient permanent magnet machine for wind power generation," *IEEE Transactions on Industry Applications*, vol. 42, no. 1, pp. 53–60, 2006.

[35] C. Liu, K. T. Chau, and X. Zhang, "An efficient wind-photovoltaic hybrid generation system using doubly excited permanent-magnet brushless machine," *IEEE Transactions on Industrial Electronics*, vol. 57, no. 3, pp. 831–839, 2010.

[36] W. S. Abu-Elhaija and A. Muetze, "Self-excitation and stability at speed transients of self-excited single-phase reluctance generators," *IEEE Transactions on Sustainable Energy*, vol. 4, no. 1, pp. 136–144, 2013.

[37] B. Singh and S. Sharma, "Voltage and frequency controllers for standalone wind energy conversion systems," *IET Renewable Power Generation*, vol. 8, no. 6, pp. 707–721, 2014.

[38] S. Sharma and B. Singh, "Asynchronous generator with battery storage for standalone wind energy conversion system," *IEEE Transactions on Industry Applications*, vol. 50, no. 4, pp. 2760–2767, 2014.

[39] A. F. Zobaa and R. C. Bansal, *Handbook of Renewable Energy Technology*, National Renewable Energy Laboratory, Golden, Colo, USA, 1996.

[40] M. M. El-Wakil, *Power Plant Technology*, McGraw-Hill, New York, NY, USA, 1984.

[41] D. L. King, W. E. Boyson, and J. A. Kratochavil, "Photovoltaic array performance model," Tech. Rep., Sandia National Laboratory, 2004.

[42] M. Hosenuzzaman, N. A. Rahim, J. Selvaraj, M. Hasanuzzaman, A. B. M. A. Malek, and A. Nahar, "Global prospects, progress, policies, and environmental impact of solar photovoltaic power generation," *Renewable and Sustainable Energy Reviews*, vol. 41, pp. 284–297, 2014.

[43] G. Zhao, H. Kozuka, H. Lin, and T. Yoko, "Solar energy conversion," in *Willy Encyclopedia of Electrical and Electronics Engineering*, pp. 638–649, John Willy & Sons, New York, NY, USA, 1999.

[44] REN21, *Renewable Global Status Report: 2009 Update*, REN21 Secretariat, Paris, France, 2009.

[45] T. Ishikawa, "Grid-connected photovoltaic power systems: survey of inverter and related protection equipments," Report of International Energy Agency, International Energy Agency, Paris, France, 2002, http://www.iea-pvps.org.

[46] Solarbuzz, "Solar cell technologies," 2010, http://www.solarbuzz.com/technologies.htm.

[47] R. Menzies, "Designing a solar power systems," in *Proceedings of the Solar Photovoltaic Energy Workshop*, Monash University, Caulfield, Australia, July 1998.

[48] Bright Green Energy, "What type of solar PV panels are available?" 2009, http://www.wirefreedirect.com/Types_Solar_PV_Panels.asp.

[49] E. D. Kosten, B. K. Newman, J. V. Lloyd, A. Polman, and H. A. Atwater, "Limiting light escape angle in silicon photovoltaics: ideal and realistic cells," *IEEE Journal of Photovoltaics*, vol. 5, no. 1, pp. 61–69, 2015.

[50] Z. N. Adamian, A. P. Hakhoyan, V. M. Aroutiounian, R. S. Barseghian, and K. Touryan, "Investigations of solar cells with porous silicon as antireflection layer," *Solar Energy Materials & Solar Cells*, vol. 64, no. 4, pp. 347–351, 2000.

[51] J. I. Hanoka, "Overview of silicon ribbon growth technology," *Solar Energy Materials and Solar Cells*, vol. 65, no. 1, pp. 231–237, 2001.

[52] J. Yang, A. Banerjee, and S. Guha, "Amorphous silicon based photovoltaics—from earth to the 'final frontier'," *Solar Energy Materials & Solar Cells*, vol. 78, no. 1–4, pp. 597–612, 2003.

[53] A. Fave, S. Quoizola, J. Kraiem, A. Kaminski, M. Lemiti, and A. Laugier, "Comparative study of LPE and VPE silicon thin film on porous sacrificial layer," *Thin Solid Films*, vol. 451-452, pp. 308–311, 2004.

[54] L. A. Dobrzański and A. Drygała, "Laser processing of multicrystalline silicon for texturization of solar cells," *Journal of Materials Processing Technology*, vol. 191, no. 1–3, pp. 228–231, 2007.

[55] P. Jiang, H. Zhang, L. Xu, X. Li, P. Zhao, and S. Zhang, "Research on a novel hybrid power system," in *Proceedings of the 9th IEEE International Conference on Mechatronics and Automation (ICMA '12)*, pp. 2494–2498, IEEE, Chengdu, China, August 2012.

[56] A. R. Prasad and E. Natarajan, "Optimization of integrated photovoltaic–wind power generation systems with battery storage," *Energy*, vol. 31, no. 12, pp. 1943–1954, 2006.

[57] A. Safdarian, M. Fotuhi-Firuzabad, and F. Aminifar, "Compromising wind and solar energies from the power system adequacy viewpoint," *IEEE Transactions on Power Systems*, vol. 27, no. 4, pp. 2368–2376, 2012.

[58] H. C. Chiang, T. T. Ma, Y. H. Cheng, J. M. Chang, and W. N. Chang, "Design and implementation of a hybrid regenerative power system combining grid-tie and uninterruptible power supply functions," *IET Renewable Power Generation*, vol. 4, no. 1, pp. 85–99, 2010.

[59] Z. Zheng, L. X. Cai, R. Zhang, and X. S. Shen, "RNP-SA: joint relay placement and sub-carrier allocation in wireless communication networks with sustainable energy," *IEEE Transactions on Wireless Communications*, vol. 11, no. 10, pp. 3818–3828, 2012.

[60] F. F. Nerini, R. Dargaville, M. Howells, and M. Bazilian, "Estimating the cost of energy access: the case of the village of Suro Craic in Timor Leste," *Energy*, vol. 79, pp. 385–397, 2015.

[61] S. Diaf, G. Notton, M. Belhamel, M. Haddadi, and A. Louche, "Design and techno-economical optimization for hybrid PV/wind system under various meteorological conditions," *Applied Energy*, vol. 85, no. 10, pp. 968–987, 2008.

[62] G. C. Bakos and N. F. Tsagas, "Technoeconomic assessment of a hybrid solar/wind installation for electrical energy saving," *Energy and Buildings*, vol. 35, no. 2, pp. 139–145, 2003.

[63] S. Diaf, M. Belhamel, M. Haddadi, and A. Louche, "Technical and economic assessment of hybrid photovoltaic/wind system with battery storage in Corsica island," *Energy Policy*, vol. 36, no. 2, pp. 743–754, 2008.

[64] S. A. Daniel and N. A. Gounden, "A novel hybrid isolated generating system based on PV fed inverter-assisted wind-driven induction generators," *IEEE Transactions on Energy Conversion*, vol. 19, no. 2, pp. 416–422, 2004.

[65] Y.-M. Chen, Y.-C. Liu, S.-C. Hung, and C.-S. Cheng, "Multiinput inverter for grid-connected hybrid PV/wind power system," *IEEE Transactions on Power Electronics*, vol. 22, no. 3, pp. 1070–1077, 2007.

[66] S.-K. Kim, J.-H. Jeon, C.-H. Cho, J.-B. Ahn, and S.-H. Kwon, "Dynamic modeling and control of a grid-connected hybrid generation system with versatile power transfer," *IEEE Transactions on Industrial Electronics*, vol. 55, no. 4, pp. 1677–1688, 2008.

[67] A. R. Reisi, M. H. Moradi, and S. Jamasb, "Classification and comparison of maximum power point tracking techniques for photovoltaic system: a review," *Renewable and Sustainable Energy Reviews*, vol. 19, pp. 433–443, 2013.

[68] K. Ishaque and Z. Salam, "A deterministic particle swarm optimization maximum power point tracker for photovoltaic system under partial shading condition," *IEEE Transactions on Industrial Electronics*, vol. 60, no. 8, pp. 3195–3206, 2013.

[69] P. Sharma and V. Agarwal, "Exact maximum power point tracking of grid-connected partially shaded PV source using current compensation concept," *IEEE Transactions on Power Electronics*, vol. 29, no. 9, pp. 4684–4692, 2014.

[70] H. A. Rub, A. Iqbal, S. M. Ahmed, F. Z. Peng, Y. Li, and G. Baoming, "Quasi-Z-source inverter-based photovoltaic generation system with maximum power tracking control using ANFIS," *IEEE Transactions on Sustainable Energy*, vol. 4, no. 1, pp. 11–20, 2013.

[71] R. Gules, J. D. P. Pacheco, H. L. Hey, and J. Imhoff, "A maximum power point tracking system with parallel connection for PV stand-alone applications," *IEEE Transactions on Industrial Electronics*, vol. 55, no. 7, pp. 2674–2683, 2008.

[72] M. A. Elgendy, B. Zahawi, and D. J. Atkinson, "Assessment of the incremental conductance maximum power point tracking algorithm," *IEEE Transactions on Sustainable Energy*, vol. 4, no. 1, pp. 108–117, 2013.

[73] M. A. Elgendy, B. Zahawi, and D. J. Atkinson, "Assessment of perturb and observe MPPT algorithm implementation techniques for PV pumping applications," *IEEE Transactions on Sustainable Energy*, vol. 3, no. 1, pp. 21–33, 2012.

[74] L. Cristaldi, M. Faifer, M. Rossi, and S. Toscani, "An improved model-based maximum power point tracker for photovoltaic panels," *IEEE Transactions on Instrumentation and Measurement*, vol. 63, no. 1, pp. 63–71, 2014.

[75] K. L. Lian, J. H. Jhang, and I. S. Tian, "A maximum power point tracking method based on perturb-and-observe combined with particle swarm optimization," *IEEE Journal of Photovoltaics*, vol. 4, no. 2, pp. 626–633, 2014.

[76] C. Konstantopoulos and E. Koutroulis, "Global maximum power point tracking of flexible photovoltaic modules," *IEEE Transactions on Power Electronics*, vol. 29, no. 6, pp. 2817–2828, 2014.

[77] A. Al Nabulsi and R. Dhaouadi, "Efficiency optimization of a dsp-based standalone PV system using fuzzy logic and dual-MPPT control," *IEEE Transactions on Industrial Informatics*, vol. 8, no. 3, pp. 573–584, 2012.

[78] B. N. Alajmi, K. H. Ahmed, S. J. Finney, and B. W. Williams, "Fuzzy-logic-control approach of a modified hill-climbing method for maximum power point in microgrid standalone photovoltaic system," *IEEE Transactions on Power Electronics*, vol. 26, no. 4, pp. 1022–1030, 2011.

[79] K. Sundareswaran, P. Sankar, P. S. R. Nayak, S. P. Simon, and S. Palani, "Enhanced energy output from a PV system under partial shaded conditions through artificial bee colony," *IEEE Transactions on Sustainable Energy*, vol. 6, no. 1, pp. 198–209, 2015.

[80] Q. Zhang, C. Hu, L. Chen et al., "A center point iteration MPPT Method with application on the frequency-modulated LLC microinverter," *IEEE Transactions on Power Electronics*, vol. 29, no. 3, pp. 1262–1274, 2014.

[81] M. O. Badawy, A. S. Yilmaz, Y. Sozer, and I. Husain, "Parallel power processing topology for solar PV applications," *IEEE Transactions on Industry Applications*, vol. 50, no. 2, pp. 1245–1255, 2014.

[82] I. R. Balasubramanian, S. I. Ganesan, and N. Chilakapati, "Impact of partial shading on the output power of PV systems under partial shading conditions," *IET Power Electronics*, vol. 7, no. 3, pp. 657–666, 2014.

[83] A. Ghaffari, M. Krstić, and S. Seshagiri, "Power optimization for photovoltaic microconverters using multivariable newton-based extremum seeking," *IEEE Transactions on Control Systems Technology*, vol. 22, no. 6, pp. 2141–2149, 2014.

[84] C. Olalla, C. Deline, and D. Maksimovic, "Performance of mismatched PV systems with submodule integrated converters," *IEEE Journal of Photovoltaics*, vol. 4, no. 1, pp. 396–404, 2014.

[85] B. Singh, C. Jain, and S. Goel, "ILST control algorithm of single-stage dual purpose grid connected solar PV system," *IEEE Transactions on Power Electronics*, vol. 29, no. 10, pp. 5347–5357, 2014.

[86] J. S. C. M. Raj and A. E. Jeyakumar, "A novel maximum power point tracking technique for photovoltaic module based on power plane analysis of $I - V$ characteristics," *IEEE Transactions on Industrial Electronics*, vol. 61, no. 9, pp. 4734–4745, 2014.

[87] M. Boztepe, F. Guinjoan, G. V. Velasco-Quesada, S. Silvestre, A. Chouder, and E. Karatepe, "Global MPPT scheme for photovoltaic string inverters based on restricted voltage window search algorithm," *IEEE Transactions on Industrial Electronics*, vol. 61, no. 7, pp. 3302–3312, 2014.

[88] M. Bansal, R. P. Saini, and D. K. Khatod, "An off-grid hybrid system scheduling for a remote area," in *Proceedings of the IEEE Students' Conference on Electrical, Electronics and Computer Science (SCEECS '12)*, pp. 1–4, IEEE, Bhopal, India, March 2012.

[89] K. Kimura and T. Kimura, "Neural networks approach for wind-solar energy system with complex networks," in *Proceedings of the IEEE 10th International Conference on Power Electronics and Drive Systems (PEDS '13)*, pp. 1–5, IEEE, Kitakyushu, Japan, April 2013.

[90] E. I. B. Gould, "Hybrid2: the hybrid system simulation model version 1.0," User Manual, National Renewable Energy Laboratory, 1996, http://www.nrel.gov/docs/legosti/old/21272.pdf.

[91] pp. 17–24, 2012, http://www.iea-pvps.org/fileadmin/dam/public/report/technical/rep11_01.pdf.

[92] E. S. Sreeraj, K. Chatterjee, and S. Bandyopadhyay, "One-cycle-controlled single-stage single-phase voltage-sensorless grid-connected PV system," *IEEE Transactions on Industrial Electronics*, vol. 60, no. 3, pp. 1216–1224, 2013.

[93] E. Mamarelis, G. Petrone, and G. Spagnuolo, "An hybrid digital-analog sliding mode controller for photovoltaic applications," *IEEE Transactions on Industrial Informatics*, vol. 9, no. 2, pp. 1094–1103, 2013.

[94] T. Senjyu, D. Hayashi, N. Urasaki, and T. Funabashi, "Optimum configuration for renewable generating systems in residence using genetic algorithm," *IEEE Transactions on Energy Conversion*, vol. 21, no. 2, pp. 459–467, 2006.

[95] X. Li, D. Hui, and X. Lai, "Battery energy storage station (BESS)-based smoothing control of photovoltaic (PV) and wind power generation fluctuations," *IEEE Transactions on Sustainable Energy*, vol. 4, no. 2, pp. 464–473, 2013.

[96] D. Weisser and R. S. Garcia, "Instantaneous wind energy penetration in isolated electricity grids: concepts and review," *Renewable Energy*, vol. 30, no. 8, pp. 1299–1308, 2005.

[97] G. E. Ahmad and E. T. El Shenawy, "Optimized photovoltiac system for hydrogen production," *Renewable Energy*, vol. 31, no. 7, pp. 1043–1054, 2006.

[98] Q. Xie, Y. Wang, Y. Kim, M. Pedram, and N. Chang, "Charge allocation in hybrid electrical energy storage systems," *IEEE Transactions on Computer-Aided Design of Integrated Circuits and Systems*, vol. 32, no. 7, pp. 1003–1016, 2013.

[99] Y. V. Makarov, P. Du, M. C. W. K. Meyer, C. Jin, and H. F. Illian, "Sizing energy storage to accommodate high penetration of variable energy resources," *IEEE Transactions on Sustainable Energy*, vol. 3, no. 1, pp. 34–40, 2012.

[100] P. Nema, R. K. Nema, and S. Rangnekar, "A current and future state of art development of hybrid energy system using wind and PV-solar: a review," *Renewable and Sustainable Energy Reviews*, vol. 13, no. 8, pp. 2096–2103, 2009.

[101] J. D. Maclay, J. Brouwer, and G. Scott Samuelsen, "Dynamic analyses of regenerative fuel cell power for potential use in renewable residential applications," *International Journal of Hydrogen Energy*, vol. 31, no. 8, pp. 994–1009, 2006.

[102] L. Xu, X. Ruan, C. Mao, B. Zhang, and Y. Luo, "An improved optimal sizing method for wind-solar-battery hybrid power system," *IEEE Transactions on Sustainable Energy*, vol. 4, no. 3, pp. 774–785, 2013.

[103] M. S. Whittingham, "History, evolution, and future status of energy storage," *Proceedings of the IEEE Special Centennial Issue*, vol. 100, pp. 1518–1534, 2012.

[104] M. Trifkovic, M. Sheikhzadeh, K. Nigim, and P. Daoutidis, "Modeling and control of a renewable hybrid energy system with hydrogen storage," *IEEE Transactions on Control Systems Technology*, vol. 22, no. 1, pp. 169–179, 2014.

[105] W. Zhou, C. Lou, Z. Li, L. Lu, and H. Yang, "Current status of research on optimum sizing of stand-alone hybrid solar–wind power generation systems," *Applied Energy*, vol. 87, no. 2, pp. 380–389, 2010.

[106] K. Nishioka, T. Takamoto, T. Agui, M. Kaneiwa, Y. Uraoka, and T. Fuyuki, "Annual output estimation of concentrator photovoltaic systems using high-efficiency InGaP/InGaAs/Ge triple-junction solar cells based on experimental solar cell's characteristics and field-test meteorological data," *Solar Energy Materials and Solar Cells*, vol. 90, no. 1, pp. 57–67, 2006.

[107] L. Marí and N. Nabona, "Renewable energies in medium-term power planning," *IEEE Transactions on Power Systems*, vol. 30, no. 1, pp. 88–97, 2015.

[108] G. Tina, S. Gagliano, and S. Raiti, "Hybrid solar/wind power system probabilistic modelling for long-term performance assessment," *Solar Energy*, vol. 80, no. 5, pp. 578–588, 2006.

[109] H. Yang, L. Lu, and W. Zhou, "A novel optimization sizing model for hybrid solar-wind power generation system," *Solar Energy*, vol. 81, no. 1, pp. 76–84, 2007.

[110] H. X. Yang, W. Zhou, L. Lu, and Z. Fang, "Optimal sizing method for stand-alone hybrid solar–wind system with LPSP technology by using genetic algorithm," *Solar Energy*, vol. 82, no. 4, pp. 354–367, 2008.

[111] A. Mellit, S. A. Kalogirou, L. Hontoria, and S. Shaari, "Artificial intelligence techniques for sizing photovoltaic systems: a review," *Renewable and Sustainable Energy Reviews*, vol. 13, no. 2, pp. 406–419, 2009.

[112] S. A. Kalogirou, "Optimization of solar systems using artificial neural-networks and genetic algorithms," *Applied Energy*, vol. 77, no. 4, pp. 383–405, 2004.

[113] R. Siddique, A. A. Faisal, M. A. H. Raihan et al., "A theoretical analysis of controlling the speed of wind turbine and assemblage of solar system in the wind energy conversion system," in *Proceedings of the IEEE International Conference on Power, Energy and Control (ICPEC '13)*, Dindigul, India, February 2013.

[114] Y. Bae, T.-K. Vu, and R.-Y. Kim, "Implemental control strategy for grid stabilization of grid-connected PV system based on German grid code in symmetrical low-to-medium voltage network," *IEEE Transactions on Energy Conversion*, vol. 28, no. 3, pp. 619–631, 2013.

[115] P. Malysz, S. Sirouspour, and A. Emadi, "An optimal energy storage control strategy for grid-connected microgrids," *IEEE Transactions on Smart Grid*, vol. 5, no. 4, pp. 1785–1796, 2014.

[116] B. O. Bilal, V. Sambou, P. A. Ndiaye, C. M. F. Kebe, and M. Ndongo, "Multi-objective design of PV-wind-batteries hybrid systems by minimizing the annualized cost system and the loss of power supply probability (LPSP)," in *Proceedings of the IEEE International Conference on Industrial Technology (ICIT '13)*, vol. 2, pp. 861–868, Cape Town, South Africa, February 2013.

[117] R. Dufo-López and J. L. Bernal-Agustín, "Multi-objective design of PV-wind-diesel-hydrogen-battery systems," *Renewable Energy*, vol. 33, no. 12, pp. 2559–2572, 2008.

[118] X. Pelet, D. Favrat, and G. Leyland, "Multiobjective optimisation of integrated energy systems for remote communities considering economics and CO_2 emissions," *International Journal of Thermal Sciences*, vol. 44, no. 12, pp. 1180–1189, 2005.

[119] J. P. Torreglosa, P. García, L. M. Fernández, and F. Jurado, "Hierarchical energy management system for stand-alone hybrid system based on generation costs and cascade control," *Energy Conversion and Management*, vol. 77, pp. 514–526, 2014.

[120] J. C. Hernández, A. Medina, and F. Jurado, "Optimal allocation and sizing for profitability and voltage enhancement of PV systems on feeders," *Renewable Energy*, vol. 32, no. 10, pp. 1768–1789, 2007.

[121] T. F. El-Shatter, M. N. Eskander, and M. T. El-Hagry, "Energy flow and management of a hybrid wind/PV/fuel cell generation system," *Energy Conversion and Management*, vol. 47, no. 9-10, pp. 1264–1280, 2006.

[122] E. Koutroulis, D. Kolokotsa, A. Potirakis, and K. Kalaitzakis, "Methodology for optimal sizing of stand-alone photovoltaic/wind-generator systems using genetic algorithms," *Solar Energy*, vol. 80, no. 9, pp. 1072–1088, 2006.

[123] A. Chaurey and T. C. Kandpal, "Carbon abatement potential of solar home systems in India and their cost reduction due to carbon finance," *Energy Policy*, vol. 37, no. 1, pp. 115–125, 2009.

[124] Q.-S. Jia, J.-X. Shen, Z.-B. Xu, and X.-H. Guan, "Simulation-based policy improvement for energy management in commercial office buildings," *IEEE Transactions on Smart Grid*, vol. 3, no. 4, pp. 2211–2223, 2012.

[125] S. M. Shaahid and M. A. Elhadidy, "Economic analysis of hybrid photovoltaic–diesel–battery power systems for residential loads in hot regions—a step to clean future," *Renewable and Sustainable Energy Reviews*, vol. 12, no. 2, pp. 488–503, 2008.

[126] S. Sarkar and V. Ajjarapu, "MW resource assessment model for a hybrid energy conversion system with wind and solar resources," *IEEE Transactions on Sustainable Energy*, vol. 2, no. 4, pp. 383–391, 2011.

[127] B. Sun, P. B. Luh, Q.-S. Jia, Z. Jiang, F. Wang, and C. Song, "Building energy management: integrated control of active and passive heating, cooling, lighting, shading, and ventilation systems," *IEEE Transactions on Automation Science and Engineering*, vol. 10, no. 3, pp. 588–602, 2013.

[128] A. Arabali, M. Ghofrani, M. Etezadi-Amoli, M. S. Fadali, and Y. Baghzouz, "Genetic-algorithm-based optimization approach for energy management," *IEEE Transactions on Power Delivery*, vol. 28, no. 1, pp. 162–170, 2013.

[129] E. Lannoye, D. Flynn, and M. O'Malley, "Evaluation of power system flexibility," *IEEE Transactions on Power Systems*, vol. 27, no. 2, pp. 922–931, 2012.

[130] S. Mei, Y. Wang, F. Liu, X. Zhang, and Z. Sun, "Game approaches for hybrid power system planning," *IEEE Transactions on Sustainable Energy*, vol. 3, no. 3, pp. 506–517, 2012.

[131] H. H. El-Tamaly and A. A. E. Mohammed, "Impact of interconnection photovoltaic/wind system with utility on their reliability using a fuzzy scheme," *Renewable Energy*, vol. 31, no. 15, pp. 2475–2491, 2006.

[132] J. V. Paatero and P. D. Lund, "Effects of large-scale photovoltaic power integration on electricity distribution networks," *Renewable Energy*, vol. 32, no. 2, pp. 216–234, 2007.

[133] X. Xu and S. V. Dessel, "Evaluation of a prototype active building envelope window-system," *Energy and Buildings*, vol. 40, no. 2, pp. 168–174, 2008.

[134] V. Quaschning, "Technical and economical system comparison of photovoltaic and concentrating solar thermal power systems depending on annual global irradiation," *Solar Energy*, vol. 77, no. 2, pp. 171–178, 2004.

[135] T. Tezuka, K. Okushima, and T. Sawa, "Carbon tax for subsidizing photovoltaic power generation systems and its effect on carbon dioxide emissions," *Applied Energy*, vol. 72, no. 3-4, pp. 677–688, 2002.

[136] D. B. Nelson, M. H. Nehrir, and C. Wang, "Unit sizing and cost analysis of stand-alone hybrid wind/PV/fuel cell power generation systems," *Renewable Energy*, vol. 31, no. 10, pp. 1641–1656, 2006.

[137] S.-J. Park, F.-S. Kang, M. H. Lee, and C.-U. Kim, "A new single-phase five-level PWM inverter employing a deadbeat control scheme," *IEEE Transactions on Power Electronics*, vol. 18, no. 3, pp. 831–843, 2003.

[138] N. A. Rahim and J. Selvaraj, "Multistring five-level inverter with novel PWM control scheme for PV application," *IEEE Transactions on Industrial Electronics*, vol. 57, no. 6, pp. 2111–2123, 2009.

[139] N. A. Rahim, K. Chaniago, and J. Selvaraj, "Single-phase seven-level grid-connected inverter for photovoltaic system," *IEEE Transactions on Industrial Electronics*, vol. 58, no. 6, pp. 2435–2443, 2011.

[140] S. Gautam and R. Gupta, "Switching frequency derivation for the cascaded multilevel inverter operating in current control mode using multiband hysteresis modulation," *IEEE Transactions on Power Electronics*, vol. 29, no. 3, pp. 1480–1489, 2014.

[141] Y. Liu, B. Ge, H. Abu-Rub, and F. Z. Peng, "An effective control method for quasi-Z-source cascade multilevel inverter-based grid-tie single-phase photovoltaic power system," *IEEE Transactions on Industrial Informatics*, vol. 10, no. 1, pp. 399–407, 2014.

[142] M. Fatu, F. Blaabjerg, and I. Boldea, "Grid to standalone transition motion-sensorless dual-inverter control of PMSG with asymmetrical grid voltage sags and harmonics filtering," *IEEE Transactions on Power Electronics*, vol. 29, no. 7, pp. 3463–3472, 2014.

[143] N. Mendis, K. M. Muttaqi, and S. Perera, "Management of battery-supercapacitor hybrid energy storage and synchronous condenser for isolated operation of PMSG based variable-speed wind turbine generating systems," *IEEE Transactions on Smart Grid*, vol. 5, no. 2, pp. 944–953, 2014.

[144] H. Quan, D. Srinivasan, and A. Khosravi, "Short-term load and wind power forecasting using neural network-based prediction intervals," *IEEE Transactions on Neural Networks and Learning Systems*, vol. 25, no. 2, pp. 303–315, 2014.

[145] A. Chauhan and R. P. Saini, "A review on integrated renewable energy system based power generation for stand-alone applications: configurations, storage options, sizing methodologies and control," *Renewable and Sustainable Energy Reviews*, vol. 38, pp. 99–120, 2014.

[146] M. Alsayed, M. Cacciato, G. Scarcella, and G. Scelba, "Multicriteria optimal sizing of photovoltaic-wind turbine grid connected systems," *IEEE Transactions on Energy Conversion*, vol. 28, no. 2, pp. 370–379, 2013.

[147] T. Zhou and W. Sun, "Optimization of battery-supercapacitor hybrid energy storage station in wind/solar generation system," *IEEE Transactions on Sustainable Energy*, vol. 5, no. 2, pp. 408–415, 2014.

[148] H.-S. Ko and J. Jatskevich, "Power quality control of wind-hybrid power generation system using fuzzy-LQR controller," *IEEE Transactions on Energy Conversion*, vol. 22, no. 2, pp. 516–527, 2007.

[149] Z. Jiang and R. Dougal, "Hierarchical microgrid paradigm for integration of distributed energy resources," in *Proceedings of the IEEE International Conference on Power Engineering Society General Meeting (PES '08)*, pp. 20–24, Pittsburgh, Pa, USA, July 2008.

[150] J. Zeng, W. Qiao, and L. Qu, "An isolated multiport DC-DC converter for simultaneous power management of multiple renewable energy sources," in *Proceedings of the 4th Annual IEEE Energy Conversion Congress and Exposition (ECCE '12)*, pp. 3741–3748, Raleigh, NC, USA, September 2012.

[151] F. Botterón, R. Carballo, R. Núñez, A. Quintana, and G. Fernandez, "High reliability and performance PWM inverter for standalone microgrids," *IEEE Latin America Transactions*, vol. 11, no. 1, pp. 505–511, 2013.

[152] K. Narula, Y. Nagai, and S. Pachauri, "The role of decentralized distributed generation in achieving universal rural electrification in South Asia by 2030," *Energy Policy*, vol. 47, pp. 345–357, 2012.

[153] G. R. Timilsina, L. Kurdgelashvili, and P. A. Narbel, "Solar energy: markets, economics and policies," *Renewable and Sustainable Energy Reviews*, vol. 16, no. 1, pp. 449–465, 2012.

[154] C.-T. Pan and Y.-L. Juan, "A novel sensorless MPPT controller for a high-efficiency microscale wind power generation system," *IEEE Transactions on Energy Conversion*, vol. 25, no. 1, pp. 207–216, 2010.

[155] K. Nishida, T. Ahmed, and M. Nakaoka, "A cost-effective high-efficiency power conditioner with simple MPPT control algorithm for wind-power grid integration," *IEEE Transactions on Industry Applications*, vol. 47, no. 2, pp. 893–900, 2011.

[156] K.-Y. Lo, Y.-M. Chen, and Y.-R. Chang, "MPPT battery charger for stand-alone wind power system," *IEEE Transactions on Power Electronics*, vol. 26, no. 6, pp. 1631–1638, 2011.

[157] N. Mendis, K. M. Muttaqi, S. Sayeef, and S. Perera, "Standalone operation of wind turbine-based variable speed generators with maximum power extraction capability," *IEEE Transactions on Energy Conversion*, vol. 27, no. 4, pp. 822–834, 2012.

[158] Y. Zou, M. E. Elbuluk, and Y. Sozer, "Stability analysis of maximum power point tracking (MPPT) method in wind power systems," *IEEE Transactions on Industry Applications*, vol. 49, no. 3, pp. 1129–1136, 2013.

[159] M. Cirrincione, M. Pucci, and G. Vitale, "Neural MPPT of variable-pitch wind generators with induction machines in a wide wind speed range," *IEEE Transactions on Industry Applications*, vol. 49, no. 2, pp. 942–953, 2013.

[160] Z. M. Dalala, Z. U. Zahid, W. Yu, Y. Cho, and J.-S. Lai, "Design and analysis of an MPPT technique for small-scale wind energy conversion systems," *IEEE Transactions on Energy Conversion*, vol. 28, no. 3, pp. 756–767, 2013.

[161] A. Urtasun, P. Sanchis, and L. Marroyo, "Small wind turbine sensorless MPPT: robustness analysis and lossless approach," *IEEE Transactions on Industry Applications*, vol. 50, no. 6, pp. 4113–4121, 2014.

[162] http://www.homerenergy.com/HOMER_pro.html.

[163] E. L. B. Gould, *Hybrid2: The Hybrid System Simulation Model*, User Manual, National Renewable Energy Laboratory, Golden, Colo, USA, 1996, http://www.nrel.gov/docs/legosti/old/21272.pdf.

[164] http://hoga-renewable.es.tl/.

[165] Ulleberg, *Standalone power systems for the future: optimal design, operation & control of solar-hydrogen energy systems [Ph.D. thesis]*, Department of Thermal Energy and Hydropower, Norwegian University of Science and Technology, Trondheim, Norway, 1998.

[166] F. Farret and M. Simões, *Micropower System Modeling with Homer*, Wiley-IEEE Press, 2006.

[167] http://www.retscreen.net/ang/centre.php.

[168] http://www.energyplan.eu/trnsys/.

[169] http://in.mathworks.com/academia/student_center/tutorials/?s_tid=acmain_st-pop-tut_gw_bod.

[170] H. Elmqvist, *Dymola: environment for object-oriented modeling of physical systems [Ph.D. thesis]*, 2009.

[171] Pvsyst contextual help, Users Guide, pp 2–13, http://files.pvsyst.com/pvsyst5.pdf.

[172] M. R. Patel, *Wind and Solar Power System*, CRC Press, New York, NY, USA, 1999.

[173] J. K. Kaldellis, *Stand-Alone and Hybrid Wind Energy Systems*, CRC Press Woodhead Publishing Series in Energy, CRC Press, 2010.

18

Load Mitigation and Optimal Power Capture for Variable Speed Wind Turbine in Region 2

Saravanakumar Rajendran and Debashisha Jena

Department of Electrical Engineering, National Institute of Technology Karnataka, Surathkal, Mangalore 575 025, India

Correspondence should be addressed to Saravanakumar Rajendran; sarrajoom@gmail.com

Academic Editor: Adnan Parlak

This paper proposes the two nonlinear controllers for variable speed wind turbine (VSWT) operating at below rated wind speed. The objective of the controller is to maximize the energy capture from the wind with reduced oscillation on the drive train. The conventional controllers such as aerodynamic torque feedforward (ATF) and indirect speed control (ISC) are adapted initially, which introduce more power loss, and the dynamic aspects of WT are not considered. In order to overcome the above drawbacks, modified nonlinear static state with feedback estimator (MNSSFE) and terminal sliding mode controller (TSMC) based on Modified Newton Raphson (MNR) wind speed estimator are proposed. The proposed controllers are simulated with nonlinear FAST (fatigue, aerodynamics, structures, and turbulence) WT dynamic simulation for different mean wind speeds at below rated wind speed. The frequency analysis of the drive train torque is done by taking the power spectral density (PSD) of low speed shaft torque. From the result, it is found that a trade-off is to be maintained between the transient load on the drive train and maximum power capture.

1. Introduction

In recent years, wind energy is one of the major renewable energy sources because of environmental, social, and economic benefits. The major classifications of wind turbines (WT) are fixed speed wind turbine (FSWT) and VSWT. Compared with FSWT, VSWT has many advantages such as improved energy capture, reduction in transient load, and better power conditioning [1]. For any kind of WT, control strategies play a major role on WT characteristics and transient load to the network [2]. In VSWT, the operating regions are classified into two major categories, that is, below and above rated wind speed. At below rated wind speed, the main objective of the controller (i.e., torque control) is to optimize the wind energy capture by avoiding the transients in the turbine components especially in the drive train. At above rated wind speed, the major objective of the controller (i.e., pitch control) is to maintain the rated power of the WT. In [3], the maximum power for VSWT is achieved by PI (proportional integral) controller, which is based on the fuzzy system. Error is taken as the input to the controller, that is, difference between the actual and optimal rotor speed,

and the output of the controller is generator torque. Fuzzy logic systems (FLS) are used for tuning the PI controller gains for various wind speed. PI gains are optimized for different wind speed by particle swarm optimization (PSO). In [4], radial bias function neural network (RBFNN) and torque observer based control algorithm are used to control the WT for optimal energy capture. RBFNN is trained online by using MPSO (modified particle swarm optimization) training algorithm. In order to achieve the maximum power, the difference between the actual and optimal rotor speed is to be minimized. In [5], a new maximum adaptive algorithm for extraction of optimal power is proposed for small WT. Perturb and absorb scheme is adapted for different wind speed to obtain optimum relationship for regulating the maximum power point. In [6], two control strategies are developed for optimal power extraction with reduced mechanical stress. The first one is tracking controller with wind speed estimator which ensures the optimal angular speed of the rotor. In the second one, a robust power tracking is developed by nonhomogenous quasicontinuous high order sliding mode controller without considering wind velocity. Maximum power extraction from VSWT is achieved by a Takagi-Sugeno-Kang (TSK) fuzzy model which is based on data driven model [7]. In TSK model, a combination of fuzzy clustering method and genetic algorithm (GA) is used for portioning the input-output space and least square (LS) algorithm is used for parameter estimation. Nonlinear static and dynamic state feedback linearization control are addressed in [8, 9], where both the single and two-mass model are taken into consideration and the wind speed is estimated by Newton Rapshon (NR) method. To accommodate the parameter uncertainty and robustness, a higher order sliding mode controller is proposed in [10], which ensures the stability of the controller in both regions, that is, below and above rated speed. Feedback torque control is applied for mathematical model FSWT for maximum power extraction [11]. In order to achieve the maximum power point in the WT, FLC tuned by GA is discussed in [12]. The width of the membership function in FLC is adjusted by GA. In [13], sliding mode controller (SMC) and integral sliding mode controller (ISMC) are designed for all the regions of variable speed variable pitch wind turbine (VSVPWT) with FAST simulator.

The objective of this paper is to prove the efficacy of nonlinear controllers which considers the dynamic aspect of the wind and aero turbine, without the wind speed measurement. Finally, the objective is to track the reference rotor speed asymptotically. This paper is organized as follows. The objective of the work is discussed in Section 2. Section 3 discusses the modeling of the two-mass model. The conventional and proposed controllers are discussed in Sections 4 and 5. In Section 6, FAST model results are analyzed. Finally, a conclusion is drawn from the obtained results in Section 7, which shows the proposed method is having better performance compared to other existing controllers.

2. Problem Formulation

Generally, WT is classified into two types, that is, fixed and variable speed WT. Variable speed WT has more advanced and flexible operation than fixed speed WT. Operating regions in variable speed WT are divided into three types. Figure 1 shows the various operating region in variable speed WT.

Region 1 represents the wind speed below the cut-in wind speed. Region 2 represents the wind speed between cut-in and cut-out. In this region, the main objective is to maximize the energy capture from the wind with reduced oscillation on the drive train. Region 3 describes the wind speed above the cut-out speed. In this region, pitch controller is used to maintain the WT at its rated power.

Figure 2 shows the WT control scheme. To achieve the above objective (Region 2), the blade pitch angle (β_{opt}) and tip speed ratio (λ_{opt}) are set to be its optimal value. In order to achieve the optimal tip speed ratio, the rotor speed must be adjusted to the reference/optimal rotor speed (ω_{ropt}) by adjusting the control input, that is, generator torque (T_g). Equation (1) defines the reference/optimal rotor speed:

$$\omega_{ropt} = \omega_{ref} = \frac{\lambda_{opt} v}{R}. \tag{1}$$

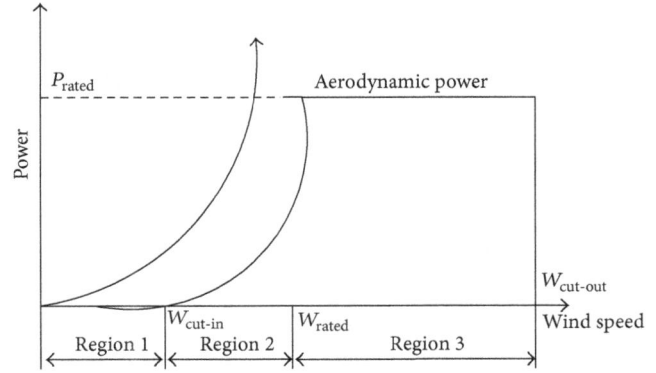

FIGURE 1: Power operating region of wind turbines.

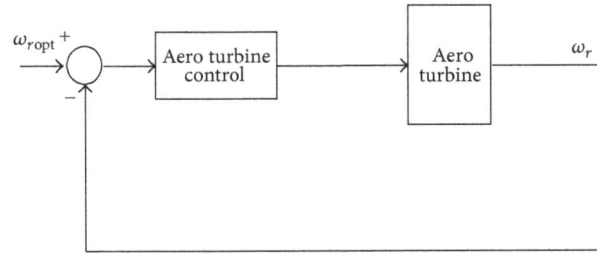

FIGURE 2: WT control scheme.

3. WT Model

A WT is a device which converts the kinetic energy of the wind into electric energy. Simulation complexity of the WT purely depends on the type of control objectives. In case of WT modelling complex simulators are required to verify the dynamic response of multiple components and aerodynamic loading. Generally, dynamic loads and interaction of large components are verified by the aeroelastic simulator. For designing a WT controller, instead of going with complex simulator, the design objective can be achieved by using simplified mathematical model. In this work, WT model is described by the set of nonlinear ordinary differential equations with limited degree of freedom. In this paper, the control law is designed based on simplified mathematical model with the objective of optimal power capture at below rated wind speed and reduced oscillation of the drive train. The proposed controllers are tested with different wind profiles. Finally, the controllers are validated for FAST WT model. The parameters of the two-mass model are given in [9]. Generally, VSWT system consists of the following components; that is, aerodynamics, drive trains, and generator are shown in Figure 3.

Equation (2) gives the nonlinear expression for aerodynamic power capture by the rotor:

$$P_a = \frac{1}{2}\rho\pi R^2 C_P(\lambda, \beta) v^3. \tag{2}$$

From (2), it is clear that the aerodynamic power (P_a) is directly proportional to the cube of the wind speed. The power coefficient C_P is the function of blade pitch angle (β)

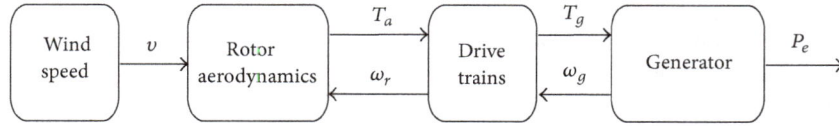

FIGURE 3: Schematic of WT.

and tip speed ratio (λ). The tip speed ratio is defined as ratio between linear tip speed and wind speed:

$$\lambda = \frac{\omega_r R}{v}.$$ (3)

Generally, wind speed is stochastic nature with respect to time. Because of this, tip speed ratio gets affected, which leads to variation in power coefficient. The relationship between aerodynamic torque (T_a) and the aerodynamic power is given in (4):

$$P_a = T_a \omega_r,$$ (4)

$$T_a = \frac{1}{2} \rho \pi R^3 C_q(\lambda, \beta) v^2,$$ (5)

where C_q is the torque coefficient given as

$$C_q(\lambda, \beta) = \frac{C_P(\lambda, \beta)}{\lambda}.$$ (6)

Substituting (6) in (5), we get

$$T_a = \frac{1}{2} \rho \pi R^3 \frac{C_P(\lambda, \beta)}{\lambda} v^2.$$ (7)

In the above equation, the nonlinear term is C_P which can be approximated by the 5th-order polynomial given in the following:

$$C_P(\lambda) = \sum_{n=0}^{5} a_n \lambda^n$$

$$= a_0 + \lambda a_1 + \lambda^2 a_2 + \lambda^3 a_3 + \lambda^4 a_4 + \lambda^5 a_5,$$ (8)

where a_0 to a_5 are the WT power coefficients.

The values of approximated coefficients are given in Table 1. Figure 4 shows the C_P versus λ curve.

Figure 5 shows the two-mass model of the WT. Equation (9) represents dynamics of the rotor speed ω_r with rotor inertia J_r driven by the aerodynamic torque (T_a):

$$J_r \dot{\omega}_r = T_a - T_{ls} - K_r \omega_r.$$ (9)

Breaking torque acting on the rotor is low speed shaft torque (T_{ls}) which can be derived by using stiffness and damping factor of the low speed shaft given in the following:

$$T_{ls} = B_{ls}(\theta_r - \theta_{ls}) + K_{ls}(\omega_r - \omega_{ls}).$$ (10)

Equation (11) represents dynamics of the generator speed ω_g with generator inertia J_g driven by the high speed shaft torque (T_{hs}) and braking electromagnetic torque (T_{em}):

$$J_g \dot{\omega}_g = T_{hs} - K_g \omega_g - T_{em}.$$ (11)

TABLE 1: Coefficients' values.

$a_0 = 0.1667$	$a_3 = -0.01617$
$a_1 = -0.2558$	$a_4 = 0.00095$
$a_2 = 0.115$	$a_5 = -2.05 \times 10^{-5}$

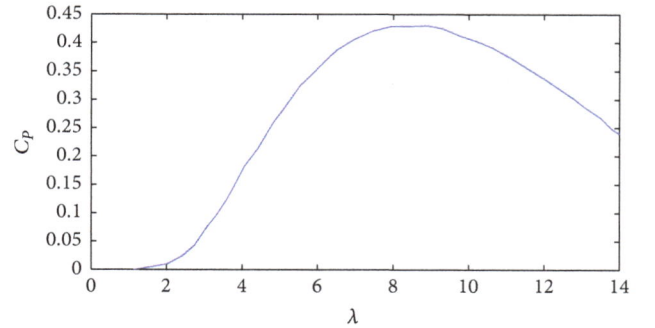

FIGURE 4: C_P versus λ curve.

Gearbox ratio is defined as

$$n_g = \frac{T_{ls}}{T_{hs}} = \frac{\omega_g}{\omega_{ls}},$$ (12)

$$T_{ls} = n_g T_{hs}.$$ (13)

From (11), the high speed shaft torque T_{hs} can be expressed as

$$T_{hs} = J_g \dot{\omega}_g + K_g \omega_g + T_{em}.$$ (14)

Putting the values of T_{hs} from (14) in (13), we get

$$T_{ls} = n_g \left(J_g \dot{\omega}_g + K_g \omega_g + T_{em} \right).$$ (15)

4. Conventional Controllers

In order to compare the results of proposed and existing conventional controllers, a brief description of the well-known control techniques, that is, ISC and ATF, is discussed in this section. In ISC, it is assumed that the WT is stable around its optimal aerodynamic efficiency curve. The two-mass model control signal is given in the following:

$$T_{em} = K_{opt_{hs}} \omega_g^2 - K_{t_{hs}} \omega_g,$$ (16)

where

$$K_{opt_{hs}} = 0.5 \rho \pi \frac{R^5}{n_g^3 \lambda_{opt}^3} C_{P_{opt}},$$

$$K_{t_{hs}} = \left(K_g + \frac{K_r}{n_g^2} \right),$$ (17)

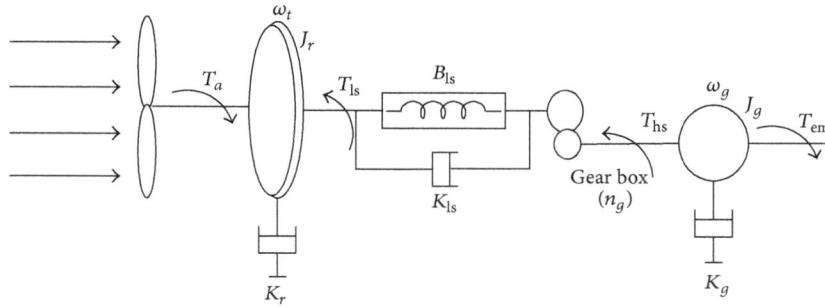

FIGURE 5: Two-mass model of the aero turbine.

where $K_{t_{hs}}$ is the low speed shaft damping coefficient brought up to the high speed shaft.

In ATF, proportional control law is used to control the WT. The rotor speed and the aerodynamic torque (T_a) are estimated using Kalman filter, which is used to control the WT [14]. The control law is given in (18):

$$T_{em} = \frac{1}{n_g}\widehat{T}_a - \left(\frac{K_r}{n_g^2} + K_g\right)\widehat{\omega}_g - \frac{K_C}{n_g^2}\left(\omega_{gref} - \omega_g\right), \quad (18)$$

$$\omega_{g_{ref}} = n_g k_w \sqrt{\widehat{T}_a}, \quad (19)$$

$$k_w = \frac{1}{\sqrt{k_{opt}}} = \sqrt{\frac{2\lambda_{opt}^3}{\rho\pi R^5 C_{Popt}}}, \quad (20)$$

$$k_{opt} = \frac{1}{2}\rho\pi\frac{R^5}{\lambda_{opt}^3}C_{Popt}. \quad (21)$$

The optimal value of proportional gain is found to be $K_c = 3 \times 10^4$.

The above existing control techniques have three major drawbacks, that is, the ATF control having more steady state error, so an accurate value of ω_{gref} is needed; in ISC, the WT has to operate at its optimal efficiency curve which introduces more power loss for high varying wind speed. Both the controllers are not robust with respect to disturbances. To avoid the above drawbacks, two nonlinear controllers, that is, MNSSFE and TSMC, are proposed.

4.1. Wind Speed Estimation. The estimation of effective wind speed is related to aerodynamic torque and rotor speed provided the pitch angle is at optimal value:

$$T_a = \frac{1}{2}\rho\pi R^3\frac{C_P(\lambda)}{\lambda}v^2. \quad (22)$$

The aerodynamic power coefficient is approximated with 5th-order polynomial as given in (8):

$$F(v) = T_a - \frac{1}{2}\rho\pi R^3\frac{C_P(\lambda)}{\lambda}v^2. \quad (23)$$

The estimated wind speed can be obtained by solving (23) using MNR. The above equation has unique solution at below rated region. With known v, the optimal rotor speed ω_{ropt} is calculated by using (1).

5. Proposed Nonlinear Controllers

5.1. Terminal Sliding Mode Control (TSMC) for Optimal Power Capture. Let us consider the linear sliding surface:

$$S = \alpha e + \dot{e}. \quad (24)$$

The first-order derivative of the above equation can be obtained as

$$\dot{S} = \alpha\dot{e} + \ddot{e}. \quad (25)$$

A nonsingular terminal sliding mode manifold [15] is first designed as

$$\sigma = S + \beta\dot{S}^{p/q}, \quad (26)$$

where $\beta > 0$ is a design constant and p and q are the positive integer, which satisfy the following condition:

$$p > q \text{ or } 1 < \frac{p}{q} < 2. \quad (27)$$

The linear sliding mode $S(t)$ is combined with nonsingular terminal manifold $\sigma(t)$. After $\sigma(t)$ reaches zero in finite time, both $S(t)$ and $\dot{S}(t)$ will also reach finite time; then, the tracking errors e and $\dot{e}, \ddot{e}$ can asymptotically converge to zero. Once σ reaches zero, it will stay on zero by using the control law. Then, the sliding surface S will converge to zero in finite time. The total time from $\sigma(0) \neq 0$ to S_{tf} can be calculated by using the equations $S + \beta\dot{S}^{p/q} = 0$, from which time taken from S_{tr} to S_{tf} is obtained as

$$t_f = t_r + \frac{(p/q)}{(p/q) - 1}\beta^{-(p/q)}\|S_{tr}\|^{(p/q)-1}. \quad (28)$$

By taking the derivative of (26),

$$\dot{\sigma} = \dot{S} + \frac{\beta p}{q}\dot{S}^{(p/q)-1}\ddot{S}. \quad (29)$$

The stability of the system is investigated by choosing the following Lyapunov function:

$$V = \frac{1}{2}\sigma^2. \quad (30)$$

By taking the derivative of the above equation, we will get

$$\dot{V} = \sigma\dot{\sigma} = \sigma\left(\alpha\left(\dot{\omega}_r - \dot{\omega}_{\text{ref}}\right) + \ddot{e} + \frac{\beta p}{q}\dot{S}^{(p/q)-1}\ddot{S}\right). \tag{31}$$

To make the controller more adaptive to uncertainty and disturbances, we introduce the parameter called F, where F contains both modelling error and external disturbances:

$$F = \Delta\left(\omega_r, \dot{\omega}_r\right) + \overline{d},$$
$$\dot{F} = 0. \tag{32}$$

From the two-mass model system equation,

$$\dot{\omega}_r = \frac{T_a}{J_r} - \frac{K_r}{J_r}\omega_r - \frac{n_g J_g}{J_r}\omega_g - \frac{n_g}{J_r}K_g\omega_g - \frac{n_g}{J_r}T_{\text{em}}$$
$$+ \frac{F}{J_r}. \tag{33}$$

By substituting (33) in (31),

$$\dot{V} = \sigma\left(\alpha\left(\frac{T_a}{J_r} - \frac{K_r}{J_r}\omega_r - \frac{n_g J_g}{J_r}\dot{\omega}_g - \frac{n_g}{J_r}K_g\omega_g\right.\right.$$
$$\left.\left. - \frac{n_g}{J_r}T_{\text{em}} + \frac{F}{J_r} - \dot{\omega}_{\text{ref}}\right) + \ddot{e} + \frac{\beta p}{q}\dot{S}^{(p/q)-1}\ddot{S}\right). \tag{34}$$

According to the above equation, the terminal sliding mode control law T_{em} is defined as

$$T_{\text{em}} = \frac{T_a}{n_g} - \frac{K_r}{n_g}\omega_r - J_g\dot{\omega}_g - K_g\omega_g + \frac{\overline{F}}{n_g}\text{sign}\left(\sigma\right)$$
$$- \frac{J_r}{n_g}\dot{\omega}_{\text{ref}} + \frac{J_r}{n_g\alpha}\ddot{e} + \frac{\beta p}{q}\frac{J_r}{n_g\alpha}\dot{S}^{(p/q)-1}\ddot{S} \tag{35}$$
$$+ \frac{J_r}{n_g\alpha}k\,\text{sign}\left(\sigma\right).$$

By substituting (35) into (34), the following equation can be obtained. Since the lumped parameter uncertainty and disturbance F are unknown in practical application and the upper bound $\overline{F}$ is very difficult to determine, the adaptive control law is adapted for lumped uncertainty $\widehat{F}$.

The Lyapunov candidate function is chosen as

$$V_1 = V + \frac{1}{2\varphi}\widetilde{F}^2, \tag{36}$$

where $\widetilde{F} = F - \widehat{F}$ and φ is a positive constant. By taking the derivative of (36),

$$\dot{V}_1 = \dot{V} + \frac{1}{2\varphi}\widetilde{F}\dot{\widetilde{F}} = \dot{V} - \frac{1}{\varphi}\widetilde{F}\dot{\widehat{F}} = \sigma\left(\alpha\left(\frac{T_a}{J_r} - \frac{K_r}{J_r}\omega_r\right.\right.$$
$$- \frac{n_g J_g}{J_r}\dot{\omega}_g - \frac{n_g}{J_r}K_g\omega_g - \frac{n_g}{J_r}\left(\frac{T_a}{n_g} - \frac{K_r}{n_g}\omega_r - J_g\dot{\omega}_g\right.$$
$$\left. - K_g\omega_g + \frac{\widehat{F}}{n_g} - \frac{J_r}{n_g}\dot{\omega}_{\text{ref}} + \frac{J_r}{n_g\alpha}\ddot{e} + \frac{\beta p}{q}\frac{J_r}{n_g\alpha}\dot{S}^{(p/q)-1}\ddot{S}\right. \tag{37}$$
$$\left. + \frac{J_r}{n_g\alpha}k\,\text{sign}\left(\sigma\right)\right) + \frac{F}{J_r} - \dot{\omega}_{\text{ref}}\right) + \ddot{e} + \frac{\beta p}{q}$$
$$\left. \cdot \dot{S}^{(p/q)-1}\ddot{S}\right) - \frac{1}{\varphi}\widetilde{F}\left(\dot{\widehat{F}} - \varphi\sigma\right).$$

The adaptive control rule can be selected as

$$\dot{\widehat{F}} = \varphi\sigma. \tag{38}$$

According to the above equation, the terminal sliding mode control law T_{em} is defined as

$$T_{\text{em}} = \frac{T_a}{n_g} - \frac{K_r}{n_g}\omega_r - J_g\dot{\omega}_g - K_g\omega_g + \frac{\widehat{F}}{n_g} - \frac{J_r}{n_g}\dot{\omega}_{\text{ref}}$$
$$+ \frac{J_r}{n_g\alpha}\ddot{e} + \frac{\beta p}{q}\frac{J_r}{n_g\alpha}\dot{S}^{(p/q)-1}\ddot{S} + \frac{J_r}{n_g\alpha}k\,\text{sign}\left(\sigma\right). \tag{39}$$

In order to make the chattering free control law, the signum function can be replaced by tanh:

$$T_{\text{em}} = \frac{T_a}{n_g} - \frac{K_r}{n_g}\omega_r - J_g\dot{\omega}_g - K_g\omega_g + \frac{\widehat{F}}{n_g} - \frac{J_r}{n_g}\dot{\omega}_{\text{ref}}$$
$$+ \frac{J_r}{n_g\alpha}\ddot{e} + \frac{\beta p}{q}\frac{J_r}{n_g\alpha}\dot{S}^{(p/q)-1}\ddot{S} + \frac{J_r}{n_g\alpha}k\,\text{tanh}\left(\sigma\right). \tag{40}$$

Substituting (38) into (37), the following can be obtained:

$$\dot{V}_1 \leq -k\sigma\tanh\left(\sigma\right). \tag{41}$$

5.2. Modified Nonlinear Static State Feedback Linearization with Estimator (MNSSFE) for Maximum Power Extraction. In [9], the authors have explained nonlinear static state feedback with estimator, where the second derivative of the rotor speed and the first derivative of the low speed shaft are considered to express the control law. In order to avoid higher derivatives and complex control law, a modified nonlinear static state feedback with estimator is proposed in this section. Rearranging the terms in (9), we will get

$$\dot{\omega}_r = \frac{1}{J_r}T_a - \frac{K_r}{J_r}\omega_r - \frac{1}{J_r}T_{\text{ls}}. \tag{42}$$

By using the relationship given in (14) and (15), we will get

$$\dot{\omega}_r = \frac{1}{J_r}T_a - \frac{K_r}{J_r}\omega_r - \frac{1}{J_r}\left(n_g\left(J_g\dot{\omega}_g + K_g\omega_g + T_{\text{em}}\right)\right). \tag{43}$$

By separating the control input T_{em}, finally, the control torque can be expressed as

$$T_{em} = \frac{T_a}{n_g} - \frac{K_r}{J_r}\omega_r - J_g\dot{\omega}_g - K_g\omega_g - \frac{J_r}{n_g}\dot{\omega}_r, \qquad (44)$$

where $\dot{\omega}_r$ is approximated by the new input w:

$$\dot{\omega}_r = w, \qquad (45)$$

$$T_{em} = \frac{T_a}{n_g} - \frac{K_r}{J_r}\omega_r - J_g\dot{\omega}_g - K_g\omega_g - \frac{J_r}{n_g}w. \qquad (46)$$

The first-order error dynamics can be written as

$$\dot{e} + a_0 e = 0, \quad a_0 > 0,$$
$$e = \omega_{ropt} - \omega_r. \qquad (47)$$

From (45), the new input w is defined as

$$w = \dot{\omega}_{ropt} + a_0 e. \qquad (48)$$

By substituting w in (46), we get the final control law for the WT two-mass model:

$$T_{em} = \frac{T_a}{n_g} - \frac{K_r}{n_g}\omega_r - J_g\dot{\omega}_g - K_g\omega_g$$
$$- \frac{J_r}{n_g}\left(\dot{\omega}_{ropt} + a_0 e\right). \qquad (49)$$

6. Validation Results

CARTs (Control Advanced Research Turbines) are located in the center of the national wind NREL (National Renewable Energy Laboratory), near Golden, Colorado. The CART3 is a three-bladed variable speed and variable pitch wind turbine and has a rating of 600 kW. It mainly consists of three parts, namely, the rotor, the tower, and the nacelle. The generator is connected to the grid through power electronics that can directly control generator torque [16]. The power electronics consist of three-phase PWM (Pulse Width Modulation) converters with a constant dc link voltage. The main objective of the grid side converter is to maintain the dc link voltage constant [17, 18].

6.1. Simulation Using FAST Model. FAST was developed by the NREL; it is used for WT aeroelastic simulator. The modelling of two- and three-blade horizontal axis wind turbines (HAWT) is obtained by FAST. This FAST code can be able to predict extreme and fatigue loads. Tower and flexible blade server are modelled by "assumed mode method." WT loads are calculated by using BEM (Blade Element Momentum) and multiple component of wind speed profile [19]. FAST code is approved by the Germanischer Lioyd (GL) WindEnergie GmbH for calculating onshore WT loads for design and certification [20]. Due to the above advantages and exact nonlinear modeling of the WT, the proposed controllers are validated by using FAST. In general,

three-blade turbines have 24 DOF (degrees of freedom) to represent the wind turbine dynamics. In this work, 3 DOF are considered for WT, that is, generator, rotor speed, and blade teeter. FAST codes are interfaced with S-function and implemented with Simulink model. FAST uses an AeroDyn file as an input for aerodynamic part. AeroDyn file contains aerodynamic analysis routine and it requires status of a WT from the dynamic analysis routine and returns the aerodynamic loads for each blade element to the dynamic routine [21]. Wind profile acts as the input file for AeroDyn. The wind input file is generated by using TurbSim which is developed by the NREL. The test wind profile with full field turbulence is generated by using TurbSim developed by NREL.

Figure 6 shows the hub height wind speed profile. In general, any wind speed consists of two components, that is, mean wind speed and turbulence component. The test wind speed consists of 10 min dataset that was generated using Class A Kaimal turbulence spectra. It has the mean value of 7 m/s at the hub height, turbulence intensity of 25%, and normal IEC (International Electrotechnical Commission) turbulence type. The above wind speed is used as the excitation of WT.

The proposed and conventional controllers are implemented using FAST interface with MATLAB Simulink. The main objectives of the controllers are to maximize the energy capture with reduced stress on the drive train. The efficiency of the controllers is compared by using the following terms, that is, aerodynamic (η_{aero}) and electrical (η_{elec}) efficiency given in the following:

$$\eta_{aero}(\%) = \frac{\int_{t_{ini}}^{t_{fin}} P_a dt}{\int_{t_{ini}}^{t_{fin}} P_{a_{opt}} dt},$$
$$\eta_{elec}(\%) = \frac{\int_{t_{ini}}^{t_{fin}} P_e dt}{\int_{t_{ini}}^{t_{fin}} P_{a_{opt}} dt}, \qquad (50)$$

where $P_{a_{opt}} = 0.5\rho\pi R^2 C_{P_{opt}}$ is the optimal aerodynamic power for the wind speed profile. The following objectives are used to measure the performance of the controllers:

(1) Maximization of the power capture is evaluated by the aerodynamic and electrical efficiency which is defined in (50).

(2) The reduced oscillation on the drive train and control torque smoothness are measured by the STD (standard deviation) and maximum value.

The abovementioned values for all the controllers are given in Table 2. The rotor speed comparisons for FAST simulator are shown in Figures 7 and 8. The conventional controllers such as ATF and ISC are not able to track the optimal reference speed. ATF has only single tuning parameter, that is, K_c, which allows reducing the steady state error. In ISC, during fast transient wind speed, it introduces more power loss. Moreover, these controllers are not robust with respect to high turbulence wind speed profile. To overcome the above drawbacks, TSMC and MNSSFE are proposed.

TABLE 2: Comparison of different control strategies based on two-mass model using FAST simulator.

Control strategy	ISC	ATF	MNSSFE	TSMC
STD (T_{ls}) (kNm)	9.629	23.03	23.13	16.00
Max (T_{ls}) (kNm)	45.62	130.8	136.7	107.81
STD (T_{em}) (kNm)	0.142	0.369	0.280	0.246
Max (T_{em}) (kNm)	1.010	2.500	1.807	1.835
η_{elec} (%)	69.73	72.87	76.23	74.81
η_{aero} (%)	85.59	85.06	94.67	94.36

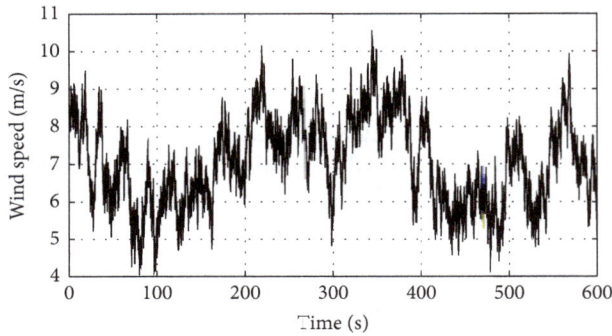

FIGURE 6: Test wind speed profile.

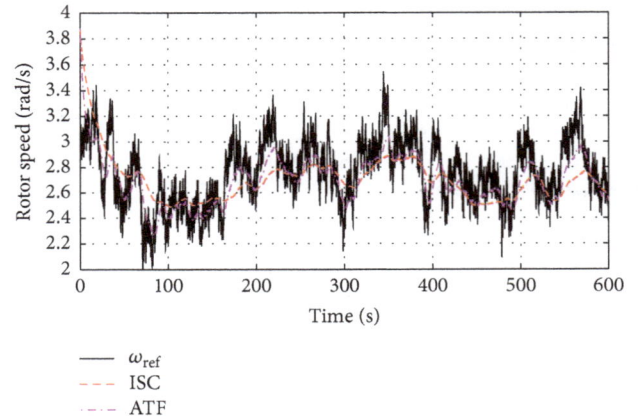

FIGURE 7: Rotor speed comparison for ATF and ISC for FAST simulator.

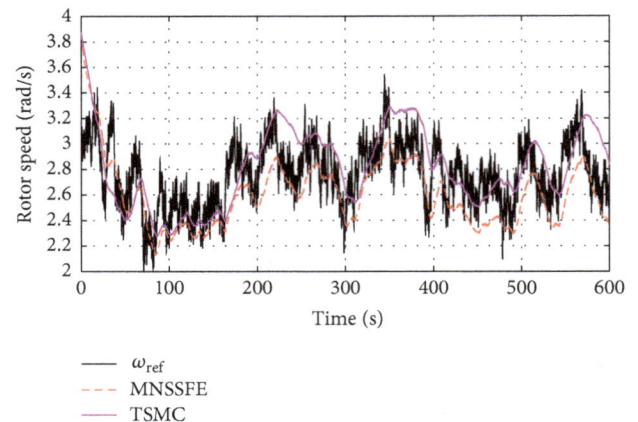

FIGURE 8: Rotor speed comparison for TSMC and MNSSFE for FAST simulator.

Figure 8 shows the rotor speed comparisons for MNSSFE and TSMC. From this figure, it is clear that at initial wind condition 0–30 sec both the controllers are not able to track the optimal rotor speed due to the initial setting in the AeroDyn input file. At high wind speed variations 220, 350, and 550 sec, the TSMC is almost tracking the reference rotor speed compared to MNSSFE. Except MNSSFE and TSMC, all the other controllers are having more tracking error in rotor speed. To achieve more power capture, the rotor speed should closely track the optimal rotor speed.

Table 2 gives the performance analysis of all the conventional and proposed controllers. From Table 2, it is clear that the STD of T_{em} and T_{ls} is the lowest for TSMC and the highest for ATF controller. This ensures that the smoothness of the control input in TSMC is better compared to other controllers. ISC has very less STD of T_{em} and T_{ls}; at the same time, the efficiency is very low compared to all the controllers. Also the ISC control only depends on the generator speed which is not accurate and for higher variation in wind speed and introduces significant power loss. So a trade-off should be made between the efficiency and the fatigue load on drive train. Except TSMC, ATF and MNSSFE are having more standard deviation which ensures more drive train transient load. Compared to ISC and ATF, the aerodynamic and electrical efficiency of the proposed TSMC and MNSSFE are better.

To analyze the controller performances in a more detailed fashion, Figures 9 and 10 show the box plot for low speed shaft torque and generator torque with the mean, median, ±25% quartiles (notch boundaries), ±75% quartiles (box ends), ±95% bounds, and the outliers. From the size of the boxes

shown, it is clear that the ISC experiences minimum variation compared to others. It ensures that ISC has the minimum transient load on the drive train; at the same time, we can find from Table 2 that the efficiency of ISC is not comparable with other controllers. Comparing the box plot of TSMC and MNSSFE, with TSMC having less variation in low speed shaft torque and generator torque, this indicates smoothness of the controller and reduction in transient load.

Figure 11 shows the box plot for rotor speed for FAST simulator. From this figure, it is clear that TSMC has almost the same variation as the variation in the reference rotor speed. It is observed that, apart from TSMC and MNSSFE, other controllers such as ATF and ISC are having more variations with respect to reference speed. This indicates that, for ATF and ISC, the obtained rotor speed is not able to track the reference rotor speed.

The frequency analysis is carried out by using the PSD on the low speed shaft torque which is shown in Figure 12. As the MNSSFE plot is completely above the TSMC plot, it is clear that low speed shaft torque variation is more

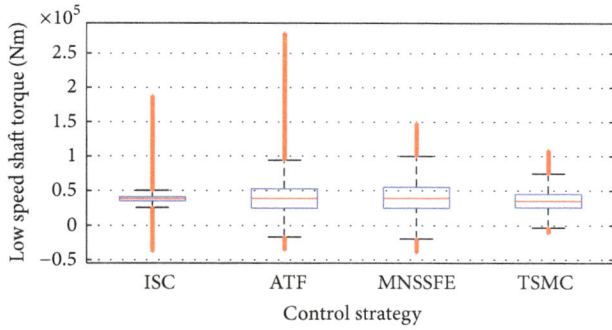

FIGURE 9: Box plot for low speed shaft torque using FAST simulator.

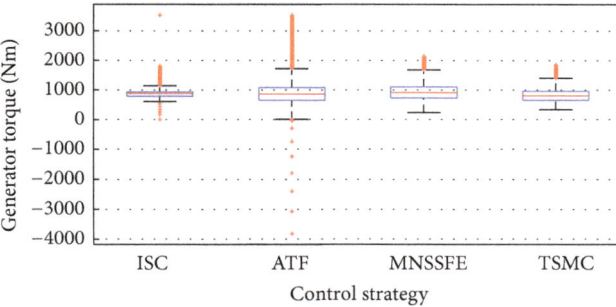

FIGURE 10: Box plot for generator torque using FAST simulator.

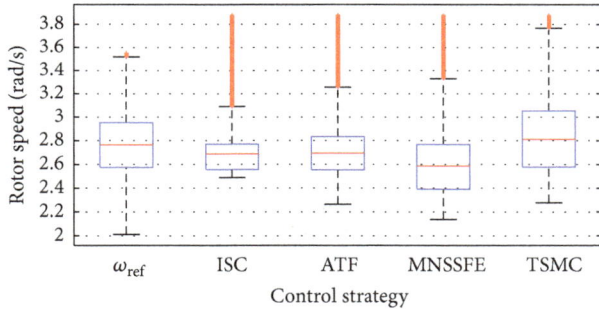

FIGURE 11: Box plot for rotor speed using FAST simulator.

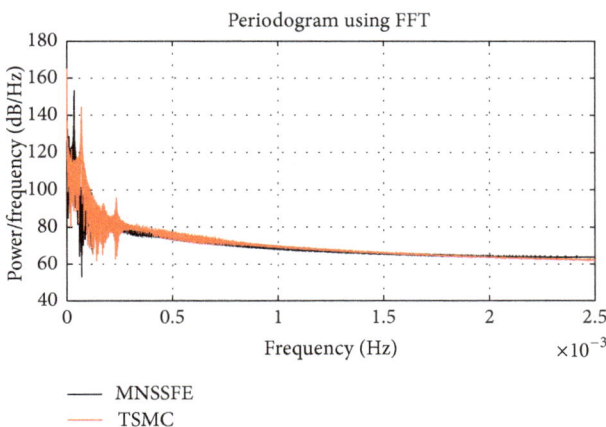

FIGURE 12: PSD for low speed shaft torque using FAST simulator.

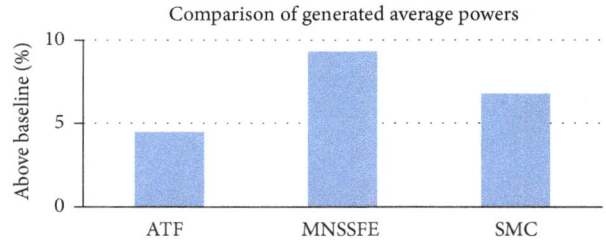

FIGURE 13: Comparison for baseline control with other controllers for generated average power.

for MNSSFE than TSMC. This ensures that TSMC gives minimum excitation to the drive train.

As shown in Figure 13, the MNSSFE controller has improved power capture by 2.53% compared to TSMC. Fast tracking introduces more variation in control input and drive train. So an intermediate tracking has to be chosen and a compromise has been made between efficiency and load mitigation. From this analysis, even though MNSSFE gives a little better efficiency than TSMC, by considering transient load on drive train and smooth control input, TSMC is found to be optimal.

In order to avoid the torsional resonance mode by choosing the proper tracking dynamics, a trade-off is made between power capture optimization and reduced transient load on low speed shaft torque. A good dynamic tracking, that is, similar to WT fast dynamics, gives better power capture but it requires more turbulence in control torque. Conversely slow tracking gives smooth control action with less power capture. Therefore, a compromise should be made between the power capture and transient load reduction. The better optimal speed tracking leads to better power capture for TSMC controller. The simulations are performed with different wind speed profiles with the mean wind speed at below rated wind speed. The results are given in Tables 3 and 4. From these tables, it is observed that, with an increase in mean wind speed, the maximum value of the control input (T_{em}) also increases. In all the cases, both TSMC and MNSSFE controllers are having almost the same efficiency but the transient load reduction is better for TSMC. As the mean wind speed increases, the standard deviation also increases for MNSSFE compared with TSMC. It is observed that when the wind speed undergoes high variation the TSMC can be able to produce better power capture with reduced transient load on the drive train.

Figure 14 shows the electrical power comparison for industrial baseline controller, with MNSSFE and TSMC. From this figure, it is observed that industrial baseline controller has more oscillation compared to MNSSFE and TSMC. Both MNSSFE and TSMC are almost having the same power. From Table 5, it is found that industrial baseline has more oscillation in control torque compared with other controllers.

Figure 15 shows the rotor speed comparison for MNSSFE and TSMC with constant additive disturbance of $1\,\mathrm{kNm}/n_g$.

TABLE 3: TSMC performance for different wind speed profiles.

Mean wind speed (m/sec)	Electrical efficiency (%)	T_{ls} standard deviation kNm	Max (T_{em}) kNm
7 (m/sec)	74.81	16.00	1.835
8 (m/sec)	73.51	16.83	1.683
8.5 (m/sec)	73.33	13.34	1.955

TABLE 4: MNSSFE performance for different wind speed profiles.

Mean wind speed (m/sec)	Electrical efficiency (%)	T_{ls} standard deviation kNm	Max (T_{em}) kNm
7 (m/sec)	76.23	23.13	1.807
8 (m/sec)	74.85	23.35	1.995
8.5 (m/sec)	74.52	23.58	2.076

TABLE 5: Performance comparison for MNSSFE and TSMC with industrial baseline controller.

	Baseline	MNSSFE	TSMC
STD (T_{em}) Nm	565.00	529.96	502.89
η_{ele} (%)	73.72	77.82	75.73

TABLE 6: Additive disturbance performance comparison for MNSSFE and TSMC.

	MNSSFE	TSMC
STD (T_{em}) kNm	0.279	0.253
STD (T_{ls}) kNm	23.06	17.97
η_{ele} (%)	75.68	75.72

From this figure, it is clear that both controllers are robust with respect to disturbance. With reference to Table 6, the STD is less for TSMC as compared to MNSSFE; at the same time, the efficiency of these controllers is similar.

7. Conclusion

This paper deals with the problem of controlling the maximum power generation at below rated wind speed of VSWT. The objective is to design a robust controller that maximizes the energy extraction from the wind while reducing the transient loads. For the above purpose, two nonlinear controllers, that is, TSMC and MNSSFE, which have the ability to reject disturbance and accommodate parameter uncertainty, are proposed in this study. Finally, it is concluded that a trade-off is to be maintained between the efficiency and mechanical stress on the drive train. The performances of these controllers are compared with the conventional ATF and ISC using FAST aeroelastic simulator. The proposed controllers are found to produce satisfactory results in achieving the control objectives.

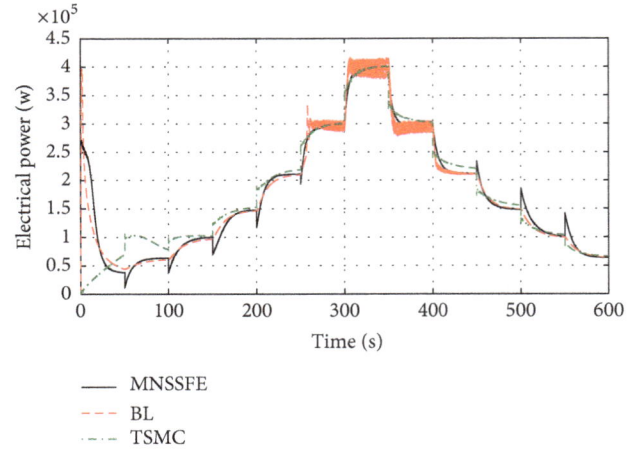

FIGURE 14: Comparison for baseline control with other controllers for electrical power.

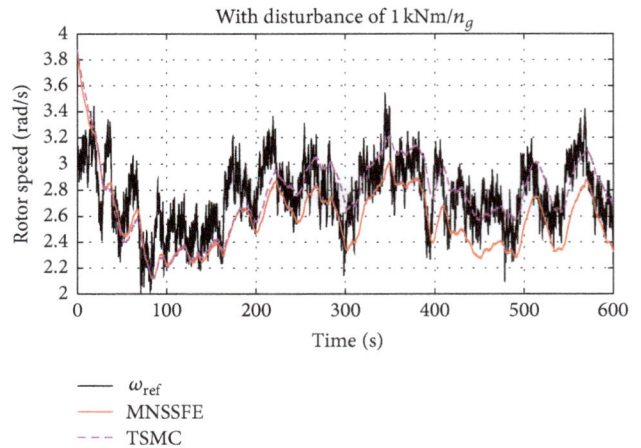

FIGURE 15: Rotor speed comparison for MNSSFE and TSMC with constant additive disturbance of $1\,\text{kNm}/n_g$.

Nomenclature

B_{ls}: Low speed shaft stiffness (N·m·rad^{-1})
$C_P(\lambda, \beta)$: Power coefficient
$C_q(\lambda, \beta)$: Torque coefficient
J_g: Generator inertia (kg·m^2)
J_r: Rotor inertia (kg·m^2)
K_g: Generator external damping (N·m·rad^{-1}·s^{-1})
K_{ls}: Low speed shaft damping (N·m·rad^{-1}·s^{-1})
K_r: Rotor external damping (N·m·rad^{-1}·s^{-1})
n_g: Gearbox ratio
P_a: Aerodynamic power (W)
P_e: Electrical power (W)
R: Rotor radius (m)
T_a: Aerodynamic torque (N·m)
T_{em}: Generator (electromagnetic) torque (N·m)
T_{hs}: High speed shaft torque (N·m)
T_{ls}: Low speed shaft torque (N·m).

Conflict of Interests

The authors declare that there is no conflict of interests regarding the publication of this paper.

References

[1] T. Burton, D. Sharpe, N. Jenkins, and E. Bossanyi, *Wind Energy Handbook*, Wiley Publications, New York, NY, USA, 2001.

[2] F. D. Bianchi, F. D. Battista, and R. J. Mantz, *Wind Turbine Control Systems: Principles, Modelling and Gain Scheduling Design*, Springer, Buenos Aires, Argentina, 2nd edition, 2006.

[3] M. Sheikhan, R. Shahnazi, and A. N. Yousefi, "An optimal fuzzy PI controller to capture the maximum power for variable-speed wind turbines," *Neural Computing and Applications*, vol. 23, no. 5, pp. 1359–1368, 2013.

[4] C.-M. Hong, C.-H. Chen, and C.-S. Tu, "Maximum power point tracking-based control algorithm for PMSG wind generation system without mechanical sensors," *Energy Conversion and Management*, vol. 69, pp. 58–67, 2013.

[5] I. Kortabarria, J. Andreu, I. M. de Alegría, J. Jiménez, J. I. Gárate, and E. Robles, "A novel adaptative maximum power point tracking algorithm for small wind turbines," *Renewable Energy*, vol. 63, pp. 785–796, 2014.

[6] J. Mérida, L. T. Aguilar, and J. Dávila, "Analysis and synthesis of sliding mode control for large scale variable speed wind turbine for power optimization," *Renewable Energy*, vol. 71, pp. 715–728, 2014.

[7] V. Calderaro, V. Galdi, A. Piccolo, and P. Siano, "A fuzzy controller for maximum energy extraction from variable speed wind power generation systems," *Electric Power Systems Research*, vol. 78, no. 6, pp. 1109–1118, 2008.

[8] B. Boukhezzar, H. Siguerdidjane, and M. Maureen Hand, "Nonlinear control of variable-speed wind turbines for generator torque limiting and power optimization," *Journal of Solar Energy Engineering*, vol. 128, no. 4, pp. 516–530, 2007.

[9] B. Boukhezzar and H. Siguerdidjane, "Nonlinear control of a variable-speed wind turbine using a two-mass model," *IEEE Transactions on Energy Conversion*, vol. 26, no. 1, pp. 149–162, 2011.

[10] B. Beltran, T. Ahmed-Ali, and M. E. H. Benbouzid, "High-order sliding-mode control of variable-speed wind turbines," *IEEE Transactions on Industrial Electronics*, vol. 56, no. 9, pp. 3314–3321, 2009.

[11] M. Liao, L. Dong, L. Jin, and S. Wang, "Study on rotational speed feedback torque control for wind turbine generator system," in *Proceedings of the International Conference on Energy and Environment Technology (ICEET '09)*, vol. 1, pp. 853–856, Guilin, China, October 2009.

[12] H. M. Amine, H. Abdelaziz, and E. Najib, "Wind turbine maximum power point tracking using FLC tuned with GA," *Energy Procedia*, vol. 62, pp. 364–373, 2014.

[13] S. Rajendran and D. Jena, "Control of variable speed variable pitch wind turbine at above and below rated wind speed," *Journal of Wind Energy*, vol. 2014, Article ID 709128, 14 pages, 2014.

[14] H. Vihriala, R. Perela, P. Makila, and L. Soderlund, "A gearless wind power drive: part 2: performance of control system," in *Proceedings of the Wind Energy for the New Millennium European Conference (EWCE '01)*, pp. 1090–1093, Copenhagen, Denmark, July 2001.

[15] S. Mondal and C. Mahanta, "Adaptive second order terminal sliding mode controller for robotic manipulators," *Journal of the Franklin Institute*, vol. 351, no. 4, pp. 2356–2377, 2014.

[16] L. J. Fingersh and K. Johnson, "Controls advanced research turbine (CART) commissioning and baseline data collection," Tech. Rep., National Renewable Energy Laboratory (NREL), 2002.

[17] R. Ottersten, *On control of back-to-back converters and sensorless induction machine drives [Ph.D. thesis]*, Chalmers University of Technology, Gothenburg, Sweden, 2003.

[18] R. Peña, R. Cardenas, R. Blasco, G. Asher, and J. Clare, "A cage induction generator using back to back PWM converters for variable speed grid connected wind energy system," in *Proceedings of the 27th Annual Conference of the IEEE Industrial Electronics Society (IECON '01)*, vol. 2, pp. 1376–1381, IEEE, Denver, Colo, USA, December 2001.

[19] M. O. L. Hansen, J. N. Sørensen, S. Voutsinas, N. Sørensen, and H. A. Madsen, "State of the art in wind turbine aerodynamics and aero elasticity," *Progress in Aerospace Sciences*, vol. 42, no. 4, pp. 285–330, 2006.

[20] A. Manjock, "Design codes FAST and ADAMS for load calculations of onshore wind turbines," Tech. Rep., National Renewable Energy Laboratory (NREL), Golden, Colo, USA, 2005.

[21] D. J. Laino and A. C. Hansen, "User's guide to the wind turbine aerodynamics computer software aerodyn," Tech. Rep., National Wind Technology Center, 2003.

19

Comparison and Optimization of Neural Networks and Network Ensembles for Gap Filling of Wind Energy Data

Andres Schmidt[1] and Maya Suchaneck[2]

[1] Department of Forest Ecosystems and Society, Oregon State University, Corvallis, OR 97331, USA
[2] Department of Geography, Ruhr University Bochum, 44780 Bochum, Germany

Correspondence should be addressed to Andres Schmidt; andres.schmidt@oregonstate.edu

Academic Editor: Shuhui Li

Wind turbines play an important role in providing electrical energy for an ever-growing demand. Due to climate change driven by anthropogenic emissions of greenhouse gases, the exploration and use of sustainable energy sources is essential with wind energy covering a significant portion. Data of existing wind turbines is needed to reduce the uncertainty of model predictions of future energy yields for planned wind farms. Due to maintenance routines and technical issues, data gaps of reference wind parks are unavoidable. Here, we present real-world case studies using multilayer perceptron networks and radial basis function networks to reproduce electrical energy outputs of wind turbines at 3 different locations in Germany covering a range of landscapes with varying topographic complexity. The results show that the energy output values of the turbines could be modeled with high correlations ranging from 0.90 to 0.99. In complex terrain, the RBF networks outperformed the MLP networks. In addition, rare extreme values were better captured by the RBF networks in most cases. By using wind meteorological variables and operating data recorded by the wind turbines in addition to the daily energy output values, the error could be further reduced to more than 20%.

1. Introduction

The Combination of climate change and the dependence on fossil fuels slowly cause changes in energy policy and trigger an increasing demand for sustainable energy sources. Global carbon dioxide emissions are ever increasing and the associated consequences for the climate are widely scientifically recognized [1–3]. Over the last decade and in particular since the release of the report of the Intergovernmental Panel on Climate Change (IPCC) in 2007, public and political awareness of renewable energy technologies has increased considerably. This is not at least due to the large and fast growing economies and the associated increase of numbers of cars and energy consumption, and therefore of CO_2 emissions [4]. Wind energy has the potential to be a vital contributor to renewable energy technologies that will substitute more and more for gas and coal [5].

In order to decrease the uncertainty of wind energy yield predictions during the planning of a single turbine or wind farm, data of nearby existing wind turbines are often used as reference for model evaluation.

In Germany, the legislation that grants priority to renewable energy sources (Renewable Energy Resources Act, EEG) states that only wind turbines in areas with sufficient wind energy potential are qualified to receive compensation for the electrical power provided for the power grid [6].

Moreover, according to the EEG, power grid owners are not required to connect turbines to the grid that does not meet or exceed 60% of the turbine type-specific reference value calculated based on reference wind conditions.

Because of these restrictions, a correct prediction for expected wind energy yield is an indispensable economic criterion for most wind farm projects. However, because of technical malfunctions, maintenance routines, or other problems, the availability of valuable comparison data from existing turbines nearby is limited. Such data limitation affects the statistical safety of model predictions of future energy yields.

Artificial neural networks are able to approximate nonlinear relationships between individual data series by adjusting network parameters in a purely data-driven, and, in our case, supervised learning process.

Here, we present a method to model the data of nearby wind turbines using different types of neural networks and network ensembles to fill in gaps in time series of wind energy outputs. Real-world operating data of six exemplary wind farms in Germany were available for this purpose.

2. Data and Methods

To provide a high planning dependability, the potential annual wind energy yields have to be estimated carefully during the planning process of a wind turbine. The accuracy of predictions becomes even more crucial when wind farms are planned due to the substantially increased financial risk. Several microscale and mesoscale wind flow models based on computational fluid dynamics are available to calculate future wind energy yields of single turbines and wind farms. Typically, terrain data such as surface roughness, orography, and existing wind obstacles for a radius of 20 km around the proposed location of the wind turbine is considered for the model of computations in combination with long-term wind statistics.

One widely used model, for instance, that has passed several stages of development over the last decades is *WAsP* (Wind Atlas Analysis and Application Program) [7]. *WAsP* was developed by the *RISØ* National Laboratory, Roskilde, Denmark, and is approved among others by the German Federal Research Ministry.

The various specific models that account for wind obstacles, orography, wake effects, and surface roughness were assembled in the model suite of *WAsP*. A detailed description of the model and its algorithms can be found in the respective literature [8].

In order to validate the predicted long-term average wind energy yields based on the model results considering the environmental conditions, the geometry of the turbine and its power curve [9] data of existing, nearby turbines are used as a reference. For that purpose, energy output values of existing wind turbines are corrected for technical availability. In addition, the wind conditions and the corresponding observed electrical energy output values of a certain year are compared and scaled to long-term data for the area of interest [10]. After correcting the observed values through linear regression to 100% long-term averages, the values can be compared to the model results that are based on long-term wind statistics and therefore also represent long-term averages [11].

2.1. Artificial Neural Networks. Artificial neural networks provide a method to map input variables on target variables by using a combination of nonlinear functions and a learning procedure that can be supervised or unsupervised [12, 13]. The ability to find mathematical functions without prior knowledge of the functional relationship makes neural networks a powerful tool to solve problems for which no analytical solution exists or if the function that relates variables to each other is unknown [14–16]. Artificial neural networks have found a growing range of applications in recent years including the field of wind energy research [17, 18]. The ability to self-adjust its parameters makes neural networks a superior fitting method compared to classical data-fitting and prediction methods [19–21].

The most popular network architecture currently used is the so-called multilayer perceptron (MLP) topology, which was presented comprehensively first by Rumelhart et al. [22].

Within this study we compare the performance of MLP networks to the performance of radial basis function (RBF) neural networks. Compared to MLP networks, the parameters of RBF networks can be adjusted faster to the data presented to the network during the training process. In addition, RBF networks are less affected by the problem of local minima [23]. MLP networks use the scalar product of the input data vector consisting of n input variables and a weight vector to calculate the neuron output usually applying hyperbolic activation functions. In contrast to that, RBF networks use the distance between the input vector and the center of the radial basis function to determine the activation value within an RBF neuron. The most commonly used activation function type with a radial basis is the Gaussian function that was also deployed in the networks used for this study. The n-dimensional vectors that determine the shape of each neuron's n-dimensional Gauss function in the input space are defined by the center vector μ and the n-dimensional variance vector σ.

The central values of μ have been optimized within this work using the k-means clustering algorithm. The input vectors were separated into K clusters and the center values of the RBF neurons were set to represent all considered input vectors while minimizing the number of clusters by finding the centers for each cluster with the smallest mean squared distance to all points in the cluster. According to Hestenes [24], this can be calculated by minimizing the function J as given in ((1) and (2)). Consider

$$J = \sum_{j=1}^{K} \sum_{i\varepsilon S_j} \left\| x_{ij} - \mu_j \right\|^2, \tag{1}$$

with

$$\mu_j = \frac{1}{N_j} \sum_{i\varepsilon S_j} x_j. \tag{2}$$

Here, μ_j is the mean of subsample S_j which the cluster j is composed of. The index i is the index over the subsample S_j. At the end of the procedure, each input vector x is assigned to the cluster center (i.e., RBF network node) to which it has the least Euclidean distance. The closer an input vector to the RBF center of a neuron is, the higher the activation value of that neuron is. Hence, the parameters of all RBF functions are adjusted during the network learning phase so that every input can be assigned to one of the RBF neurons in the hidden layer and the weighted sum of the activation functions can be transformed to satisfyingly match the target values. More details about the large field of learning algorithms for neural networks can be found in the literature [12, 13, 16].

For the network training and evaluation process, each available learning dataset was divided into 3 different dataset sections: training data, test data, and validation data. The training data is transformed by the network functions. After the output of the network is compared to the real available measured outputs, the error is determined. The error value used for this performance check is the final RMSE (root of the mean squared error) summing up the differences between modeled output and known measured output according to

$$E = \sqrt{\frac{1}{N}\sum_{j=1}^{N}\left(m_j - y_j\right)^2}. \tag{3}$$

Here, m_j is the measured result within the available learning dataset (i.e., data that contains all values for all input variables and measured samples of the corresponding target value) and y_i is the value estimated by the neural network. N is the number of data records used to determine the error E for the training data and test or validation data, respectively.

The parameters of the network are, then, iteratively adjusted applying the conjugate gradient descent method for the training algorithm [12, 24] until an acceptable error is reached.

In a second step the test data is used. Test data was not used for the network parameter adjustment during the training. The trained network is now applied to the test dataset and the RMSE is determined again.

If the error for the training data is small and the error for the test data is large, this indicates that the network parameters have been over adjusted to the training data. Hence, the training has to be started again until minima for both, the training error and the test error, have been reached.

Once the optimal balance between training data error and test data error is reached, the network has the ability to sufficiently generalize the functional relationship between input and target variable well and is not over-fitted to a specific training dataset.

The third dataset section, the validation data, has not been used for the learning process of the network at all and is processed with the adjusted neural network in a final step. Hence, the validation error is the only of the three error values that is admissible to assess the goodness of a neural network model. The division of the data into the training data, test data, and validation data was conducted according to values presented in the literature [25, 26]. The validation error indicates the final performance of the trained network when applied on new data and is therefore given in the results presented.

In order to scale all input variables to values between 0 and 1 in a preliminary step we applied min-max normalization to prevent the network results from being biased by the stronger numerical influence of a variable measured in units with larger numbers than a variable that is limited to a smaller scale by default [27]. For each dataset, we trained specific neural networks and determined the three best networks. For that purpose, a batch algorithm was conducted to test 20,000 networks for each dataset while varying the number of neurons, number of input variables, network type, and activation functions in the nodes.

To demonstrate the potential of the method, three pairs of turbine sites were selected with different large-scale wind conditions, whereupon the wind conditions amongst the pairs are similar due to the relative spatial adjacency.

2.2. Real-World Case Studies with Operating Data of Exemplary Wind Farms. The daily energy output W of existing wind turbines provide important information and are used as input data for the neural networks. Also, the mean values, minima and maxima of wind speed, wind direction, instantaneous power, and number of rotor revolutions recorded by the operating and surveillance software were taken into account for the neural network calculation. The three exemplary test areas are located in Germany where the prevailing wind conditions are dominated by the global West wind drift of the temperate latitudes in the Northern hemisphere.

The locations were selected with respect to their different orographic features to validate the robustness of the method presented.

It is a coastal location, a site in the *Muensterland* lowland area in Northwest Germany, and an area located in the mountainous forested uplands of West central Germany (Figure 1).

Overall, the data from 21 modern gearless wind turbines of the manufacturer Enercon (ENERCON GmbH, 26605 Aurich, Germany) recorded over a period of four years were available for the calculations.

3. Results and Discussion

In the following section the results for 3 case studies are presented, consecutively with increasing topographic complexity. In each case, we provide detailed information about turbine types, wind conditions, and topographic circumstances that affect the local wind fields for the distinct types of terrain.

In each case, specific neural networks were trained to reproduce the electrical energy output using the data of existing wind turbines with various distances to the respective target turbines and wind farms. The performances of the best trained networks are presented and errors values are given and compared.

3.1. The Coastal Sites. Due to the flat landscape with no significant changes in elevation, orographic effects on the wind speed are negligible around the coastal sites of *Hinte* and *Jennelt.*

The site belongs to the district of the city of Aurichin the East Frisia region in Lower Saxony, Germany. The North Sea is located in a distance of about 12 km to the West of the site *Hinte.* The nearest city Emden is located at a distance of 5.8 km South of the wind farm *Hinte.* The immediate vicinity of the site is characterized by meadows and agriculturally used land. The terrain is flat and mostly free of wind obstacles.

The wind farm is located 1.7 km Northwest of *Hinte.* The wind farm consists of a total of 15 wind turbines operated by different companies with an average annual energy output of

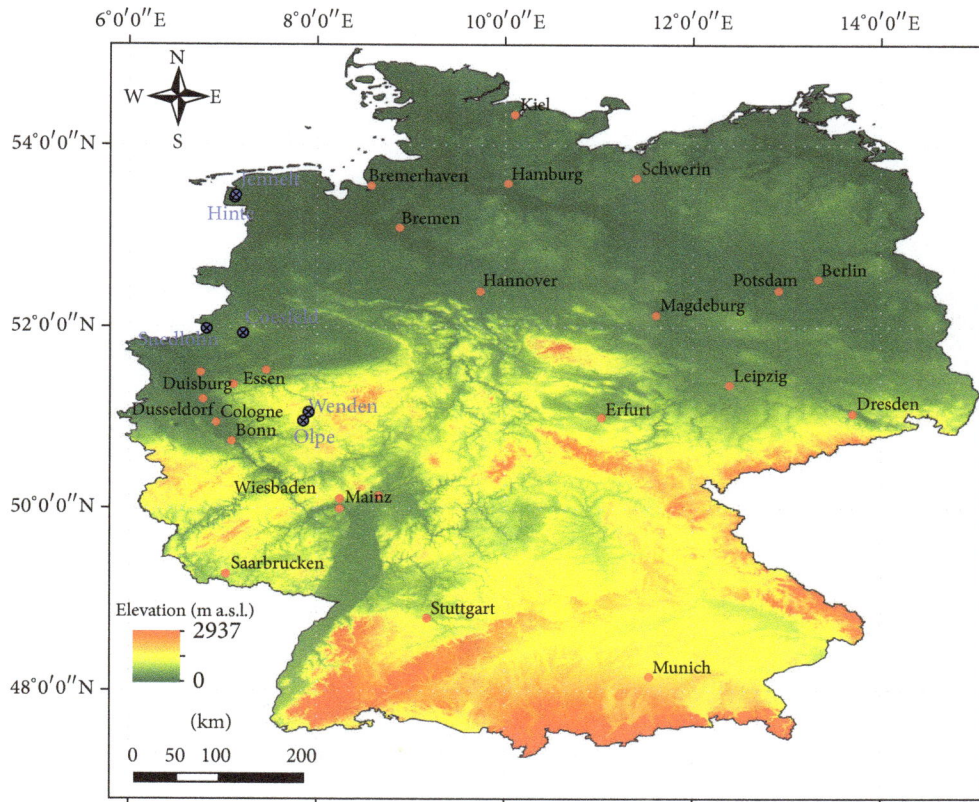

FIGURE 1: Shaded relief overview map of the incorporated wind farm sites in Germany. The elevations shown are based on the NASA Shuttle Radar Topography Mission data.

TABLE 1: Coordinates and data of the wind turbines at the site *Hinte*.

Turbine	Type	Hub height [m]	Rotor Ø [m]	Nominal power [kW]	Longitude (WGS 84)	Latitude	Elevation (m a.s.l.)
HI 1	E 66/18.70	65.0	70.0	1800	7.170441E	53.420212N	0
HI 2	E 66/18.70	65.0	70.0	1800	7.166404E	53.424987N	0
HI 3	E 66/18.70	65.0	70.0	1800	7.166359E	53.422247N	0

TABLE 2: Coordinates and data of the wind turbines at the site *Jennelt*.

Turbine	Type	Hub height [m]	Rotor Ø [m]	Nominal power [kW]	Longitude (WGS 84)	Latitude	Elevation (m a.s.l.)
JE 1	E 66/18.70	65.0	70.0	1800	7.137170E	53.448060N	2
JE 2	E 66/18.70	65.0	70.0	1800	7.132560E	53.456434N	2
JE 3	E 66/18.70	65.0	70.0	1800	7.139583E	53.456116N	1
JE 5	E 66/18.70	65.0	70.0	1800	7.139436E	53.452882N	1
JE 5	E 66/18.70	65.0	70.0	1800	7.125192E	53.452810N	2

47 GWh. The specifications and geographical coordinates of the turbines used for the calculations are given in Tables 1 and 2, respectively.

Due to ongoing expansions and changes, the wind farm information given in this study always refer to the turbines with data incorporated in the analyses and do not necessarily represent the current total number of installed turbines.

The wind farm *Jennelt* is located at 4.1 km to the Northwest of the wind farm *Hinte*. The North Sea lies West of the site in a distance of 7.4 km.

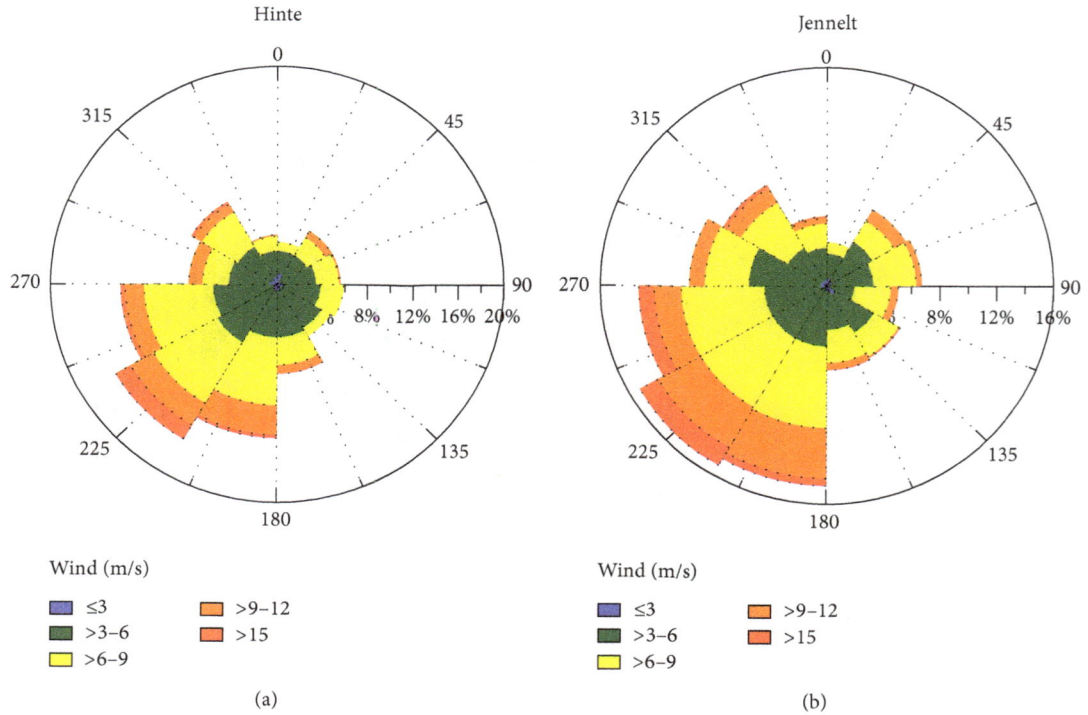

FIGURE 2: Average wind conditions at the coastal sites in *Hinte* (target site) and *Jennelt* (input site) during the observation period from January 2006 to January 2008.

The data collection period for the site pair *Hinte/Jennelt* is two years, from January 1, 2006, to December 31, 2008. Thus, 732 data records were available for the two sites.

The wind conditions at the sites *Hinte* and *Jennelt* measured by anemometers on the nacelle of the wind turbines (i.e., 67 m above ground level) were similar with regard to the wind speeds and the distribution of wind directions during the data collection period (Figure 2). Even though the wind measurements on the nacelle of the plants are subjected to certain errors by flow distortions, they are still sufficiently accurate to capture the speed and direction for most applications [28].

Comparisons between the wind speeds measured on the nacelle anemometer and an undisturbed measurement of wind in front of the rotor show only minor deviations of 2% and less [29].

600 daily data records were used for the training of the neural networks. The remaining records were used to the same parts for the test dataset and the validation dataset.

The nonparametric Spearman rank correlation coefficient r_s was used for the comparison of the measured electrical energy output W and the results reproduced by the neural networks. The RMSE values given refer to the validation data set that was composed of values that were randomly distributed over the entire dataset. In Table 3, the results for the three best networks are shown, that is, the networks with the lowest validation errors and highest correlation coefficients.

TABLE 3: Summary of results for the energy output of the target turbines at the coastal site *Hinte*.

Target variable	Input variables	Network type	Network topology	RMSE (kWh)	r_S
$W_{HI\,1}$	$W_{(JE\,1-JE\,5)}$	MLP	5-6-1	1326	0.98
$W_{HI\,1}$	$W_{(JE\,1-JE\,5)}$	RBF	5-44-1	1312	0.99
$W_{HI\,1}$	$W_{(JE\,1-JE\,5)}$	RBF	5-80-1	1458	0.99
$W_{HI\,2}$	$W_{(JE\,1-JE\,5)}$	MLP	5-10-1	1794	0.99
$W_{HI\,2}$	$W_{(JE\,1-JE\,5)}$	RBF	5-51-1	1911	0.99
$W_{HI\,2}$	$W_{(JE\,1-JE\,5)}$	RBF	5-53-1	2208	0.98
$W_{HI\,3}$	$W_{(JE\,1-JE\,5)}$	MLP	5-6-1	2256	0.98
$W_{HI\,3}$	$W_{(JE\,1-JE\,5)}$	RBF	5-42-1	2244	0.98
$W_{HI\,3}$	$W_{(JE\,1-JE\,5)}$	RBF	5-80-1	2330	0.98

All correlations given are significant on a 95% confidence level ($P \leq 0.05$). The network topology refers to the number of input variables, the number of hidden nodes, and the number of output nodes of the neural networks used. The number of hidden neurons within RBF networks is usually higher than for the MLP type networks to accomplish the same ability of generalization [13, 30].

The differences in network performance are overall small. All networks achieved very high correlations of 0.98 and 0.99 and the statistics for the measured and reproduced power output data are similar (Tables 3 and 4).

Figure 3 shows the observed and network reproduced frequency distributions of the daily values of W for the

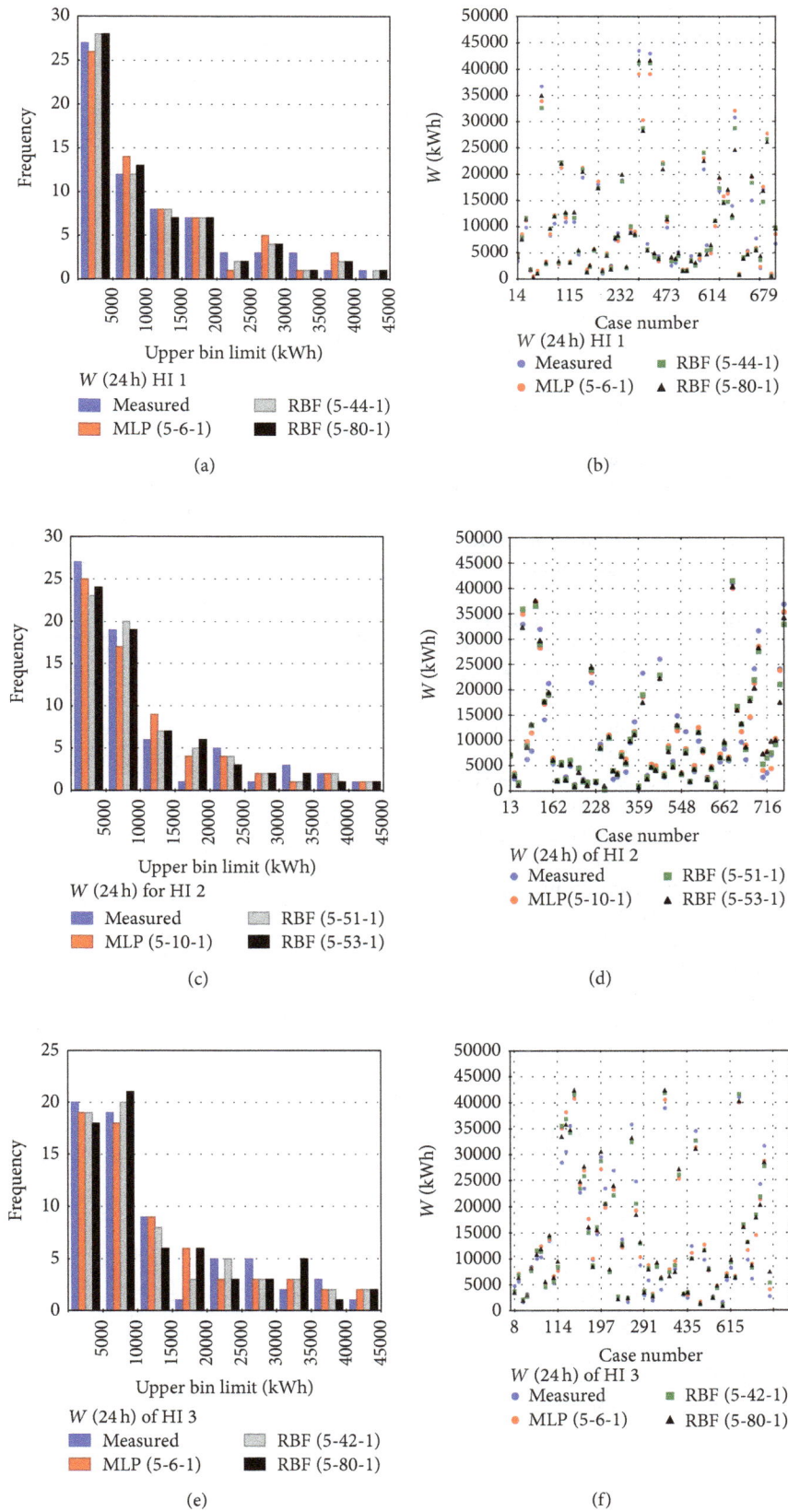

FIGURE 3: Results for the 3 best networks for the dataset of the coastal wind farms *Hinte* and *Jennelt*. The case numbers in the right panels represent individual data records of the complete time series available that were chosen randomly and assigned to the validation dataset.

TABLE 4: Descriptive statistics of the measured MES time series and the time series modeled by the best artificial neural network ANN calculated over all available daily values.

Target variable	Average		Standard deviation		Maximum		Minimum	
	Measured	ANN	Measured	ANN	Measured	ANN	Measured	ANN
$W_{HI\,1\,(kWh)}$	11137	11104	9918	9743	43627	42695	186	90
$W_{HI\,2\,(kWh)}$	11477	11452	10358	10138	44491	43017	168	205
$W_{HI\,3\,(kWh)}$	11220	11213	10123	9895	43589	41752	216	251

validation data set by the three best neural networks for the three target turbines H1 (a), H2 (c), and H3 (e). The frequency distributions show that all networks produce a similar value range of the energy output for the three turbines indicating that the functional relationships between the input and output variables are well captured by all 3 best networks (Table 3). It is noteworthy that, for the rare peak values of over 40 MWh of turbine HI 1, the RBF networks exhibit smaller differences to the measured data than the MLP networks (Figures 6(a) and 6(b)). These extreme values are not outliers in the statistical sense but rare, yet physically meaningful, values.

The fact that RBF networks outperform MLP networks while generalizing the underlying functional relation between input and output also covering extreme values has also been observed in other studies [31, 32].

Also the direct comparisons of individual nonconsecutive samples of the validation dataset (Figures 3(b), 3(d), and 3(f)) show that the two RBF networks estimate values on either side of the end of the scale are the best. The first and second order statistic moments of the measured and modeled data are in good agreement (Table 4).

Prior to the learning process, three continuous periods of approximately three weeks each were extracted from the dataset. This data was then calculated with the best neural network for each turbine of the target wind farm (Figure 4). In contrast to the actual validation data (Figure 3) that were also excluded from the training processes, this artificially created data gap simulates real-world situations when continuous gaps in the records for several days due to technical problems or for several hours due to machine maintenance may occur. The energy output values of the system of wind turbines could be reproduced closely with the neural networks.

One reason for the very reliable and accurate calculation ($0.98 \leq r_S \leq 0.99$) is given by the similarity of the data. The wind conditions at the locations of the wind turbines are not compromised by a complex terrain or large topographic elements. Furthermore, the turbines of both wind farms are of the same type with the same hub height. This assumption about the initial similarity of the data of the two wind farms is supported by the linear regression matrix for the measured data of the two wind farms (Figure 5).

3.2. Sites in the Westphalian Basin. With regard to the orography, the *Muensterland* lowland area around the sites of *Coesfeld* and *Suedlohn* can be classified as a transitional landscape between the flat area in the very Northwest of

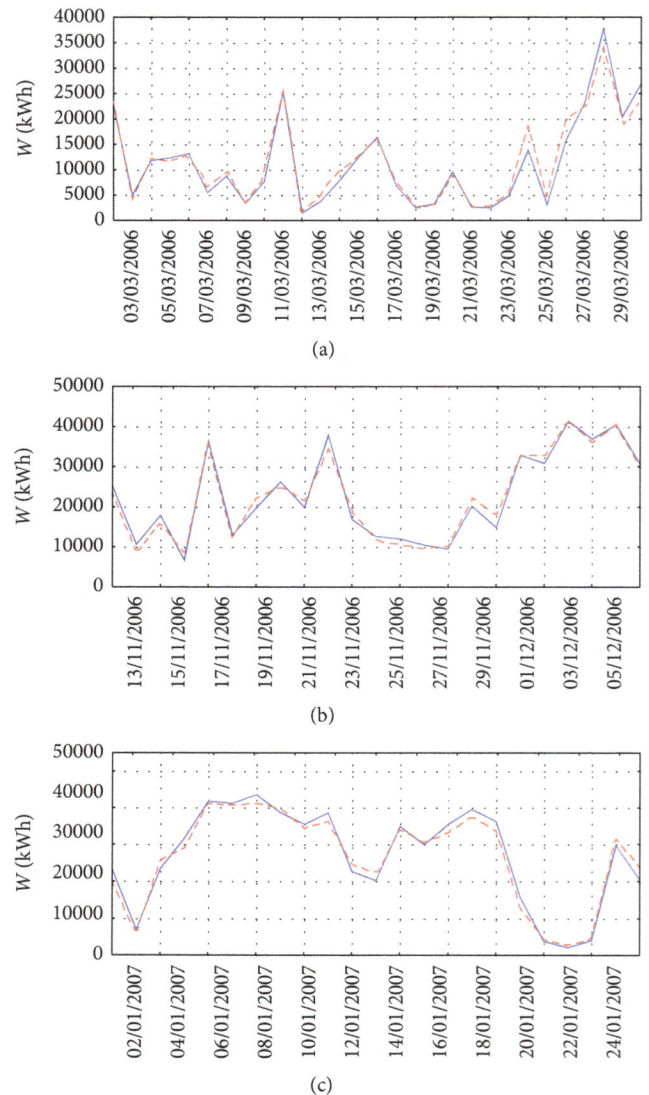

FIGURE 4: Comparison of the measured (blue continuous line) and network reproduced (red dashed line) daily energy values of the turbines at the coastal site.

Germany close to the North Sea and the low mountain range to the Southeast.

Despite some slight elevations with weak slopes, the landscape is relatively flat as typical for the Westphalian Basin in the Southern marginal area of the North German

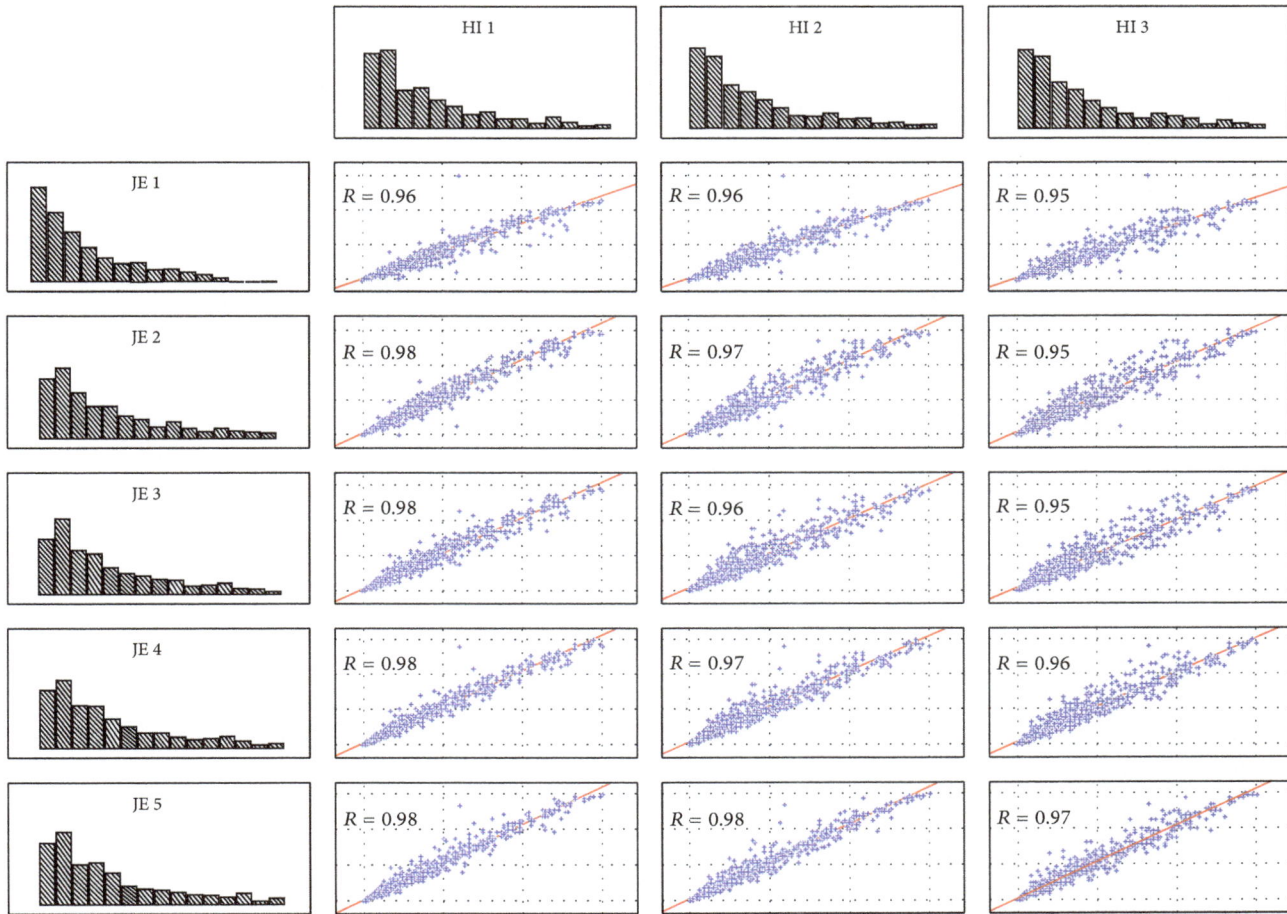

FIGURE 5: Graphic correlation matrix with linear regression coefficients R for the daily energy yields at the coastal sites for the entire data period from the beginning of 2006 through the end of 2008. The frequency distributions are equally divided into 15 classes for all distributions.

Plain. The surface roughness is generally higher compared to the coastal sites *Hinte* and *Jennelt* due to many small towns and some small forests that disrupt the predominantly agricultural region.

The wind farm in *Coesfeld* consisting of five wind turbines is located at a distance of 24 km South-East of the town *Coesfeld*. The farm is composed of two different types of wind turbines, two turbines of the type E-58/10.58, and three larger turbines of the type E-66/10.70. The topography in the immediate vicinity around the wind farm is slightly undulating. The specifications and geographical coordinates of the turbines in the Westphalian Basin area used for the calculations are given in Tables 5 and 6, respectively.

The target wind farm *Suedlohn* is located at a distance of about 1.5 km East of the border to the Netherlands. The wind farm is located Southwest of the community of *Suedlohn*, 24 km away from the wind farm *Coesfeld*. As for the wind farm *Coesfeld*, the topography in *Suedlohn* is expected to have some effects on the wind regime due to upwind obstruction effects at hills [8, 33]. The near surrounding of the site is dominated by arable land, pasture, and widespread farm buildings.

At the site *Suedlohn*, noise emissions restrictions produce legal reasons to reduce the sound level of the turbines affecting the farm houses in the vicinity of the turbines during nighttime. Therefore, the turbines SU 2 and SU 3 are operated with sound reduced performance characteristics from 22:00 to 6:00 which reduces the nominal power output from 800 kW to 600 kW.

The data collection period for the two sites in the Westphalian Basin covers two years from January 1, 2006, to January 1, 2008, providing 732 daily data records. The wind is dominated by winds from West and Southwest at both wind farms because no larger topographic structures are affecting the superimposed West wind drift.

However, the Southwesterly wind directions exhibit a slightly higher frequency at the site *Coesfeld* (Figure 6). In order to achieve the best comparability anemometer data from turbine CO 1 was used for the comparison of the wind conditions as the hub height of 70.5 m is closest to the hub height of 75.6 of the turbines at the target site *Suedlohn*.

The correlation coefficients (Table 7) indicate that the combination of increased topographic structure and the distance of 25 km between the wind farm delivering the input

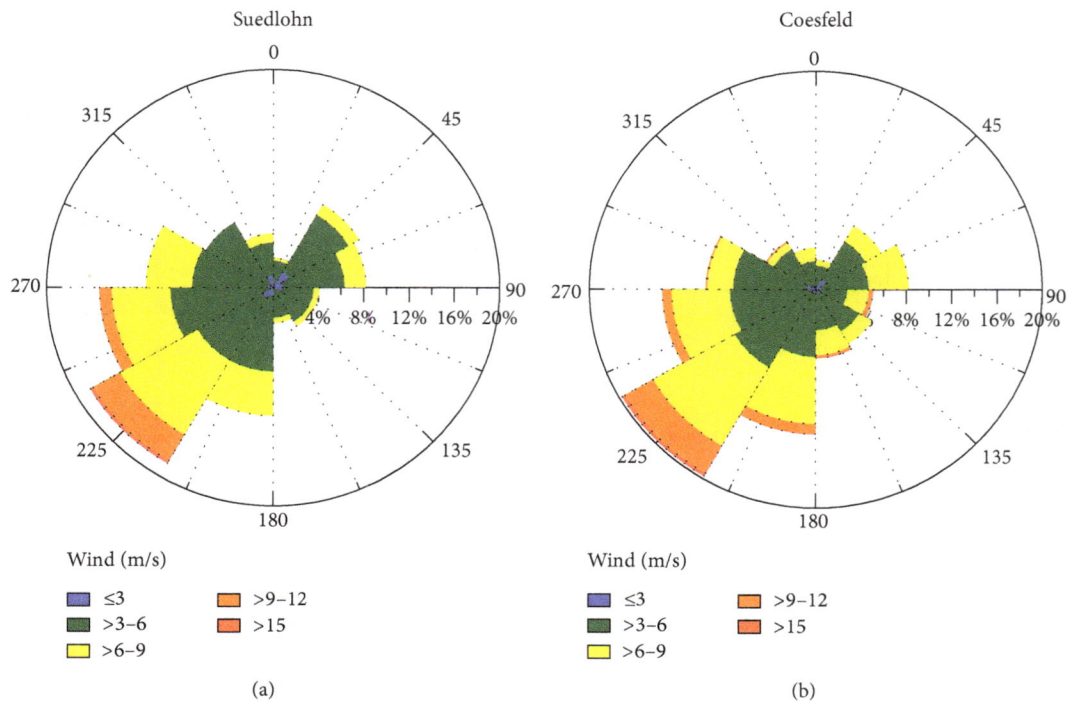

FIGURE 6: Average wind conditions at the locations *Suedlohn* (target site) and *Coesfeld* (input site) during the data collection period from January 2006 to January 2008.

TABLE 5: Coordinates and data of the wind turbines at the site *Coesfeld*.

Turbine	Type	Hub height [m]	Rotor Ø [m]	Nominal power [kW]	Longitude (WGS 84)	Latitude	Elevation (m a.s.l.)
CO 1	E-58/10.58	70.5	58.6	1000	7.2321950E	51.9316590N	136
CO 2	E 66/18.70	98.0	70.0	1800	7.2198140E	51.9311870N	101
CO 3	E 66/18.70	86.0	70.0	1800	7.2113600E	51.9261410N	121
CO 4	E 66/18.70	98.0	70.0	1800	7.2247980E	51.9358810N	114
CO 5	E-58/10.58	89.0	58.6	1000	7.1940630E	51.9213830N	91

TABLE 6: Coordinates and data of the wind turbines at the site *Suedlohn*.

Turbine	Type	Hub height [m]	Rotor Ø [m]	Nominal power [kW]	Longitude (WGS 84)	Latitude	Elevation (m a.s.l.)
SU 1	E 48/8.48	75.6	48.0	800	6.828237E	51.950485N	47
SU 2	E 48/8.48	75.6	48.0	800/600	7.086687E	51.952653N	48
SU 3	E 48/8.48	75.6	48.0	800/600	7.140800E	51.952483N	53

TABLE 7: Correlation matrix for the daily energy output at the locations of *Coesfeld* and *Suedlohn* for the analyzed period.

	CO 1	CO 2	CO 3	CO 4	CO 5
SU 1	0.52	0.53	0.51	0.52	0.56
SU 2	0.54	0.53	0.52	0.53	0.57
SU 3	0.52	0.53	0.48	0.53	0.54

All correlations shown are significant on a 95% confidence level.

data (*Coesfeld*) and the target wind farm (*Suedlohn*) lead to a functional relation between, then, input and target variables that cannot be described well by a simple linear approach.

Thus, due to their ability to emulate any nonlinear functions, neural networks provide an ideal tool for such situations. The available energy output data were divided randomly in the training data, test data, and validation data (602 training data records, 65 test data records, and 65 validation data records).

TABLE 8: Summary of the results for the target site *Suedlohn* in the Westphalian Basin within the area of the North German Plain.

Target variable	Input variables	Network type	Network topology	RMSE (kWh)	r_S
$W_{SU\,1}$	$W_{(CO\,1-CO\,5)}$	MLP	5-6-1	1154	0.95
$W_{SU\,1}$	$W_{(CO\,1-CO\,5)}$	MLP	5-13-1	1092	0.95
$W_{SU\,1}$	$W_{(CO\,1-CO\,5)}$	RBF	5-33-1	1153	0.95
$W_{SU\,2}$	$W_{(CO\,1-CO\,5)}$	RBF	5-25-1	513	0.97
$W_{SU\,2}$	$W_{(CO\,1-CO\,5)}$	RBF	5-21-1	588	0.96
$W_{SU\,2}$	$W_{(CO\,1-CO\,5)}$	MLP	5-8-1	627	0.96
$W_{SU\,3}$	$W_{(CO\,1-CO\,5)}$	MLP	5-8-1	917	0.97
$W_{SU\,3}$	$W_{(CO\,1-CO\,5)}$	RBF	5-15-1	1179	0.95
$W_{SU\,3}$	$W_{(CO\,1-CO\,5)}$	RBF	5-17-1	1026	0.96

TABLE 9: Descriptive statistics of the measured MES time series and the time series modeled by the best artificial neural network ANN calculated over all available daily values.

Target variable	Average		Standard deviation		Maximum		Minimum	
	Measured	ANN	Measured	ANN	Measured	ANN	Measured	ANN
$W_{SU\,1\,(kWh)}$	3980	3995	3402	3296	17188	15327	363	264
$W_{SU\,2\,(kWh)}$	3582	3637	3229	3327	16103	15234	300	229
$W_{SU\,3\,(kWh)}$	3631	3619	3215	3062	16247	15560	287	592

The neural network modeling results for the energy yield values of the turbines SU 1 to SU 3 are given in Table 8 considering all results of the respective three best networks.

Only the power output values of all five turbines were used as input vectors for all neural networks.

The correlations of the network reproduced values with the corresponding measured values are all high and statistically significant ($r_S = 0.95$, $P \leq 0.05$) for the validation data sets of all target turbines (Table 8).

Since turbine SU 2 was affected most by the noise reduction restrictions, the absolute RMSE values are smallest for that turbine which has to be taken into account while interpreting the relatively low RMSE values. Nevertheless, the associated Spearman rank correlation coefficient shows that, despite altering operating conditions during nighttime, the neural networks produce satisfactory results (Tables 8 and 9). Nevertheless, the scattering of differences between modeled and observed values when considering multiple networks is an unwanted effect.

Since the accumulated RMSE is calculated from the squared residual values, the direction of deviation is not considered when constraining the length of the network learning process as shown in (3). Network ensembles provide a way to reduce these variations among a group of neural networks trained for the same purpose.

By weighted averaging of the output values of individual networks, the variations are smoothed and deliver improved results [34, 35]. The weight factors for the averaging were determined using the RMSE values of the incorporated networks.

It is noteworthy in the context of the present work that the results from the individual networks are satisfactory. For the application in practice, there is at least no urgent need for improvement. Nevertheless, since this study outlines the usefulness of neural networks for energy output modeling for wind turbines in order to close data gaps, the performance of a network ensemble was analyzed for turbine SU 1.

As shown in Table 10, the results could be improved using an ensemble composed of three neural networks. In comparison to the values gained through the best single RBF network (Table 8), r_s improved from 0.95 to 0.97, while the RMSE could be reduced by 82 kWh.

The distance between the two wind farms in the *Muensterland* lowland area is more than 20 km. Moreover, the more complex topography leads to a relationship between the power output values of the two wind farms that can hardly be captured by simple linear functions (Table 7). Nevertheless, the results show that the neural network approach delivers sound results using data that were collected over a period of 2 years only.

3.3. Mountainous Sites with Complex Topography. The region around the third pair of sites is located in the mountainous area within the Rhine Massif near the Southern rim of the Ebbe Mountains in central Germany (Figure 1) in the densely forested *Sauerland* region.

Accordingly, the topographic and in particular the orographic conditions will have a significant influence on regional scale and local scale wind fields [36]. Caused by the vast forested areas with different stand ages and scattered towns, the surface roughness is significantly higher compared to the previously presented sites causing higher shear stress on the airmass. These topographic conditions make the wind field modeling challenging and make a simple

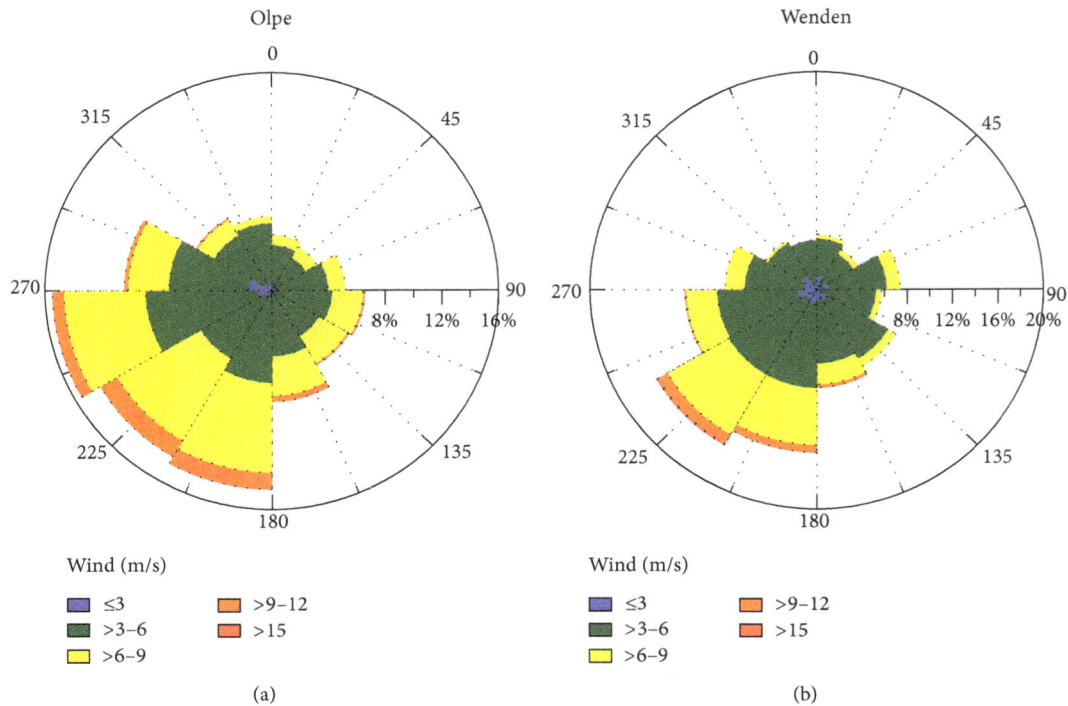

FIGURE 7: Average wind conditions at the sites *Olpe* and *Wenden* during the period from January 2004 through January 2008.

TABLE 10: Summary of the results for the calculation of the income of investment.

Target variable	Input variables	Network type	RMSE (kWh)	r_S
$W_{SU\,1}$	$W_{(CO\,1-CO\,5)}$	Ensemble $(2 \times MLP + 1 \times RBF)$	1010	0.97

SU 1 using an ensemble from the previously optimized individual networks.

spatial interpolation of wind velocities and wind energy yields, respectively, unfeasible. The technical specifications and geographical coordinates of the turbines used for the calculations for the mountainous site are given in Tables 11 and 12, respectively.

The area is mostly used for forestry and grazing. The wind farm *Wenden* is located 1.2 km South of the village *Wenden*. The wind farm *Olpe* delivering the input data for the neural network training and calculations is located 11 km North-Northeast of the wind farm *Wenden* and about 3.6 km Northeast of the town *Olpe*. The topography in the vicinity of both wind farms is undulating.

The data collection period for the turbines in *Olpe* and *Wenden* exceeds four years from January 1, 2004, until April 1, 2008. However, because the input vector is mathematically mapped on the output value by the neural networks, the number of variables in the input vector cannot be altered from the dimensionality used for the training and adjustment

procedure when applying the trained network to new data. Hence, 1430 daily records that contain the synchronal operating data of all five wind turbines were available.

The wind conditions measured at 80 m above ground show significant differences between the two wind farm sites. At both locations the Southwestern wind directions exhibit the highest wind speeds. The West-Southwesterly and South-Southwesterly wind directions are more frequent in *Olpe* compared to the site in *Wenden* (Figure 7). The distributions of the wind direction indicate the effect of the orography in the complex terrain creating pronounced local scale wind fields with distinct distributions of wind velocity and direction at the two sites.

A training dataset of 1100 data records was used for the calculations on the site pair *Olpe/Wenden*. The test dataset and the validation dataset consisted of 165 data records each.

The Spearman rank correlation coefficients r_S are lower compared to the coastal sites and the sites in the Westphalian Basin area presumably caused by the more complex, nonlinear relationship between wind energy yields of the two wind farms. Nevertheless, correlations are still high reaching values from 0.90 to 0.93 (Table 13).

The comparison of the frequency distributions show that the energy output values in the most occupied class of less than 1000 kWh per day are better captured by the RBF networks in both cases (Figure 8). Also, for the other bins, the frequencies calculated by the best RBF agree best with the measured frequencies with the exception of a single value above 9000 kWh calculated by the RBF (topology 3-22-1) that

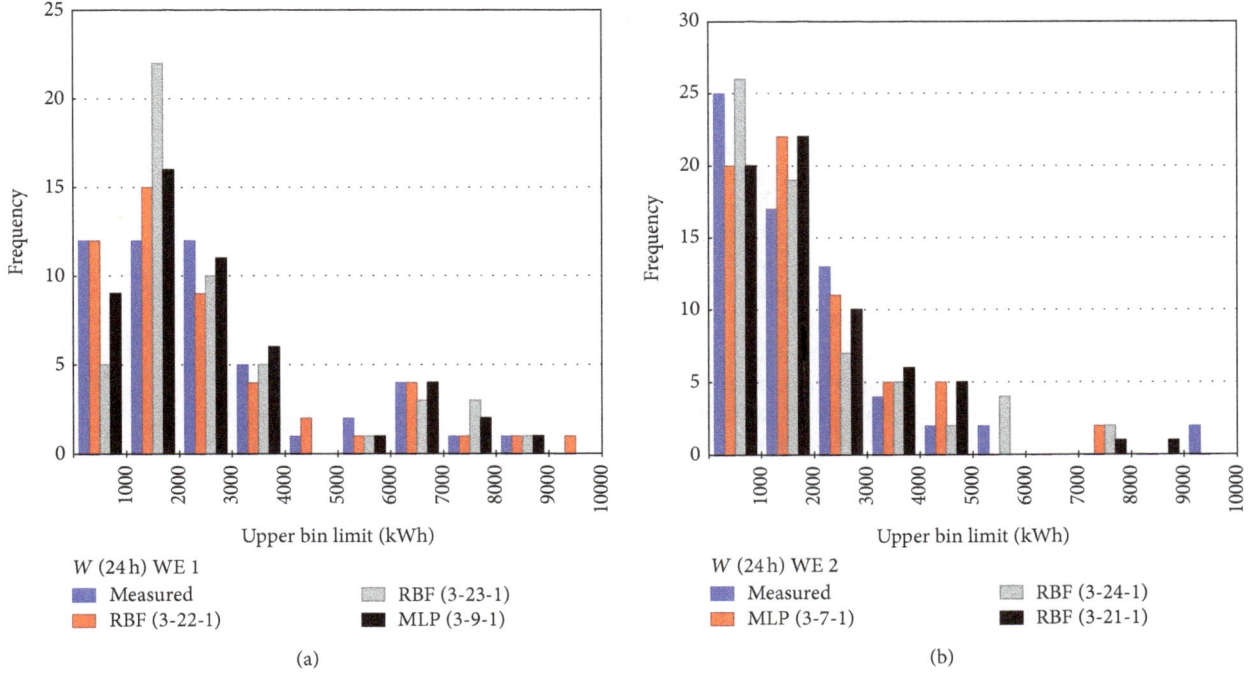

FIGURE 8: The frequency distributions for the measured data and the data calculated by the respective 3 best networks for the turbines WE 1 (a) and WE 2 (b).

TABLE 11: Coordinates and data of wind turbines at the site *Olpe*.

Turbine	Type	Hub height [m]	Rotor Ø [m]	Nominal power [kW]	Longitude (WGS 84)	Latitude	Elevation (m a.s.l.)
OL 1	E 40/6.44	78.0	44.0	600	7.913383	51.054727	520
OL 2	E 66/18.70	86.0	70.0	1.800	7.917154	51.055103	528
OL 3	E 66/18.70	98.0	70.0	1.800	7.910023	51.053212	512

TABLE 12: Coordinates and data of wind turbines at the site *Wenden*.

Turbine	Type	Hub height [m]	Rotor Ø [m]	Nominal power [kW]	Longitude (WGS 84)	Latitude	Elevation (m a.s.l.)
WE 1	E 40/6.44	78.0	44.0	600	7.871504E	50.960112N	423
WE 2	E 40/6.44	78.0	44.0	600	7.873792E	50.959684N	424

TABLE 13: Summary of results from the 3 best neural networks out of 20000 iteratively tested networks for the target turbines in the complex terrain of the *Sauerland* region.

Target variable	Input variables	Network type	Network topology	RMSE (kWh)	r_S
$W_{WE\,1}$	$W_{(OL\,1-OL\,3)}$	MLP	3-9-1	640	0.90
$W_{WE\,1}$	$W_{(OL\,1-OL\,3)}$	RBF	3-22-1	602	0.91
$W_{WE\,1}$	$W_{(OL\,1-OL\,3)}$	RBF	3-22-1	588	0.92
$W_{WE\,2}$	$W_{(OL\,1-OL\,3)}$	MLP	3-7-1	747	0.92
$W_{WE\,2}$	$W_{(OL\,1-OL\,3)}$	RBF	3-24-1	630	0.93
$W_{WE\,2}$	$W_{(OL\,1-OL\,3)}$	RBF	3-21-1	668	0.90

was not observed in the corresponding validation dataset (Figure 8(a)).

To account for the more complex topographic conditions, we tested the addition of more input variables that potentially carry the information needed to better map the input values onto the energy yields of the target turbines. While choosing additional input variables, one must avoid that redundant variables are chosen as an increased number of variables will affect the ability to find the global minimum of the error function during the network training process [13, 37]. This effect is also known as the "curse of dimensionality" and affects all multivariate optimization algorithms [38].

Sensitivity analyses are one method to ensure that only variables are used as input that are important to map the input vectors on the target values. Sensitivity analyses determine the influence of each variable on the minimum of RMSE by, first, determining the minimal error, while taking all input variables into account. In the next steps, the values of the input variables will be partially replaced by random values, while the other input variables remain unchanged. This is done consecutively through all variables. The minimum error E_r applying the random valuesis then compared to the original minimum error (E_0) achieved with the original values of the respective variable. This is done by simply calculating the ratio $R_E = E_r/E_0$. Thus, an error ratio R_E equal to 1 or less indicates that the variable does not add additional information. In that case, the variable can be considered disruptive or at least redundant and should not be used as input variable.

In Table 14, the error ratios for the network that was optimized for the reproduction of the energy output values of the turbine WE 1 are shown using the operating data recorded by the three turbines of the *Olpe* wind farm.

The ranks reflect the influence of each input variable on the result. The higher a specific variable is ranked, the more it contributes to the error minimization. Only variables that exhibit an error ratio R_E > 1.1 were used for the neural networks during further analyses [32]. The error ratios confirm that the energy output values have a large impact on the quality of the results. It is also shown that additional wind meteorological variables and operating data such as the instantaneous power contain important information to model the target variable. The fact that the wind direction exhibits an error ratio of 2.233 also underlines the importance of the topographic effects on the wind energy yields in complex terrain. Since the dimensionality of the input vector was increased, more neurons in the hidden layers were needed for the two RBF networks used [12].

Through application of the additional important variables the performance of the neural network could be increased (Table 15). The correlation between the energy output calculated with the neural network utilizing the additional input variables and the measured energy output could be increased from a maximum of 0.92 (Table 12) to 0.97 (Table 15). Furthermore, The RMSE accumulated over 165 validation data records was reduced by 20.4% from 599 on 477 kWh. Using an RBF network ensemble did not increase the correlation but slightly reduced the RMSE by another 1.9% (Table 15).

TABLE 14: Ranked error ratios for available input variables of the wind farm *Olpe*.

Rank	Variable	R_E
1	$W_{(OL\,2)}$	2.790
2	$W_{(OL\,1)}$	2.258
3	Average wind direction (OL 1)	2.233
4	Average instantaneous power (OL 1)	1.889
5	$W_{(OL\,3)}$	1.808
6	Average instantaneous power (OL 2)	1.550
7	Minimum instantaneous power (OL 1)	1.507
8	Daily hours of operation (OL 2)	1.296
9	Minimum instantaneous power (OL 2)	1.278
10	Maximum instantaneous power (OL 2)	1.274
11	Maximum wind speed (OL 1)	1.216
12	Average of wind speed (OL 3)	1.208
13	Average of wind speed (OL 2)	1.204
14	Minimum wind speed (OL 3)	1.088
15	Minimum number of revolutions (OL 3)	1.082

Only the 15 variables with R_E > 1 are shown.

4. Conclusion

We presented a neural network approach to model the daily energy yields of wind turbines by training neural networks using the data of other wind farms. The method was deployed on three examples with different spatial setups and distances between the input sites and the target sites in exemplary regions covering a variety of topographic complexity. The results show that artificial neural networks provide a capable mathematical tool to deliver reliable results. The data modeled by the trained neural networks are highly correlated to the corresponding data measured by the operating and surveillance system of the turbines with coefficients of 0.9 and higher.

Differences between the predictions of the best networks are small for the coastal sites as well as for the sites in the mostly flat region of the Westphalian Basin. Both network types tested allow a sound and accurate filling of the data gaps. However, the RBF networks turned out to better capture extreme values compared to the respective best MLP networks.

The biggest advantage of the method of artificial neural networks to fill data gaps is the fact that no information about the relationships between the variables or the statistical distributions of the individual quantities must be assumed prior to the network training.

In combination with the fast learning procedure for RBF networks, this makes them a suitable approach for wind energy yield predictions in practice.

The operating data of wind turbines are recorded today by default with high temporal resolution and many important technical parameters as well as basic wind meteorological measurements.

Our results show that these additional measurements can significantly increase the performance of the neural networks

TABLE 15: Summary of results for the modeled energy output of turbine WE 1 using 13 input variables from the site *Olpe*.

Destination variable	Input variables	Network type and topology	RMSE (kWh)	Correlation (r_S)
$W_{WE\,1}$	$W_{(OL\,1)}$ average instantaneous power (OL 1) Minimum instantaneous power (OL 1) Average wind direction (OL 1) Maximum wind speed (OL 1)	Ensemble $2 \times$ RBF (15-92-1) (15-101-1)	477	0.97
	$W_{(OL\,2)}$ Daily hours of operation (OL 2) Minimum instantaneous power (OL 2) Average instantaneous power (OL 2) Maximum instantaneous power (OL 2) Average wind speed (OL 2)	RBF (15-95-1)	494	0.97
	$W_{(OL\,3)}$ Average wind speed (OL 3)	RBF (15-93-1)	486	0.97

especially in areas with complex topography. Therefore, the approach presented helps to reduce the uncertainty of future wind energy predictions conducted with wind flow models. The work presented delivers the methodical framework to be deployed anywhere on the globe where input and related output data is available to train the artificial neural networks.

Due to the increasing demand for energy the application of specifically trained networks for turbines in complex terrain is of great interest in practice. Distances between wind turbines used as input and the target locations can be significantly larger than the distances considered in our case studies. Modeled power output values could be used as input for a different ANN that is applied on turbines even further away from the turbines that deliver the primary input values. Exploring the capability of the ANN approach to work on larger spatial scales by using consecutive stages of input and output datasets is a goal for future research in order to utilize the inherent flexibility of artificial neural networks and increase the planning dependability for wind energy projects.

Conflict of Interests

The authors declare that there is no conflict of interests regarding the publication of the paper.

Acknowledgments

The operating data of the wind turbines was kindly provided by ENERCON GmbH, SOLVENT GmbH, and SL-wind energy GmbH. The authors thank these companies for their support. The authors also would like to thank Joshua Baur for editing the language of the manuscript.

References

[1] IPCC, "Climate Change 2013: The Physical Science Basis," in *Contribution of Working Group I To the Fifth Assessment Report of the Intergovernmental Panel on Climate Change*, T. F. Stocker, D. Qin, G. K. Plattner et al., Eds., p. 1535, Cambridge University Press, Cambridge, United Kingdom, 2013.

[2] P. Friedlingstein, R. A. Houghton, G. Marland et al., "Update on CO_2 emissions," *Nature Geoscience*, vol. 3, no. 12, pp. 811–812, 2010.

[3] G. P. Peters, G. Marland, C. le Quéré, T. Boden, J. G. Canadell, and M. R. Raupach, "Rapid growth in CO_2 emissions after the 2008-2009 global financial crisis," *Nature Climate Change*, vol. 2, no. 1, pp. 2–4, 2012.

[4] G. P. Peters, C. L. Weber, D. Guan, and K. Hubacek, "China's growing CO2 emissions - A race between increasing consumption and efficiency gains," *Environmental Science and Technology*, vol. 41, no. 17, pp. 5939–5944, 2007.

[5] X. Lu, M. B. McElroy, and J. Kiviluoma, "Global potential for wind-generated electricity," *Proceedings of the National Academy of Sciences of the United States of America*, vol. 106, no. 27, pp. 10933–10938, 2009.

[6] EEG, "Renewable energy sources Act of 25 October 2008," last amended by the Act of 20 December 2012, Art. 5G I, 2730, 2012.

[7] I. Troen and E. L. Petersen, *European Wind Atlas*, Risø National Laboratory, Roskilde, Denmark, 1989.

[8] N. G. Mortensen, L. Landberg, I. Troen, and E. L. Petersen, "Wind Atlas Analysis and Application Program (WAsP), Vol. 1: Getting Started. Vol. 2: User's Guide," Risø National Laboratory, Roskilde, Denmark, 1993.

[9] R. Gasdch and J. Twele, *Wind Power Plants: Fundamentals, Design, Construction and Operation*, Springer, Berlin, Germany, 2nd edition, 2012.

[10] W. Winkler, M. Strack, and A. Westerhellenweg, "Scaling and evaluation of wind data and wind farm energy yields," *DEWI Magazine*, vol. 23, pp. 76–84, 2003.

[11] FGW, "Technical Guidelines for Wind Turbines. Part 6. Determination of wind potential and energy yields," Revision 8, Berlin, Germany, 2011.

[12] C. M. Bishop, *Neural Networks For Pattern Recognition*, Oxford University Press, Oxford, UK, 1996.

[13] D. W. Patterson, *Artificial Neural Networks. Theory and Applications*, Prentice Hall, Singapore, 1996.

[14] D. S. Broomhead and D. Lowe, "Multi variable functional interpolation and adaptive networks," *Complex Systems*, vol. 2, pp. 321–355, 1988.

[15] B. D. Ripley, *Pattern Recognition and Neural Networks*, Cambridge University Press, Cambridge, UK, 1996.

[16] S. Haykin, *Neural Networks: A Comprehensive Foundation*, Prentice Hall, Upper Saddle River, NJ, USA, 2nd edition, 1999.

[17] S. Li, D. C. Wunsch, E. A. O'Hair, and M. G. Giesselmann, "Using neural networks to estimate wind turbine power generation," *IEEE Transactions on Energy Conversion*, vol. 16, no. 3, pp. 276–282, 2001.

[18] Z. Liu, W. Gao, Y. H. Wan, and E. Muljadi, "Wind power plant prediction by using neural networks," in *Proceedings of the IEEE Energy Conversion Conference and Exposition*, Raleigh, NC, USA, September 2012.

[19] W. A. Oost and E. M. Oost, "An alternative approach to the parameterization the momentum flux over the sea," *Boundary-Layer Meteorology*, vol. 113, no. 3, pp. 411–426, 2004.

[20] S. Ayoubi and K. L. Sahrawat, "Comparing multivariate regression and artificial neural network to predict barley production from soil characteristics in Northern Iran," *Archives of Agronomy and Soil Science*, vol. 57, no. 5, pp. 549–565, 2011.

[21] H. Z. Hamid Zare Abyaneh, "Evaluation of multivariate linear regression and artificial neural networks in prediction of water quality parameters," *Journal of Environ Health Science and Engineering*, vol. 12, article 40, 2014.

[22] D. E. Rumelhart, G. E. Hinton, and R. J. Williams, "Learning internal representations by error propagation," in *Parallel Distributed Processing: Explorations in the Microstructure of Cognition*, D. E. Rumelhart, J. L. McClelland, and PDP Research Group, Eds., vol. 1, pp. 318–362, Cambridge University Press, Cambridge, Mass, USA, 1986.

[23] E. Metcalfe, J. Teteh, and S. L. Howells, "Optimisation of radial basis and backpropagation neural networks for modelling autoignition temperature by quantitative-structure property relationships," in *Chemometrics and Intelligent Laboratory Systems*, vol. 32, pp. 177–191, 1996.

[24] M. Hestenes, *Conjugate Direction Methods in Optimization*, Springer, New York, NY, USA, 1980.

[25] G. J. Bowden, H. R. Maier, and G. C. Dandy, "Optimal division of data for neural network models in water resources applications," *Water Resources Research*, vol. 38, no. 2, pp. 2–11, 2002.

[26] R. Khosla, "Knowledge-Based Intelligent Information and Engineering Systems," in *Proceedings of the KES 9th International Conference*, Lecture Notes in Computer Science, p. 933, Melbourne, Australia, September 2005.

[27] A. Schmidt, C. Hanson, J. Kathilankal, and B. E. Law, "Classification and assessment of turbulent fluxes above ecosystems in North-America with self-organizing feature map networks," *Agricultural and Forest Meteorology*, vol. 151, no. 4, pp. 508–520, 2011.

[28] B. Smith and H. Link, "Anemometer measurements for use in turbine power performance tests," Preprint American Wind Energy Association (AWEA) Windpower Conference Portland, Oregon, USA, June 2002.

[29] A. Albers and H. Sacman, "Wind speed and turbulence evaluation for performance and gondola anemometer measurements," *DEWI Magazine*, vol. 10, pp. 51–62, 1997.

[30] J. A. Anderson, *An Introduction To Neural Networks*, MIT Press, Boston, Mass, USA, 1995.

[31] M. Qu, F. Y. Shih, J. Jing, and H. Wang, "Automatic solar flare detection using MLP, RBF, and SVM," *Solar Physics*, vol. 217, no. 1, pp. 157–172, 2003.

[32] A. Schmidt, T. Wrzesinsky, and O. Klemm, "Gap filling and quality assessment of CO_2 and water vapour fluxes above an urban area with radial basis function neural networks," *Boundary-Layer Meteorology*, vol. 126, no. 3, pp. 389–413, 2008.

[33] Y. N. Maharani, S. Lee, and Y. K. Lee, "Topographical effects on wind speeds over various terrains: a case study for Korean peninsula," in *Proceedings of the 7th Asia-Pacific Conference on Wind Engineering*, Taipei, Taiwan, November 2009.

[34] M. P. Perrone and L. N. Cooper, "When networks disagree: ensemble methods for hybrid neural networks," in *Artificial Neural Networks For Speech and Vision*, R. J. Mammone, Ed., pp. 126–147, Chapman & Hall/CRC, London, UK.

[35] B. Bakker and T. Heskes, "Clustering ensembles of neural network models," *Neural Networks*, vol. 16, no. 2, pp. 261–269, 2003.

[36] K. J. Eidsvik, "A system for wind power estimation in mountainous terrain. prediction of askervein hill data," *Wind Energy*, vol. 8, no. 2, pp. 237–249, 2005.

[37] J. M. Zurada, A. Malinowski, and I. Cloete, "Sensitivity analysis for minimization of input data dimension for feedforward neural networks," in *Proceedings of the IEEE International Symposium on Circuits and Systems*, pp. 447–450, IEEE Press, London, UK, February 1994.

[38] R. E. Bellman, *Adaptive Control Processes*, Princeton University Press, Princeton, NJ, USA, 1961.

20

Performance Analysis of Savonius Rotor Based Hydropower Generation Scheme with Electronic Load Controller

Rajen Pudur[1] and Sarsing Gao[2]

[1]*Electrical Engineering, National Institute of Technology, Yupia, Arunachal Pradesh 791112, India*
[2]*Department of Electrical Engineering, North Eastern Regional Institute of Science and Technology (NERIST), Nirjuli, Arunachal Pradesh 791 109, India*

Correspondence should be addressed to Rajen Pudur; rajenpudur1977@gmail.com

Academic Editor: Pallav Purohit

This paper describes the performance of electronic load controller (ELC) of asynchronous generator (AG) coupled to an uncontrolled Savonius turbine and variable water velocity. An AC-DC-AC converter with a dc link capacitor is employed to maintain the required frequency. The ELC which is feeding a resistive dump load is connected in parallel with the generating system and the power consumption is varied through the duty cycle of the chopper. Gate triggering of ELC is accomplished through sinusoidal pulse width modulation (SPWM) by sensing the load current. A MATLAB/Simulink model of Savonius rotor, asynchronous generator, ELC, and three-phase load is presented. The proposed scheme is tested under various load conditions under varying water velocities and the performances are observed to be satisfactory.

1. Introduction

Of late, distributed generation schemes are becoming popular in developed as well as developing countries to augment the existing power scenario. Induction generators are best suited for such applications due to their many advantages over other generators available [1]. The conventional sources of energy from coal, oil, natural gas, and uranium are observed to pollute air and water and produce ecological imbalances and, therefore, are hostile to both human beings and other plants and animals in the long run. Therefore, alternative sources of energy, like wind, solar, hydro, tidal, geothermal, and so forth, need to be explored to bring about a change in the energy scenario, in terms of both capacity and quality. Hydropower schemes generate a clean and environment friendly form of energy. However, large hydropower plants with generating capacity above 25 MW are often considered to endow with drawbacks such as high initial cost and environmental impacts [2]. Therefore, distributed generation scheme with run-of-river scheme is considered to be a benign source of energy. The recent survey conducted by Energy Next (EN) in collaboration with the Ministry of New and Renewable Energy (MNRE), Govt. of India, in April 2015 reveals that the States of Arunachal Pradesh, Himachal Pradesh, and Uttarakhand alone have small hydropower (SHP) potential of 5447.16 MW with harnessed capacity of 917.63 MW [3]. This figure may go up with installation of more run-of-river schemes in these areas. In smaller capacity hydropower plants, use of uncontrolled turbines is preferred, thus allowing the water flow as it comes and driving the turbine-generator set. Use of costly governors in such applications is unwise; thus use of load control mechanism is usually adopted. In this paper, a mechanism is evolved to regulate the consumer's load by diverting additional load to a dump load, thereby maintaining the power output of the machine by the use of an electronic load controller (ELC).

The use of ELC in the proposed scheme is realized by incorporating an AC-DC-AC converter similar to that of a wind energy conversion system, which produces a constant voltage and constant frequency from variable water velocity. Once a stable voltage and frequency are achieved, the total output power remains constant and hence the power switching between the main load and the dump load becomes feasible.

The turbine selected in the present study is a vertical axis Savonius rotor, which is commonly used as wind turbines

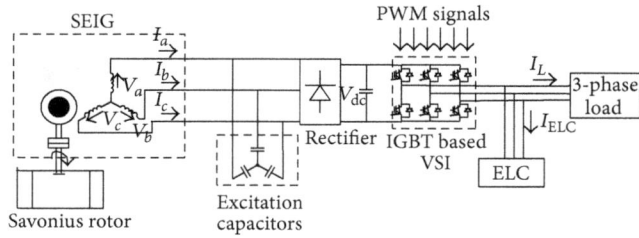

FIGURE 1: Schematic diagram of power generating scheme.

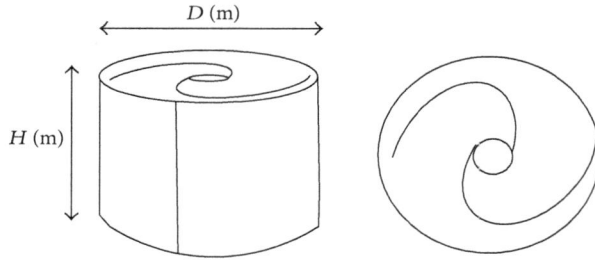

FIGURE 2: Savonius rotor.

[4, 5]. It is a unique fluid mechanical device which works on drag effect mechanism rather than lift mechanism as in the case of the rest of the wind turbines. It is well proven that Savonius rotor can be effectively used for generating power using hydrodynamics in addition to aerodynamics principles [6]. It is a high solidity rotor. It is simple and easy to construct and can be implemented in any suitable locations [7, 8]. A typical Savonius rotor is shown in Figure 2. The drag coefficient of the concave surface is larger than the convex surface, thus forcing the rotor to rotate. It generates much higher starting torque compared to other vertical axis turbines [9]. This type of turbine is self-starting and provides high torque at low speeds. Implementation of ELC in conventional small scale hydropower generating schemes has been detailed in [10–13]. In these systems, reaction type of turbines which are ideally suited for low head hydropower plants is advocated.

The schematic diagram of the proposed scheme is shown in Figure 1.

2. System Configuration

In the proposed scheme, a Savonius rotor coupled with a 7.5 kW, 3-phase, 415 V, 14.5 A, 50 Hz, Y-connected, 4-pole, squirrel cage induction machine is employed. The excitation of the AG is achieved by means of a delta connected capacitor bank of 140 μF. The dimension of Savonius rotor so selected is given in the appendix.

The power output from Savonius turbine is given by

$$P = 0.5 C_p A \rho V^3, \tag{1}$$

where P is power output (W), ρ is the density of water (kg/m^3), A is the swept area of rotor (m^2), V is the velocity of water (m/s), and C_p is the power coefficient.

Tip speed ratio is given by

$$\text{TSR} = \frac{\omega D}{2V}, \tag{2}$$

where ω is the angular velocity and D is rotor diameter (m).

Coefficient of torque (C_t) is given by

$$C_t = \frac{C_p}{\text{TSR}}. \tag{3}$$

Shaft torque (T_{sh}) is given by

$$T_{sh} = \frac{P}{\omega} = \frac{0.5 C_p A \rho V^3}{2\pi N/60}. \tag{4}$$

The generated voltage and frequency of the machine are likely to vary with the varying velocity of river water. This, in turn, would result in unbalanced voltage and frequency at the load end. Thus, for uncontrolled turbine with varying input power, it is essential to determine the real and reactive power requirements of the generating machine in order to arrest the changes in terminal voltage and frequency. To maintain the rated voltage and frequency at the load end therefore, an AC-DC-AC converter realized with the help of an uncontrolled rectifier and insulated gate bipolar transistor (IGBT) based current controlled voltage source inverter (CC-VSI) is used. The triggering pulses to the three-legged IGBTs are varied in accordance with the varying input power.

The variation in the DC link capacitor voltage presents the direct-axis component of current from the machine. The peak value of the line-to-line voltage from the machine is computed and compared with the reference peak value ($415\sqrt{2}$ V). The difference between these two quantities is the reactive power required by the machine or is the amount of quadrature-axis component of current to be supplied to the machine. These two axes reference currents, namely, I_{ds}^*, I_{qs}^*, are converted into three-phase form by inverse Park's transformation [14]. The "cos(ωt)" and "sin(ωt)" terms needed for Park's transformation are derived with the help of a phase locked loop (PLL) which is fed with unit templates of line voltages from the machine. The three-phase reference currents thus obtained are compared with the actual load currents in a hysteresis current controller to yield the firing signals for the six devices in the voltage source inverter (VSI). The fluctuation in capacitance voltage is due to power consumed by the devices in the VSI and filter resistance.

2.1. Generation of Unit Voltage Templates. The line voltages (V_{ab}, V_{bc}, and V_{ca}) of the generator terminals are considered sinusoidal and therefore, their amplitudes are computed as

$$V_{\text{tactual(peak)}} = \sqrt{\frac{2}{3} \left(V_{ab}^2 + V_{bc}^2 + V_{ca}^2 \right)}. \tag{5}$$

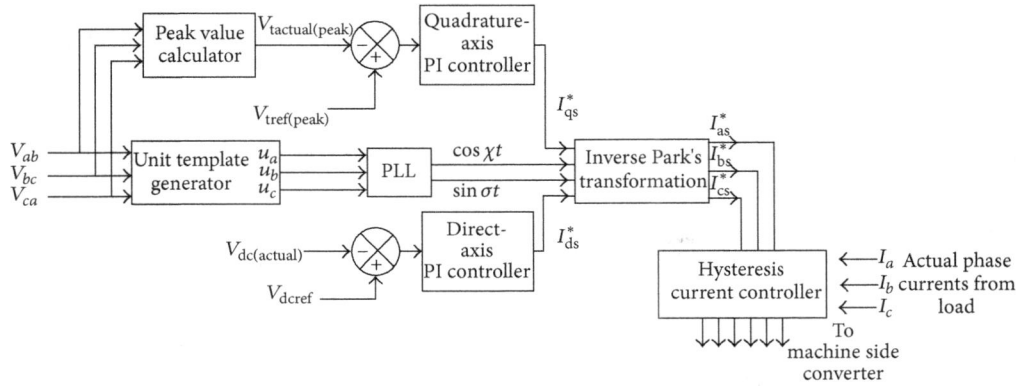

FIGURE 3: Schematic diagram of control scheme.

The unit template voltages are derived as

$$u_{\text{a}} = \frac{V_{ab}}{V_{\text{tactual(peak)}}},$$ (6a)

$$u_{\text{b}} = \frac{V_{bc}}{V_{\text{tactual(peak)}}},$$ (6b)

$$u_{\text{c}} = \frac{V_{ca}}{V_{\text{tactual(peak)}}}.$$ (6c)

2.2. Quadrature-Axis Component of Reference Source Currents. The ac voltage error $V_{\text{err}(n)}$ at the nth sampling instant is given by

$$V_{\text{err}(n)} = V_{\text{tref(peak)}(n)} - V_{\text{tactual(peak)}(n)},$$ (7)

where $V_{\text{tref(peak)}(n)}$ is the peak value of the three-phase ac voltage being sensed at the generator terminal at the nth instant. The output of the PI controller ($I^{*}_{\text{qs}(n)}$) for maintaining the ac terminal voltage constant at the nth instant is expressed as

$$I^{*}_{\text{qs}(n)} = I^{*}_{\text{qs}(n-1)} + K_{\text{pa}}\left\{V_{\text{err}(n)} - V_{\text{err}(n-1)}\right\} + K_{\text{ia}}V_{\text{err}(n)},$$ (8)

where K_{pa} and K_{ia} are the proportional and integral gain constants of the PI controller, $V_{\text{err}(n)}$ and $V_{\text{err}(n-1)}$ are voltage errors in the nth and $(n - 1)$th instants, and $I^{*}_{\text{qs}(n-1)}$ is the amplitude of quadrature component of the reference source current at the $(n - 1)$th instant.

2.3. Direct-Axis Component of Reference Source Currents. The error in dc bus voltage ($V_{\text{dcerr}(n)}$) of the VSI at the nth sampling instant is given by

$$V_{\text{dcerr}(n)} = V_{\text{dcref}(n)} - V_{\text{dc(actual)}(n)},$$ (9)

where $V_{\text{dcref}(n)}$ is the reference dc voltage and $V_{\text{dc(actual)}(n)}$ is the DC link voltage of the VSI being sensed at the nth instant. The output of the PI controller ($I^{*}_{\text{ds}(n)}$) for maintaining dc bus voltage at the nth instant is expressed as

$$I^{*}_{\text{ds}(n)} = I^{*}_{\text{ds}(n-1)} + K_{\text{pd}}\left\{V_{\text{dcerr}(n)} - V_{\text{dcerr}(n-1)}\right\}$$
$$+ K_{\text{id}}V_{\text{dcerr}(n)},$$ (10)

where $I^{*}_{\text{ds}(n)}$ is considered as the amplitude of active source current at the nth instant while K_{pd} and K_{id} are the proportional and integral gain constants of dc voltage PI controller.

2.4. Reference Source Currents. The reference source currents (I^{*}_{as}, I^{*}_{bs}, and I^{*}_{cs}) are obtained with the help of inverse Park's transformation as shown in

$$I_{dq0} = TI_{abc}$$
$$= \sqrt{\frac{2}{3}}\begin{bmatrix} \cos(\omega t) & \cos\left(\omega t - \dfrac{2\pi}{3}\right) & \cos\left(\omega t + \dfrac{2\pi}{3}\right) \\ \sin(\omega t) & \sin\left(\omega t - \dfrac{2\pi}{3}\right) & \sin\left(\omega t + \dfrac{2\pi}{3}\right) \\ \dfrac{\sqrt{2}}{2} & \dfrac{\sqrt{2}}{2} & \dfrac{\sqrt{2}}{2} \end{bmatrix}\begin{bmatrix} I_{\text{a}} \\ I_{\text{b}} \\ I_{\text{c}} \end{bmatrix}.$$ (11)

2.5. Current Controller. The load currents (I_{a}, I_{b}, and I_{c}) are compared with the reference source currents (I^{*}_{as}, I^{*}_{bs}, and I^{*}_{cs}) and error signals are passed through hysteresis band to generate the firing pulses, which are operated to produce output voltage in manner to reduce the current error. Figure 3 shows the three-phase CC-VSI with hysteresis current controller.

The output of the current controller decides the switching patterns to be given to the IGBTs in the VSI. The current errors are computed as

$$I_{\text{aserr}} = I^{*}_{\text{as}} - I_{\text{a}},$$
$$I_{\text{bserr}} = I^{*}_{\text{bs}} - I_{\text{b}},$$ (12)
$$I_{\text{cserr}} = I^{*}_{\text{cs}} - I_{\text{c}}.$$

3. Design and Control of ELC

Although the input power is varying, with the help of AC-DC-AC converter, the generator terminal voltage and frequency are observed to be constant. Thus, the total power generated remains constant, so a power diverter circuit such as an ELC may be connected in parallel to the main load. This circuit diverts the unused power to an auxiliary load or dump load thus helping in achieving balance of the power system. The ELC is designed using IGBT based chopper switch.

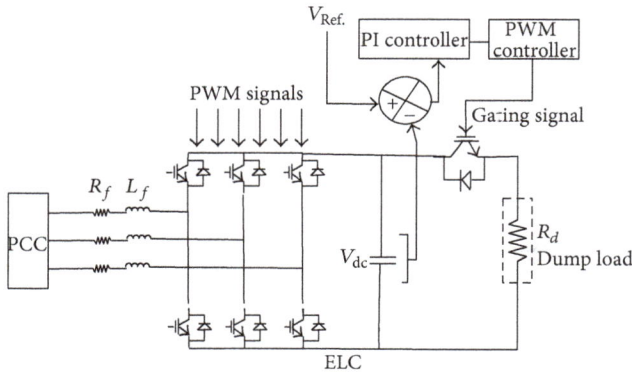

FIGURE 4: Schematic diagram of ELC and control scheme for chopper.

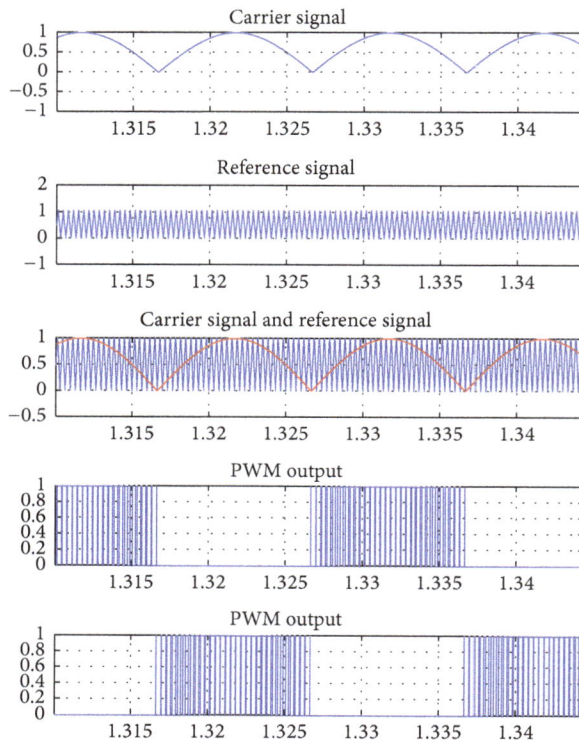

FIGURE 5: Output pulses from SPWM method.

Figure 4 shows the control scheme of ELC and chopper circuit. The ELC is connected to main circuit with point of common coupling (PCC). The three-phase load currents are compared with the reference carrier current generated with 4 kHz. The error signal is passed through sinusoidal pulse width modulation (SPWM), and pulses are then fed to IGBT of the voltage source inverter, as shown in Figure 5.

The rating of IGBT switches depends on the rated voltage and power of AG. The DC output voltage of VSI corresponding to rated voltage of AG is given by

$$V_{dc} = \frac{V_L \times 3\sqrt{2}}{\pi} = 1.35 \times 415 = 560 \, \text{V}, \quad (13)$$

where V_L is rms line voltage of AG.

For feeding reactive power in case of 0.8 pf lagging reactive load it is found that AGs require 130%–160% of rated generated power [10–12]. Therefore, for 7.5 kW generators the VAR rating of the controller should be around 9.975 kVAR (133%).

Then, the apparent power S is given by

$$S = \sqrt{7.5^2 + 9.98^2} = 12.48 \, \text{kVA}. \quad (14)$$

So the current rating of the converter is

$$3VI_c = 12.48, \\ I_c = 17.36 \, \text{A}. \quad (15)$$

Now, average current flowing through ELC can be taken as 90% of converter rms current, considering the worst case of load unbalancing:

$$I_{(average)} = 0.90 \times 17.36 = 15.62 \, \text{A}. \quad (16)$$

Considering voltage ripple in V_{dc} in the order of 2% then

$$V_{dc(ripple)} = 2\% \text{ of } 560 = 11.2 \, \text{V}. \quad (17)$$

Taking the values of $V_{dc(ripple)}$, $I_{(average)}$, and $\omega = 314 \, \text{r/s}$, we have

$$C_{dc} = \frac{I_{(average)}}{(2\omega V_{dc})} = 2220.7 \, \mu\text{F} \approx 3000 \, \mu\text{F}. \quad (18)$$

3000 μF is selected which is available in market.

Rating of dump load is

$$R_{dp} = \frac{V_{dc}^2}{P_R} = \frac{560^2}{7500} = 48 \, \Omega \approx 50 \, \Omega. \quad (19)$$

Value of R_{dp} is selected as 50 Ω for giving wide range of control to the controller, where R_{dp} is resistance of dump load and P_R is rated power of generator.

The control signal for chopper is generated through the error signal generated by comparing voltage across DC capacitor and the reference voltage of 560 V. Chopper regulations maintain the machine terminal voltage constant by feeding additional load to dump load.

The output power of the AG is held constant at varying consumer loads. Thus, the generated power is given by

$$P_{gen} = P_{ELC} + P_{Load}, \quad (20)$$

where P_{gen} is generated power by the AG, P_{Load} is consumer's load, and P_{ELC} is the power absorbed by the ELC [13].

4. Results and Discussion

A complete hydropower generation scheme consisting of a Savonius rotor, asynchronous machine, and AC-DC-AC converter connected to three-phase load is modeled and simulated in MATLAB/Simulink environment. Varying water velocities ranging from 1.96 m/s to 2.02 m/s are considered. Table 1 shows the variation of active power with the velocity of water.

TABLE 1: Variation of output power with velocity of water.

S/number	Velocity of water (m/s)	Active power output (kW)
1	1.0	0.5
2	1.3	1.2
3	1.8	1.35
4	1.96	2.0
5	2.0	2.5

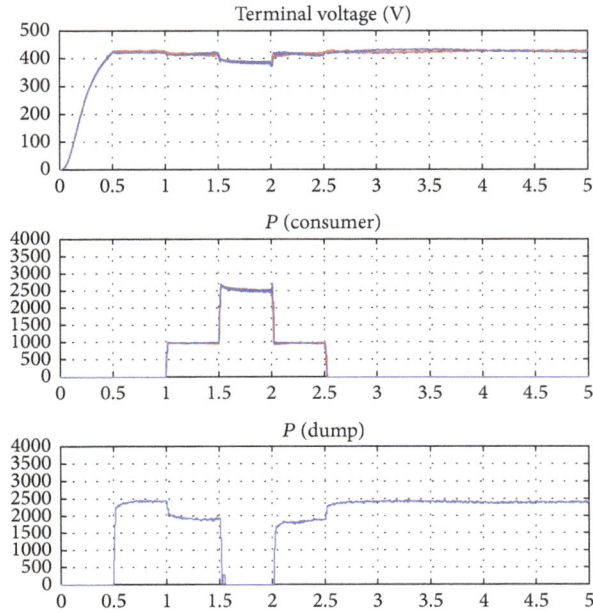

FIGURE 6: Variation of consumer power and dump load power.

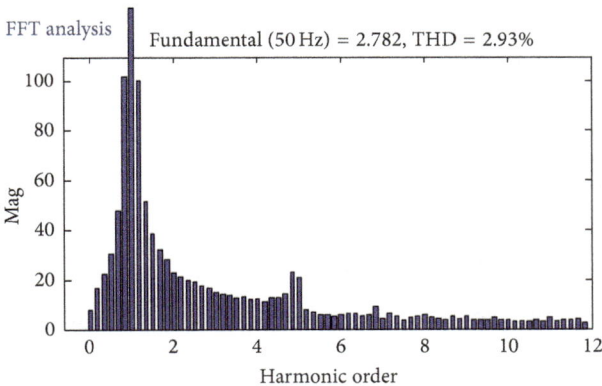

FFT analysis   Fundamental (50 Hz) = 2.782, THD = 2.93%

FIGURE 7: FFT analysis of load current at full load.

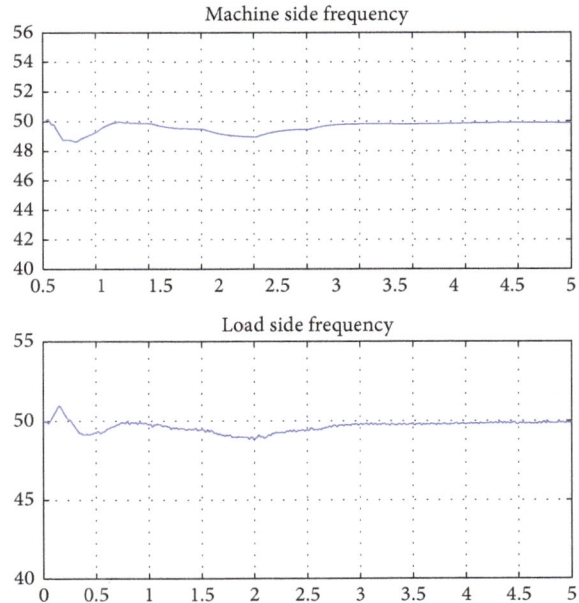

FIGURE 8: System frequencies.

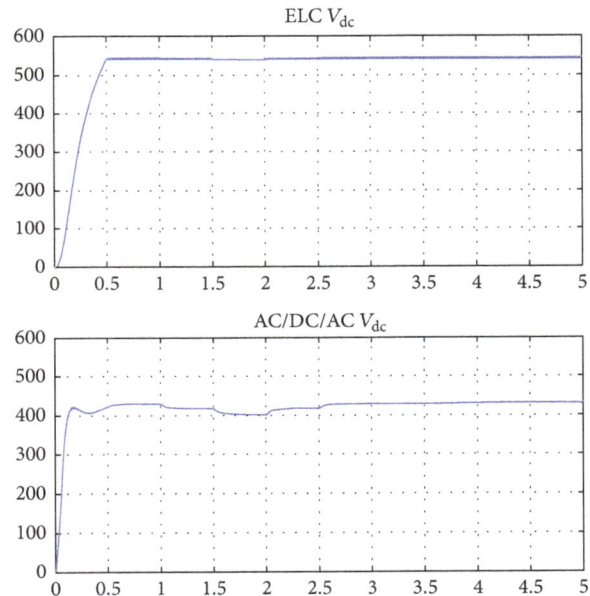

FIGURE 9: DC link voltage of AC-DC-AC converter and ELC.

Three-phase resistive loads are switched in at different instants; three-phase loads of 1 kW, 2.5 kW, and 1 kW, respectively, are connected and then disconnected at 1.0 s and 1.5 s, 1.5 s and 2.0 s, and 2.5 s and 3.0 s, respectively, as shown in Figure 6. As seen from the figure, the rated terminal voltage is maintained while the machine is loaded and there is a precise load sharing between the main load and the dump load. The generator is operated to produce a maximum power output of 2.5 kW. When the consumer's load is zero, the dump load takes the total generated power. On the other hand,

when the consumer's load is equal to the maximum generator output, no power is diverted to dump load. This act of power switching between the two ensures a constant total power output from the machine at constant terminal voltage.

The Fast Fourier Transforms (FFT) analysis of load current at full load indicates a total harmonic distortion (THD) of 2.93%. This is shown in Figure 7. The ELC starts working at 0.5 s when the voltage across the DC link capacitor builds up to 560 volts. Figures 8 and 9 show the system frequencies and DC link voltages of AC-DC-AC converter and ELC while Figure 10 shows the MATLAB/Simulink model of the

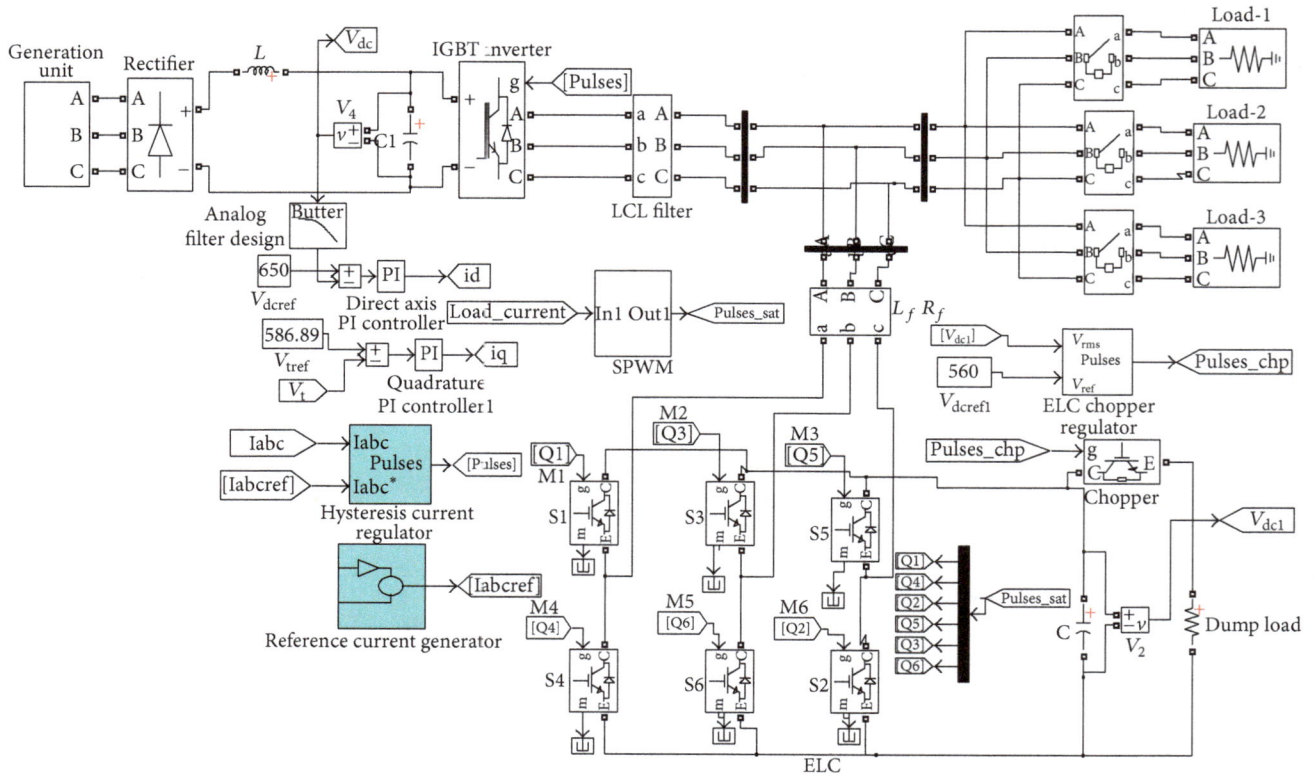

FIGURE 10: MATLAB model of AG-ELC system.

complete system. In this paper, variable water velocity is considered, as the analysis becomes easier when the input power remains constant, which means velocity of water remains the same, but in actual practice, it will never be constant; hence the use of dual converters plays a vital role here.

5. Conclusions

The work presented in this paper demonstrates a successful power switching between the main load and the dump load while maintaining a constant terminal voltage. The performance of ELC is satisfactory. The AC-DC-AC converter performs satisfactorily with independent control of active and reactive power thus giving a constant voltage and frequency output at the machine terminal. The proposed system may be adopted in rural areas where distributed power generation is a suitable option and the consumers are less. The power diverted to the dump load may also be gainfully utilized in battery charging and supplying power to some auxiliary circuit or equipment.

Appendices

A. Machine Parameters

Three-phase generating unit was as follows: 7.5 kW, 415 V, 14.5 A, 50 Hz, Y-connected, 4-pole, squirrel induction machine:

$R_s = 0.9\,\Omega$, $R_r = 0.66\,\Omega$, $X_{ls} = X_{lr} = 1.437\,\Omega$, and X_{ml} (sat.) $= 35.74\,\Omega$.

B. Controller Parameters for AC/DC/AC Controller

Consider $C_{dc} = 2000\,\mu F$, $K_{pa} = 0.0118$, $K_{ia} = 0.0018$, $K_{pd} = 0.036$, and $K_{id} = 0.0008$.

C. Savonius Rotor Parameters

The Savonius rotor parameters are as follows: D_r (rotor diameter) = 2 m, H_r (rotor height) = 2 m, C_p (power coefficient) = 0.25, ρ (water density at 25°C) = 997.0479 kg/m³, and inner diameter = 1.8 m, with thickness = 100 mm, no overlapping.

D. LCL Filter Parameters

Consider $L_f = 80$ mH, $C_f = 40\,\mu F$, $R_f = 30\,\Omega$.

E. ELC Parameters

The ELC parameters are as follows: current rating = 17.36 A, DC capacitor rating (selected) = 3000 μF, rating of dump load (selected) = 50 Ω, controller parameter, $K_p = 0.5$, and $K_i = 8$.

Conflict of Interests

The authors declare that there is no conflict of interests regarding the publication of this paper.

References

[1] Md. Orai, T. Ahmed, M. Nakaoka, and M. Z. Youssef, "Efficient performances of induction generator for wind energy utilization," in *Proceedings of the 30th Annual Conference of IEEE Industrial Electronics Society*, Busan, South Korea, November 2004.

[2] T. Abbasi and S. A. Abbasi, *Renewable Energy Sources: Their Impact on Global Warming and Pollution*, PHI Publication, 3rd edition, 2013.

[3] Energy Next, "Mission possible: potential and installed capacity," *Monthly Magazine, Ministry of New and Renewable Energy (MNRE), Government of India*, vol. 5, no. 6, pp. 14–21, 2015.

[4] S. Gao and R. Pudur, "Harnessing hydroelectric power using Savonius rotor coupled with asynchronous generator connected to grid," in *Proceedings of the IEEE PES Asia-Pacific Power and Energy Engineering Conference (APPEEC '13)*, pp. 1–4, Kowloon, Hong Kong, December 2013.

[5] R. Pudur and S. Gao, "Savonius rotor based hydropower generation," in *Proceedings of the International Symposium on Aspects of Mechanical Engineering & Technology for Industry (AMETI '14)*, vol. 1, pp. 104–111, Itanagar, India, December 2014.

[6] M. N. I. Khan, T. Iqbal, M. Hinchey, and V. Masek, "Performance of savonius rotor as a water current turbine," *The Journal of Ocean Technology*, vol. 4, no. 2, pp. 71–83, 2009.

[7] U. K. Saha and M. Jaya Rajkumar, "On the performance analysis of Savonius rotor with twisted blades," *Renewable Energy*, vol. 31, no. 11, pp. 1776–1788, 2006.

[8] Md. Hadi Ali, "Experimental comparison study for Savonius wind turbine of two & three blades at low wind speed," *International Journal of Modern Engineering Research*, vol. 3, no. 5, pp. 2978–2986, 2013.

[9] G. Kailash, T. I. Eldho, and S. V. Prabhu, "Performance study of modified savonius water turbine with two deflector plates," *International Journal of Rotating Machinery*, vol. 2012, Article ID 679247, 12 pages, 2012.

[10] B. Singh, G. K. Kasal, A. Chandra, and K.-A. Haddad, "Electronic load controller for a parallel operated isolated asynchronous generator feeding various loads," *Journal of Electromagnetics Analysis and Applications*, vol. 3, no. 4, pp. 101–114, 2011.

[11] B. Singh, G. K. Kasal, and S. Gairola, "Power quality improvement in conventional electronic load controller for an isolated power generation," *IEEE Transactions on Energy Conversion*, vol. 23, no. 3, pp. 764–773, 2008.

[12] B. Singh, S. S. Murthy, and S. Gupta, "Transient analysis of self-excited induction generator with electronic load controller (ELC) supplying static and dynamic loads," *IEEE Transactions on Industry Applications*, vol. 41, no. 5, pp. 1194–1204, 2005.

[13] S. Gao, G. Bhuvaneswari, S. S. Murthy, and U. Kalla, "Efficient voltage regulation scheme for three-phase self-excited induction generator feeding single-phase load in remote locations," *IET Renewable Power Generation*, vol. 8, no. 2, pp. 100–108, 2014.

[14] J. M. D. Murphy and F. G. Turnbull, *Power Electronic Control of AC Motors*, Pergamon Press, New York, NY, USA, 1990.

21

Techno-Economic Feasibility of Small Scale Hydropower in Ethiopia: The Case of the Kulfo River, in Southern Ethiopia

Zelalem Girma[1,2]

[1]*Electrical and Computer Engineering Department, Arba Minch University, P.O. Box 21, Arba Minch, Ethiopia*
[2]*University of Kassel, Mönchebergstraße 19, 34109 Kassel, Germany*

Correspondence should be addressed to Zelalem Girma; zelalem.girma@amu.edu.et

Academic Editor: Shuhui Li

This paper presents the technical and economic feasibility of grid connected small scale hydropower construction in selected site of the Kulfo River in southern Ethiopia. In doing so the paper presents the general overview of Ethiopia electric power situation; small scale hydropower situation and barriers and drivers for its development; site assessment and cost estimation methods and at the end presents techno-economic analysis of small scale hydropower development on the Kulfo River in southern Ethiopia. The technical and economic feasibility of the site have been studied by using HOMER, RETscreen, and SMART Mini-IDRO software. The result of simulation shows that the construction of small scale hydropower in the Kulfo River is technically and economically feasible with total net present cost of $13,345,150, cost of energy $0.028/kWh, simple payback period of 12.4 year, and internal rate of return 12.9%. The result also shows that construction of hydropower curtails greenhouse gas emissions such as carbon dioxide by 96,685,45 kg/year, sulfur dioxide by 4,1917 kg/year, and nitrogen dioxide by 20,500 kg/year.

1. Introduction

Ethiopia is located in east Africa with total area of 1.1 million sq. kilometres and a population of more than 90 million and is endowed with enormous renewable energy resources that include 45,000 MW hydropower, 10,000 MW geothermal power, 1,350,000 MW wind, and massive solar and biomass potential [1]. Biomass covers 90% of the total energy consumption, mainly used for cooking in the household. Hydropower contributes significantly to electric generation; the current installed electrical capacity reached 2268 MW and two big hydropower projects with capacity 1870 MW (Gilgel Gibe III) and 6250 MW (Grand Renaissance dam) are under construction. The installed capacity is expected to jump to about 8,000–10,000 MW by the end of the growth and transformation plan (2015) [2].

The country power generation dominated by large hydropower. The mountainous landscape feature coupled with hydrological condition enables the country to generate electricity from hydropower at relatively lower cost when compared to other energy sources. The energy consumption of the country is 45 kWh/capita which is the lowest when compared to averages of 578 and 2752 kWh/capita for Africa and the world, respectively [3, 4]. The total electric access rate is around 41% and less than 10% of the rural people connected to the national grid. The government has taken different measures to increase electrification access in the country of which formulation of energy policy in 1994 is one of the positive drives [5]. The policy encourages the use of indigenous resources and renewable energy to secure energy supply and reduce use and dependency on fossil fuel. The policy puts hydropower resource development as top priority due to availability of high potential site suitable to generate electricity at relatively lower cost. Furthermore, the revised policy in 1997 and 2013 encourages private independent power producer (IPP) to participate in energy generation by formulating necessary incentives and feed in tariff law [6, 7]. The revised policy also gave due attention for rural electrification by using renewable energy based off-grid technology.

Ethiopian electric power corporation (EEPCO) and Ethiopian rural energy development and promotion centre (EREDPC) are the implementing agencies of grid expansion and off-grid electrification for rural area, respectively, under

Ministry of Water, Irrigation and Energy [8]. EREDPC is mandated for off-grid access expansion by promoting private sector led off-grid rural electrification through participation of the private sector, cooperatives, community based organization, and local government where EEPCO cannot cover them due to economic terms. According to a 25-year master plan, EEPCO focused on the development of medium and large hydropower plant [9] even though the country has substantial rivers and streams suitable for small scale hydropower development.

The country generates around 91% of its power from large scale hydropower and small scale hydropower development gets little attention from the government side and contributes a small portion in the energy pool of the country. The total generation potential of hydropower is estimated to be 45 GW of which only 2% is taped to date [10, 11]. The government five-year (2010–2015) growth and transformation plan mainly focus on the development of large hydropower, whereas small and micro hydropower development have been left to private sector and NGO who are willing to support rural electrification program. As a result the contribution of small and micro hydropower in the energy pool of the country is insignificant. However, there are numerous potential sites identified by the government to generate electric power in small, mini, and micro hydropower capacity. Currently there are few small hydropower plants operational; most of them built by the German Cooperation Organization (GIZ). According to [6], the potential of small and micro hydropower development of the country is estimated from 1500 to 3000 MW or about 10% of the overall hydropower potential. If this potential is exploited and put into operation, it could provide a considerable contribution to the energy mix of the country by meeting the power deficit in the national grid, substituting diesel generators in main and isolated grid and electrifying remote rural area.

In recent times the country has registered remarkable economic performance with average annual growth of 10% over the past 10 years, which is double the sub-Saharan Africa and triple the world average growth over this period [12]. The fast growing economy demands a high energy with annual consumption rate increment of 25%. In recent times, the imbalance between demand and supply of electricity coupled with the inefficiency of electric utility service created huge gap and also negatively affected the economy of the country. The development of small hydropower in potential rivers in the country with low construction and commissioning time will alleviate the power imbalance.

Therefore, this paper examines techno-economic feasibility of small hydropower development on the Kulfo River in the Gamo Gofa zone, near to Arba Minch town in the southern part of Ethiopia to give insight to government, private sector investors, and interested NGO who are willing to contribute to small scale power generation development of the country.

The paper is organized in eight sections. Section 1 is an introduction; Section 2 describes the situation of small scale hydropower development in Ethiopia, its classification, barriers, and drivers; the working principle is described in Section 3; Section 4 discusses site assessment and cost

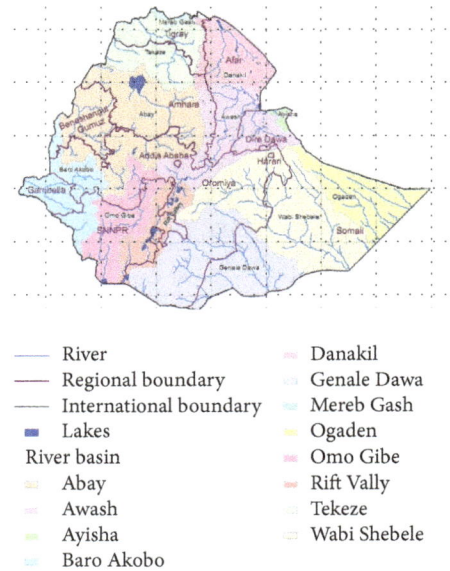

FIGURE 1: Location of major river basin in Ethiopia [15].

estimation method; Section 5 discusses the background of the study site and load profile; Section 6 discusses methodologies; the simulation result will be discussed in Section 7; and conclusion is put in Section 8.

2. Small Scale Hydropower Development in Ethiopia

Small scale hydropower is estimated to be 10% of the total hydropower potential of the country. However, in terms of technical feasibility, the potential could be reduced by more than half to about 5% due to inaccessibility, and proximity to grid and service centres [13]. The available potential of small scale hydropower in the country has hardly been exploited so far due to government focus on large scale hydropower development to meet the energy demand of the country.

As feasibility study, the government identified around 299 hydropower potential sites within eleven river basins with a total potential of 7877 MW including both large and small hydropower. Figure 1 shows major location of river basin in Ethiopia. The Abay river basin is the largest basin in terms of hydropower potential site estimated about 79000 Gwh/yr which cover about 49% of all river basins [14]. The potential for small scale hydropower lies in western and southwestern Ethiopia, where annual rainfall ranges from 300 mm to over 900 mm especially in Omo Gihbe basin and Abay basin.

2.1. Classification of Hydropower Plant. The hydropower plant is classified broadly into different classes based on quantity of water available, available head, and nature of the load [16]. However, classifications vary from country to country as there is currently no internationally agreed standard. Ethiopia uses a classification of hydropower systems which differs from other countries as shown in Table 1.

TABLE 1: Hydropower classification in Ethiopia [17].

Terminology	Capacity	Unit
Large	>30	MW
Medium	10–30	MW
Small	1–10	MW
Mini	501–1000	kW
Micro	11–500	kW
Pico	≤10	kW

In the past majority of small scale hydropower schemes in the country were abandoned due to the encroachment of the national grid with cheaper and more reliable electricity. Currently only one small and two mini hydropower (MHP) schemes are functional under EEPCOs Self-Contained System (SCS), namely, Sor (5 Mw), Yadot (350 kW), and Dembi (800 kW), with a cumulative installed capacity of 6.15 MW. Moreover, another four new small hydropower schemes (Gobecho I = 7 kW, Gobecho II = 30 kW, Hagara Sodicha = 55 kW, and Ererte = 33 kW) have been installed in the southern part of Ethiopia in Sidama zone with the help of the German Cooperation Organization (GIZ) as pilot project in 2011 [17].

To facilitate and support the financing of small scale hydropower scheme the government has also set aside rural energy development and promotion centre under Ministry of Water, Irrigation and Energy, mandated to

(i) promote small scale hydropower and other renewable energy sources,

(ii) provide financial support to develop SHP and other renewable energy sources by setting rural electrification fund.

Furthermore, feed in tariffs is under review to encourage private sector participation in power sector development. Therefore, the government incentives, policy, and regulations put SHP business in favourable condition in Ethiopia in recent times.

2.2. Drivers and Barriers of Small Scale Hydropower Development in Ethiopia. There are several pull and push mechanisms set by the government in order to spur the market of SHP despite considerable barriers for market development.

2.2.1. Drivers

(i) Favourable renewable energy policy: the policy favours the development of electric power from renewable energy sources and established Ethiopian energy agency to be mandated to regulate the electricity market, electricity price regulation, power purchase agreement (PPA), licensing of independent power producer (IPP), and regulating access to the grid by private power producer.

(ii) Establishment of Ethiopian rural energy development and promotion centre (EREDPC): it is established at the federal level with a mandate to promote renewable energy technology for rural electrification by setting aside rural energy fund by collecting donation from different organization and government and give soft loans with low interest rate for private power producer.

(iii) Feed in tariff: the government of Ethiopia announced feed in tariff for powers purchased from IPP for different types of renewable sources which encourages IPPs to enter into power generation business.

(iv) Introduction of climate resilient green economy strategy (CRGE): Ethiopia initiated and implemented this policy strategy to participate in global climate change mitigation campaign and protect the country from climate change and as a result planned to develop 25 GW of electricity from renewable energy source (22 GW from hydro + 1 GW from geothermal power + 2 GW from wind)

2.2.2. Barriers

(i) Absence of expertise to fabricate parts, work, and maintain small hydro power plant in the country is one of the barriers.

(ii) Inaccessibility of small and micro hydro power spare parts in local market is another barrier.

(iii) Low proposed feed in tariff results in low return on investment for IPP discouraging the private investment.

(iv) Expansion of irrigation projects in small hydrostreams may prevent hydropower development in downstream.

3. Working Principle of Small Scale Hydropower

The working principle of small hydropower is not different from that of large scale hydropower. It captures the energy of falling water to generate electricity. The water turbine, which is different type depending upon the head and flow rate, converts the energy of falling water into mechanical energy [18]. The electric alternator or generator coupled with turbine converts mechanical energy of rotating shaft into electrical energy according to Faradays' law of electromagnetic induction. The amount of electricity produced mainly depends upon the two factors [19]: a) head: the distance that the water falls; b) flow rate: the volume of water that pass through a given point per second usually measured in meter cube per second. For fixed head the more the water is falling per second on the turbine, the more the power will be produced and vice versa. The flow rate of a given stream may vary seasonally depending upon the location of the site. Different types of water turbine can be used to convert kinetic energy of the flowing water into mechanical energy (rotation of the shaft). The selection of the turbine depends upon head and flow rate as explained in [20, 21]. Furthermore, care has to be taken in terms of constructability, cost, efficiency, maintenance and serviceability, portability, and scope of modularity during turbine selection.

4. Site Assessment and Cost Estimation

Assessment of the site is a prerequisite in any hydropower development [22–24]. From the result of site assessment one can decide whether the given site is a viable option for hydropower development or not [25]. The key parameters during the assessment are the pressure head, the flow rate of the given river, and wire to water efficiency of the overall system. This parameter can be easily found through measurement and manufacturer specification. Then the power which can be generated at a specific site can be calculated by using the following formula:

$$P = 9.81 \times Q \times H \times \eta, \qquad (1)$$

where P is power output in kW, Q is turbine flow in m^3/sec, H is net head in meter (elevation between intake at the river and out late at the turbine less head loss along the power channel), 9.81 is acceleration due to gravity (m/sec^2), and η is overall efficiency of the system.

As seen from the above equation the power generated from the turbine depends upon the discharge rate Q, the net head H, and overall efficiency of the system since other variables are constant in the equation. For the same power output one can either increase head or discharge rate. Usually the head is site-dependent and could not be varied. However, the flow rate can be varied by controlling the water entering into the penstock. However, the turbine should have a capacity to accommodate the increased discharge.

Furthermore, the head, the discharge, and the desired rotational speed of the generator determine the type of turbine to be used. More head or faster flowing water means more power.

Design flow is the maximum flow for which the hydrosystem is designed. It will likely be less than the maximum flow of the stream (especially during the rainy season), more than the minimum flow, and a compromise between potential electrical output and system cost [26]. The flow duration curve (FDC) provides means of selecting the right design discharge by taking into account reserved (residual) flow for environmental and aquatic life purpose. Usually the design flow is assumed to be the difference between the mean annual flow and the residual flow [27]:

$$Q_{\text{design}} = Q_{\text{mean}} - Q_{\text{residual}}. \qquad (2)$$

Once the design flow and net head are estimated, suitable head can be selected from turbine selection chart and also note that every turbine has a minimum technical flow under which the turbine cannot operate or has very low efficiency.

In general, planning a hydropower project is a complex and iterative process, where consideration is given to the environmental impact, technological options, economic evaluation, and other constraints. Even though it is difficult to provide a detailed guide on how to evaluate a hydropower scheme, it is possible to provide a short feasibility study of

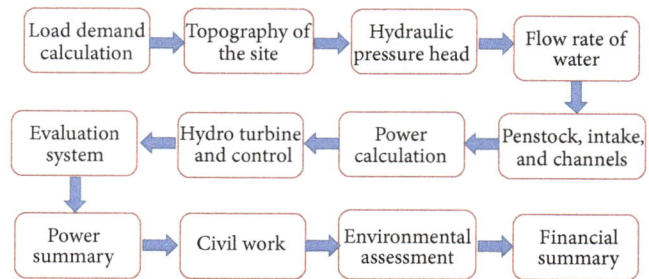

FIGURE 2: Planning and evaluation of a small hydropower plant [30].

a given site configuration in order to develop the project [28, 29]. Figure 2 shows the steps of developing and planning a micro hydropower project [30].

4.1. Cost Estimation. The geographical and geological features along with the effective head, available flow, equipment (turbines, generators, etc.), and civil engineering works determine the capital required for any small hydropower project [31]. In general the cost of hydropower project highly depends upon the site and the location of the project, whether the parts are manufactured locally or imported, and the availability of local skilled manpower to construct and maintain the plant.

Among the many factors that affect the cost of a project are site topography, rock quality, availability of access roads, and the distance to the interconnected grid, earthquake risk, and sediment load in the river [32]. Of course, hydrology and local cost of labor, cement, steel, and explosives also must be factored into the cost equation. In order to grasp the cost structure of hydropower plant around the world and Ethiopia search and review of literatures have been carried out from relevant published papers and reports [33, 34]. Several studies have been carried out to analyze the cost of small hydropower development depending upon the hydraulic characteristics of a given site and a number of cost estimation equations were developed to suite the site specific condition. The researchers on [35–37] developed empirical equations to estimate the cost of hydropower projects based on cost of electromechanical equipment, installed power, hydraulic head, location factors, and so forth. However, developed equations have limitation to apply for all countries in the world since the assumptions used were not inclusive of the nature in all countries. Therefore the World Bank group and IEA [38] studied extensively the project cost of different hydropower projects in the globe and come out with the cost range table depending upon the hydropower type (Table 2) [38].

A recent study of International Renewable Energy Agency (IRENA-2012) [39] also shows that the investment cost of large hydropower plants with storage typically ranges from as low as USD 1050/kW to as high as USD 7650/kW while the range of small hydropower projects is between USD 1300/kW and 8000/kW depending upon the site condition. Figure 3 shows the investment cost in different country, including Ethiopia, and confirms the investment cost report by the Ethiopia's Ministry of Water, Irrigation and Energy which

TABLE 2: World Bank and IEA cost estimate of hydropower [38, 40].

Project cost $/kW	Head range (m)	Remark
Estimate of World Bank group 2012		
1800–8000	2.3–13.5	Low head
1000–3000	27–350	High head
Estimate of IEA 2010		
2000–7500	Small scale hydropower	
2500–10000	Mini hydropower	
1500–2500	Low head hydropower	

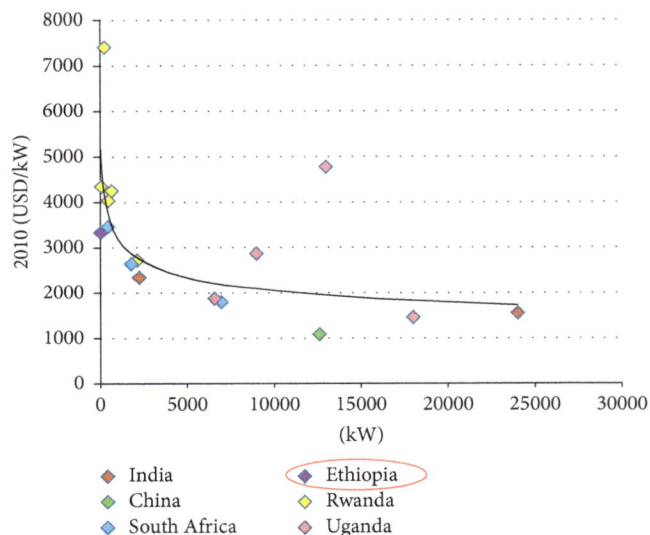

FIGURE 3: Installed capital costs for small hydro in developing countries by capacity [39].

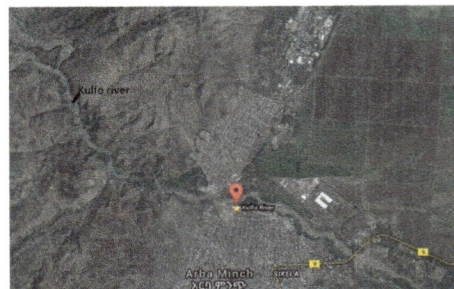

FIGURE 4: Google satellite map image of the site.

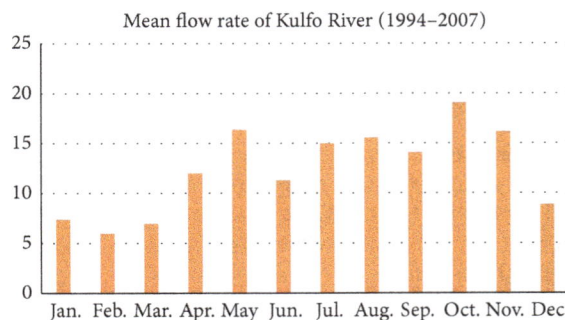

FIGURE 5: Average stream flow of the Kulfo River (1994–2007).

ranges from 3500 to 4000 $/kW [39]. In the economic assessment of proposed hydropower $3500/kW was used to estimate the capital cost of the plant.

5. Background of the Study Site

The Kulfo River basin is situated in relatively dry southern area of the Ethiopia in Gamo Gofa zone near to Arba Minch town at latitude 6°N and 37.5°E and is still under geographical modification with hilly topography and impervious soil texture as shown in Figure 4. The river flows through Arba Minch forest and drains into Lake Chamo. The site was selected due to the fact that the river flows throughout the year; it is near to national grid so that it can be easily connected to national grid with low grid interconnection charge; the train of the site is very suitable for hydropower development; and the construction of the power plant does not have social and environmental impact.

As seen from the eleven-year (1994–2007) daily flow data of the river in Figure 5 which was collected from Arba Minch University gauging station, the river has a minimum flow rate of 6 m³/sec. on February and maximum flow of 19.1 m³/sec. on October. Its average flow rate is 12.4 m³/sec. The river has high daily and intermonth variability and low interyear

variability. In cases where there were gaps within the data, due to temporary failures of the measuring equipment, the record system, or any other reason, the gaps were noted and the average data of the previous day and the day after the missed data was taken. Furthermore, the data sets were carefully screened for anomalies. Figure 5 shows the monthly flow rate of the Kulfo River.

Prefeasibility study has been done on the site in order to get basic information on the situation of the site, the variability of the river, and the demographic and topographic nature of the site and also to analyze the suitability of the topography for hydropower generation. Form feasibility study it is noted that, in downstream of the river, there is agricultural land owned by private investor which uses part of the river for irrigation. Part of the river is also used by people settled along the water shed of the river. As a result, the location selected for construction of small scale hydropower is above the agricultural land and does not affect the operation of farming in downstream.

As shown in Figure 5 the river has high variability and is not suitable to construct run of river scheme without diversion. The diversion also helps to settle the debris and to control flow of water during rain and dry seasons. Due to hydraulic head limitation and flow constraints the maximum economical potential of the river has calculated as 2.2 MW by taking 50% of available flow rate (9522 L/sec.) at the gross head of 25 meters.

SMART Mini-IDRO software [41, 42] has been used to draw flow duration curve (FDC) and to analyze the preliminary electric generation potential of the river. SMART

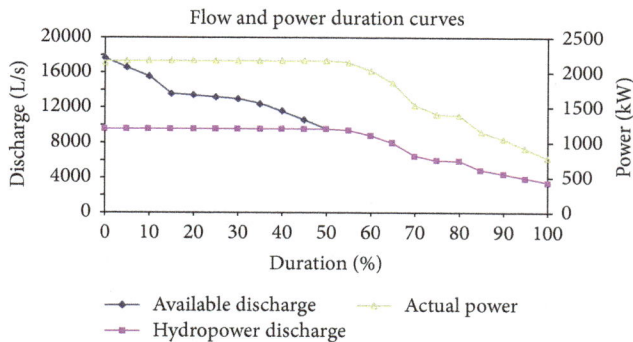

FIGURE 6: FDC and power curve.

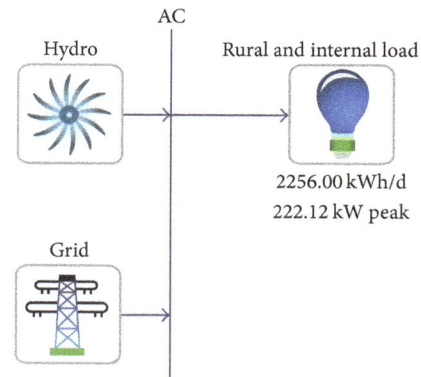

FIGURE 7: HOMER system configuration.

Mini-IDRO is a tool for technical and economical evaluation of mini hydropower plants and evaluates the energy production, benefits, and financial aspects and assesses the discharge availability. From the SMART Mini-IDRO software analysis, the river has theoretical potential of 4.5 MW, the technical potential of 4 MW, and economic potential of 2.2 MW. Figure 6 shows FDC and power curve of the site.

6. Methodology

Extensive literature review has been done to grasp the status of electrification and its challenge in the country by giving particular attention on small scale hydropower development to know its past and present status, drivers, barriers, and deployment. The site assessment and cost estimation method in hydropower development have also been reviewed. After getting overall situation on electrification status, small and large scale hydropower development, site assessment, and cost estimation methods, case study site (the Kulfo River) has been selected in southern Ethiopia near Arba Minch town with the following assumptions:

(i) The developed small hydropower is intended to be owned by the private power producer (IPP).

(ii) The hydropower first supplies the rural village nearby and supplies surplus power to the national grid at Low voltage.

(iii) EEPCO is the only buyer of surplus electricity with agreed feed in tariff.

(iv) Small hydropower station can purchase power from EEPCO during dry season when hydropower fails to supply full load to the rural village.

With the above assumptions techno-economic analysis of the small hydropower constructed on the Kulfo River has been done by using HOMER, RETscreen, and SMART Mini-IDRO software. HOMER is micro power optimization model developed by U.S. National Renewable Energy Laboratory (NREL) to assist the design of micro power system and to facilitate the comparison of different technologies [43, 44]. The software can model off-grid and grid connected power system. It performs three principal tasks: simulation, optimization, and sensitivity analysis. In the simulation process,

the software models the performance of a micro power system configuration each hour of the year to determine its technical feasibility and life cycle cost. In the optimization process it searches among feasible options the one that satisfies technical constraints at the lowest life cycle cost. In the sensitivity analysis process it assesses the effect of uncertainty or change in the variable over which the designer has no control such as change in flow rate, interest rate, and inflation rate. HOMER uses net present cost (NPC) method to represent the life cycle cost of the system and rank the optimal feasible one according to total net present cost and present the feasible one with lowest total net present cost as the optimal system.

HOMER software has been used to find the optimal total net present cost (TNPC), generation cost of the power plant, to do sensitivity analysis on determinant but uncontrollable variables (flow rate, inflation rate, load change, and grid sale capacity), to compute the total amount electricity purchased from and sold to the grid in kilowatt hour (kWh). RETscreen software has been used to compute simple payback period, internal rate of return and to draw cumulative cash flow within project life time. SMART Mini-IDRO software is used to draw flow duration curve, to determine design flow, and to compute theoretical, technical, and economic potential of proposed hydropower.

7. System Configuration and Simulation Result in HOMER Environment

Figure 7 shows the configuration of the proposed grid connected small scale hydropower with local and internal load in HOMER simulation and optimization environment. The configuration contains hydro, grid, and load as main component.

7.1. Hydro Component. The hydro component in the simulation needs equipment capital cost, replacement cost, maintenance, and operation cost as input variable; the system lifetime for economic evaluation and available head, design flow rate, percentage of minimum and maximum flow rate, efficiency, and pipe friction losses as turbine parameter. After inserting input variables HOMER calculates the electrical

TABLE 3: Turbine and economic input parameter in HOMER software.

Turbine parameter		Description
Available head (m)	25	Net head
Design flow rate (L/sec)	9,522	50% duration
Minimum flow ratio (%)	10	10% of design flow rate which is limited by the turbine to start generation of power
Maximum flow ratio (%)	100	100% design flow
Efficiency (%)	96	Turbine efficiency
Penstock pipe loss	1.4%	
Economic parameter ($)		Description
Capital cost	7,847,000	$3500/kW
Replacement cost	7,800,000	Assumed
O&M cost	313,880	4% of capital cost
Lifetime of the project	30 years	Project lifetime

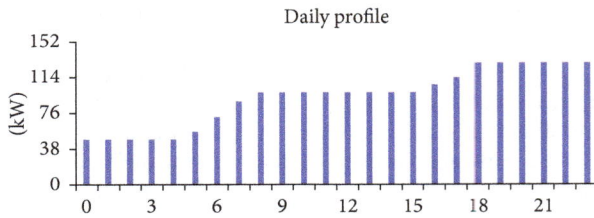

FIGURE 8: Daily load profile of hypothetical village.

power output of the hydro turbine using the following equation:

$$P_{\text{hyd}} = \frac{\eta_{\text{hyd}} \cdot \rho_{\text{water}} \cdot g \cdot h_{\text{net}} \cdot \dot{Q}_{\text{turbine}}}{1000 \text{ W/kW}}. \tag{3}$$

Table 3 shows the input value used in hydro component for proposed hydropower design in HOMER software. The calculated electric power by using (3) is 2,241.86 kW.

7.2. Load Component and Analysis of the Site.

For load analysis, hypothetical nearby villages with a peak electrical load of 222 kW, with average energy consumption of the 2256 kWh/day, and with load factor 0.42 were assumed. The assumed load composed of the household appliance, the small enterprise electric machines such as a saw mill, electric welding machine, and other machines used by small enterprises. The daily load profile is as shown in Figure 8.

The consumers grouped into low, medium, and high income class according to yearly income they can generate. The communal services such as school, administrative building, and religious institutions are also considered in load estimation. The assumption was based on the survey made

TABLE 4: Rate schedule (Step 1: define and select a rate).

Rate	Price ($/kWh)	Sellback ($/kWh)	Demand ($/kW/mo)
Rate 1	0.048	0.060	0.000
Rate 2	0.048	0.080	0.000

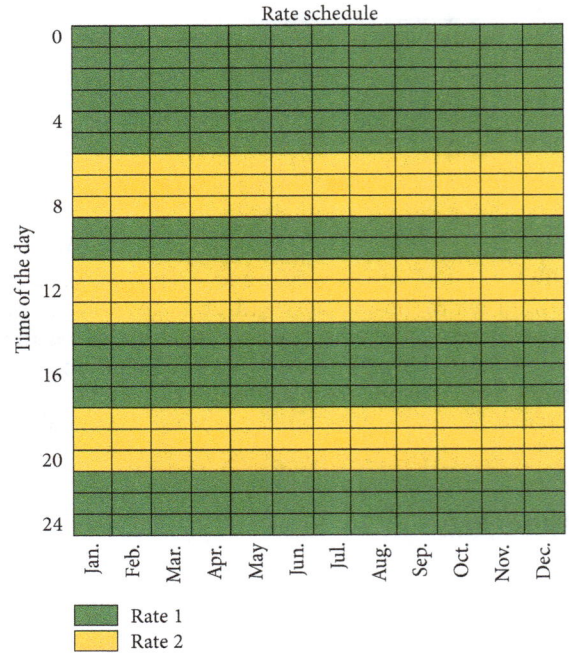

FIGURE 9: Grid sale and purchase rate schedule (see Table 4).

on the grid connected village with the same socioeconomic condition of the hypothetical village.

In order to simulate the load in a more realistic way 10% day-to-day and 20% time-to-time random variable were added in load profile. Furthermore to include future load growth sensitivity analysis has been done on the total load. The result of simulation shows that the proposed hydropower can supply energy of 11,086 kWh without purchasing the power from national grid, above which it starts to purchase power from the grid in order to bridge supply shortage.

7.3. Grid Component.

HOMER software has the capacity to simulate grid connected power generation and in doing so it takes as input purchase and sellback rate of electricity. Figure 9 and Table 4 show the rate and grid schedule used in simulation. According to draft feed in tariff document [45], the EEPCO purchases power from IPP with a rate of 0.06 $/kWh during off-peak period (rate 1) and 0.08 $/kWh during peak hours (rate 2) for hydropower based generations and sale with rate 0.048 $/kWh irrespective of peak hours. Furthermore, in the draft feed in tariff proposal the utility requests the IPP to cover grid connection cost and this is assumed to be 4% of the total investment cost in simulation. The maximum power that IPP can purchase from grid is limited to 500 kW and with this scenario IPP can sale up to 2200 kW of the power to the grid after covering the internal

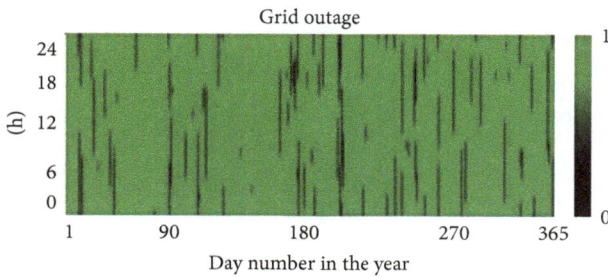

FIGURE 10: Reliability of national grid.

and village load. The simulated rate and schedule are shown in Table 4 and Figure 9.

Since the national grid in Ethiopia is not reliable, grid reliability issue was also included during simulation. The mean failure frequency is taken as 80, repair time variability is taken 90%, and mean repair time was assumed as 5 hours after grid failure. The result of reliability analyses is shown in the following by using random grid outage.

The black lines in Figure 10 show grid outage during which the proposed hydro power could not sell to the grid. Therefore, during this time some way of frequency control needed in power stations.

7.4. Sensitivity Analysis. In order to accommodate the uncertainty of some variables during simulation sensitivity analysis on some essential variables has been done. Sensitivity analysis is used to evaluate the effects of uncertainty on selected input parameters. It is used to quantify the economic consequences of a potential, but uncontrollable changes in important parameters in the future [46, 47]. The sensitivity analysis is very essential during simulation in HOMER software since it gives answers to the designer what if questions. In simulation the following sensitivity variables have been used:

   (i) Total load.

  (ii) The designed flow rate.

 (iii) Inflation.

 (iv) Grid sale capacity.

These variables have the most uncertainty factor in the design. For example, the load may increase in the future as new consumers connected to the local grid and the power consumption of existing user may rise due to usage of electricity for income generation. The flow rate of the river varies throughout the year and accordingly the design flow rate. From flow duration curve it has been seen that there is high flow time and low flow time and two or three turbines may be used to efficiently utilize the available flow.

The inflation is one of the volatile variables in Ethiopia even if the government puts several measures in order to control it. Therefore, this variable has also been used in sensitivity analysis. Figures 11 and 12 show the impact of inflation and the design flow rate on the energy cost of the overall system. Figures 11 and 12 show impact of flow rate and inflation on cost of energy.

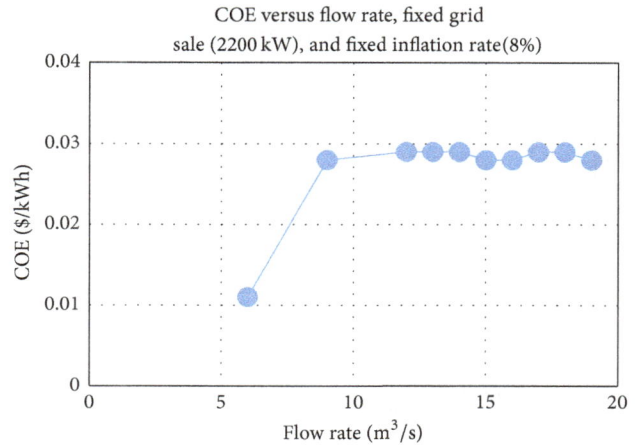

FIGURE 11: Relation between costs of energy versus variation in design flow rate.

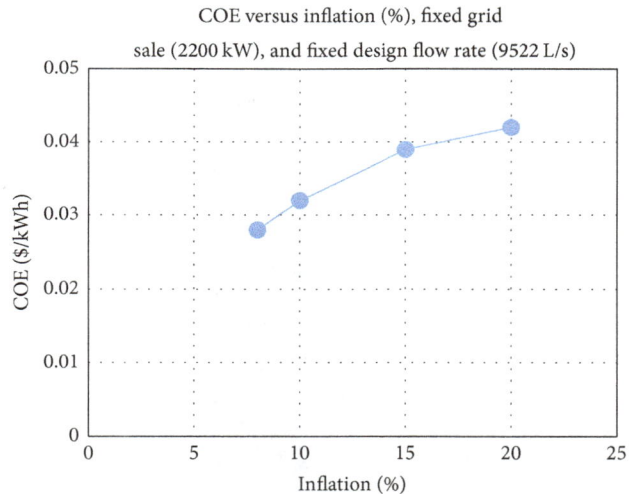

FIGURE 12: Relation between costs of energy versus inflation.

The local load demand as shown in Figure 13 is 818,695 kWh/yr which is 5.1% of the total generation of proposed hydropower (16,116,005 kWh/yr). This indicates that the proposed hydropower can cover the local load without buying the power from the national grid until the local load reaches its production capacity. Even in the worst month, month of February, with a flow rate of 6 m^3/sec. and even if this flow rate persists throughout the year the hydropower can generate 11,373,948 kWh/yr and can cover its local load demand.

As shown in Figure 14 the proposed hydropower has significant capacity to sell the surplus electricity to the grid. The lowest energy sold occurred on February which is the dry season and the higher sales occurred on the months of May, July, and August. The highest sale occurs on October. According to simulation the total amount of energy that can be sold to the grid was around 15,298,333 kWh/year.

Comparison of energy generated from hydropower and load demand per year

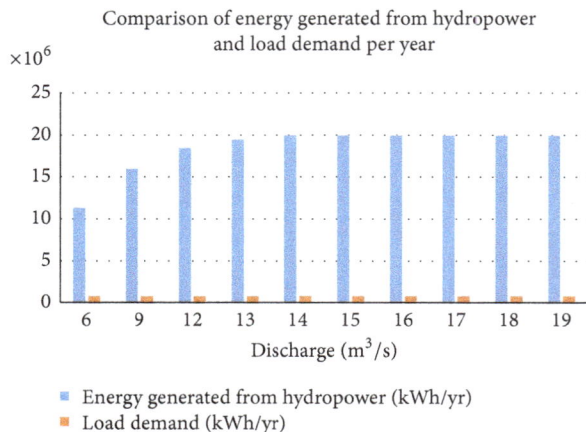

FIGURE 13: Energy generated versus load demand per year.

Energy sold to the grid year in kWh

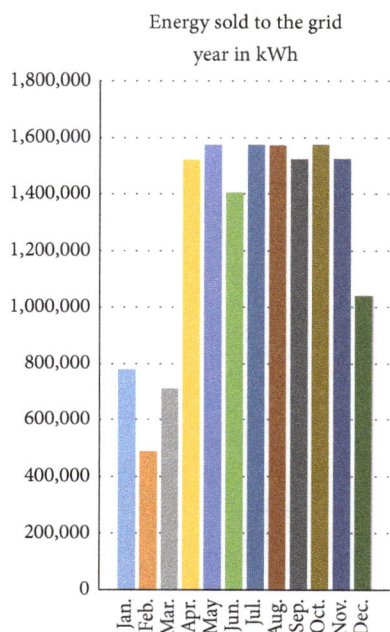

FIGURE 14: Energy sold to the grid in each month of the year.

The total earning from grid sale was around $1,031,914 per year. Figure 15 shows the monthly income from grid sell. The lowest sell. occurred in the month of February and the highest sale occurs in the month of October.

8. Conclusion

Ethiopia has immense potential for small scale hydropower development. However, to tap these potential active government engagement in facilitating policy and regulatory reform regrading small hydropower is needed. After decades of powers sector reform in the country which allows IPP to produce and sale electricity to national grid, no active participation is seen from private sector. The main bottleneck is the feed in tariff law which is not finalized yet. In addition,

Earning from grid sale in US dollar per year

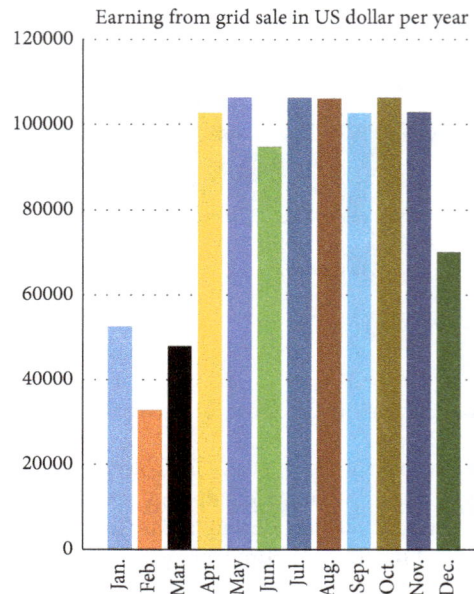

FIGURE 15: Monthly earning from grid sale.

Financial viability		
Pretax IRR-equity	(%)	12.9%
Pretax IRR-assets	(%)	4.9%
Simple payback	(yr)	12.4%
Equity payback	(yr)	13.4%

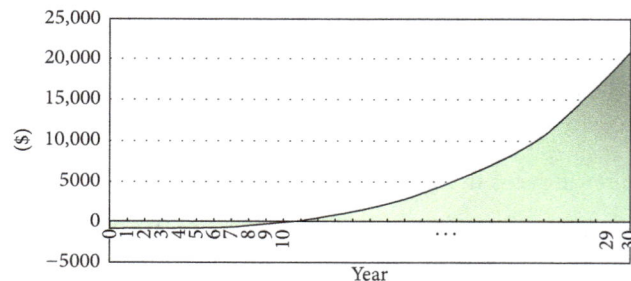

FIGURE 16: Cumulative cash flow of the project with RETscreen simulation software.

TABLE 5: System architecture.

Hydro	Hydroelectric	2,242	kW
Grid	Grid	500	kW
Dispatch strategy	Load following		

TABLE 6: Cost summary.

Total net present cost	−13345150	$
Levelized cost of energy	−0.028	$/kWh

government has to use various push and pull mechanisms to promote and motivate IPP in power generation market. Moreover the required data regarding small hydropower

TABLE 7: Net present costs.

Component	Capital	Replacement	O&M	Fuel	Salvage	Total
Hydro	7,847,000	0	9,416,400	0	0	17,263,400
Grid	313,880	0	−30,957,424	0	0	−30,643,544
Other	5,000	0	30,000	0	0	35,000
System	8,165,880	0	−21,511,026	0	0	−13,345,146

TABLE 8: Annualized costs.

Component	Capital	Replacement	O&M	Fuel	Salvage	Total
Hydro	261,567	0	313,880	0	0	575,447
Grid	10,463	0	−1,031,914	0	0	−1,021,451
Other	167	0	1,000	0	0	1,167
System	272,196	0	−717,034	0	0	−444,838

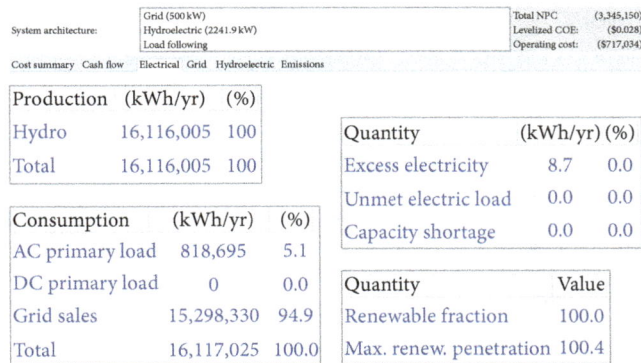

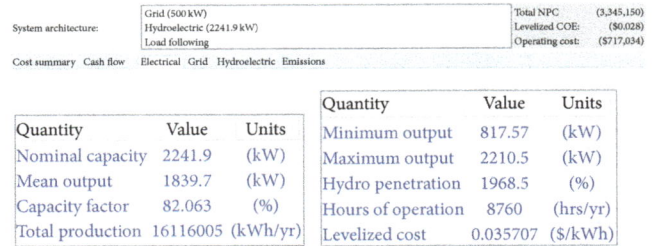

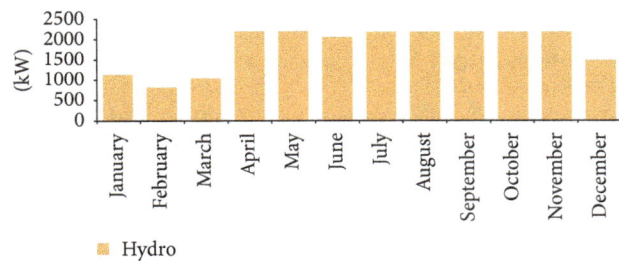

FIGURE 17: Electrical output.

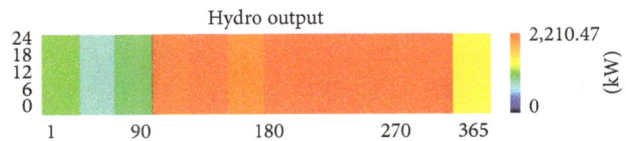

FIGURE 18: Hydroelectric output.

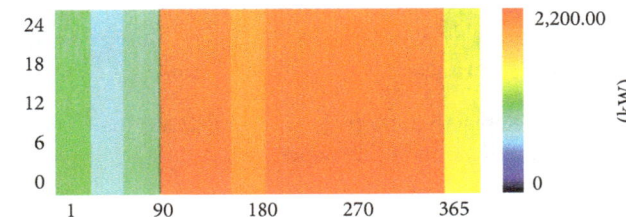

FIGURE 19: Energy sold to grid.

development has to be gathered and put into database so that interested IPP can access and do informed decisions.

In this work overall electrification status in Ethiopia and small scale hydropower development situation with its drivers and barriers have been reviewed in the first few sections. The policy and regulatory changes in powers sector reform have also been dealt. Then techno-economic feasibility study on selected site in southern region on the Kulfo River has been done in order to assess and study technical and economic feasibility of the project. Techno-economic analysis has been done by using HOMER and RETscreen software has been used to calculate payback period and IRR and also used to draw cumulative cash flow (Figure 16). Overall potential (theoretical, technical, and economic) of

proposed hydropower has been computed by using SMART Mini-IDRO software. The objective is to show the overall situation of small hydropower and its technical and economic feasibility by using simulation.

The result of HOMER simulation software shows that small hydropower development is profitable in the proposed specific site. It has very low levelized cost of energy (COE) around $0.028/kWh for proposed local load. It has also least total net present cost of $13,345,150 and can deliver 95% of the generated power to the grid after covering the local load. It is also seen from the result of RETscreen software that the project has simple payback time of 12.4 years with IRR of

TABLE 9: Electrical.

Quantity	Value	Units
Excess electricity	9	kWh/yr
Unmet load	0	kWh/yr
Capacity shortage	0	kWh/yr
Component	Production (kWh/yr)	Fraction (%)
Total	16,116,005	100
Load	Consumption (kWh/yr)	Fraction (%)
AC primary load	818,695	5
DC primary load	0	0
Grid sales	15,298,330	95
Total	16,117,025	100

TABLE 10: Emissions.

Pollutant	Emissions	Units
Carbon dioxide	−9668545	kg/yr
Carbon monoxide	0	kg/yr
Unburned hydrocarbons	0	kg/yr
Particulate matter	0	kg/yr
Sulfur dioxide	−41917	kg/yr
Nitrogen oxides	−20500	kg/yr

12.9%. As shown in the simulation of this particular site, small scale hydropower is a technical and economical feasibility in this specific selected site.

Appendix

Sample HOMER Simulation Output

System Report. See Tables 5, 6, 7, 8, 9, and 10 and Figures 17, 18, and 19.

Conflict of Interests

The author declares that there is no conflict of interests regarding the publication of this paper.

References

[1] D. Derbew, "Ethiopia's Renewable Energy Power Potential and Development Opportunities," 2013, http://www.irena.org/DocumentDownloads/events/2013/July/Africa%20CEC%20session%203_Ministry%20of%20Water%20and%20Energy%20-Ethiopia_Beyene_220613.pdf.

[2] Ethiopian Energy Authority, June 2015, http://www.ethioenergyauthority.gov.et/attachments/article/63/DRAFT%20FINAL-%20MARCH%206%202015%20%20%20Version%20IIIcomment.pdf.

[3] D. Chandrasekharam and V. Chandrasekhar, "Clean development mechanism through geothermal: Ethiopian scenario," in *Proceedings of the 4th African Rift Geothermal Conference*, Nairobi, Kenya, November 2012.

[4] A. Evans, *Resources, Risk and Resilience: Scarcity and Climate Change in Ethiopia*, Center on International Cooperation, New York University, 2012, http://cic.nyu.edu/sites/default/files/evans_security_ethiopia_2012.pdf.

[5] Small scale Hydro power for rural development, Nile basin capacity building network NBCBN, 2005, http://www.nbcbn.com/Project_Documents/Progress_Reports/Hp-G1.pdf.

[6] M. Abebe, "The Ethiopian power sector reform and its role for enhancing IPP investment opportunities," April 2015, http://www.engerati.com/hubs/articles/ethiopian-power-sector-reform-and-its-role-enhancing-ipps-investment-opportunities.

[7] M. Teferra, "Power sector reforms in Ethiopia: options for promoting local investments in rural electrification," *Energy Policy*, vol. 30, no. 11-12, pp. 967–975, 2002.

[8] USAID, "InvestmentBrief for energy sector in Ethiopia," http://www.usaid.gov/powerafrica.

[9] Ethiopian Power System Expansion Master Plan update (EPSEMPU), http://www.bgr.de/geotherm/ArGeoCl/pdf/06%20Energy%20master%20plan.pdf.

[10] M. Abebe, "Hydropower development in Ethiopia vis-a-vis perspective regional power trade," in *Proceedings of the International Information System for the Agricultural Science and Technology (AGRIS '05)*, pp. 155–180, 2005.

[11] Ethiopia Energy Situation, https://energypedia.info/wiki/Ethiopia_Energy_Situation#Hydropower.

[12] Country Economic Brief—UNDP in Ethiopia, http://www.et.undp.org/content/dam/ethiopia/docs/Country%20Economic%20Brief%201%20final%20for%20web.pdf.

[13] Shanko, Melessaw Target Market Analysis: Ethiopia's Small Hydro Energy Market, https://www.giz.de/static/themen_umleitung/index.html.

[14] W. Abtew and A. M. Melesse, "Enhancing water productivity in crop-livestock systems of the Nile Basin: improving systems and livelihoods," in *Proceedings of the Workshop on Hydrology and Ecology of the Nile River Basin under Extreme Conditions*, pp. 16–19, Aardvark Global, Addis Ababa, Ethiopia, June 2008.

[15] Major River Basins of Ethiopia, February 2015, http://www.idp-uk.org/Resources/Maps/TopographicThematic%20Maps/Major-River-Basins-Map.pdf.

[16] W. J. Klunne, "Small hydropower in Southern Africa—an overview of five countries in the region," *Journal of Energy in Southern Africa*, vol. 24, no. 3, pp. 14–25, 2013.

[17] K. Meder, *Application of environment assessment related to GIZ ECO micro hydropower plants in the Sidama Zone/Ethiopia [M.S. thesis]*, Heidelberg University, Heidelberg, Germany, 2011.

[18] O. Paish, "Small hydro power: technology and current status," *Renewable and Sustainable Energy Reviews*, vol. 6, no. 6, pp. 537–556, 2002.

[19] A. Omojolaa and O. A. Oladejib, "Small hydro power for rural electrification in Nigeria," *American Journal of Science and Engineering*, vol. 1, no. 2, 2012.

[20] M. Saikia, "Remote village electrification by small hydropower project in Assam," *Indian Journal of Energy*, vol. 3, no. 1, pp. 140–147, 2014.

[21] H. Sharma and J. Singh, "Run off river plant: status and prospects," *International Journal of Innovative Technology and Exploring Engineering*, vol. 3, no. 2, pp. 210–213, 2013.

[22] B. C. Kusre, D. C. Baruah, P. K. Bordoloi, and S. C. Patra, "Assessment of hydropower potential using GIS and hydrological modeling technique in Kopili River basin in Assam (India)," *Applied Energy*, vol. 87, no. 1, pp. 298–309, 2010.

[23] C.-S. Yi, J.-H. Lee, and M.-P. Shim, "Site location analysis for small hydropower using geo-spatial information system," *Renewable Energy*, vol. 35, no. 4, pp. 852–861, 2010.

[24] L. Kosnik, "The potential for small scale hydropower development in the US," *Energy Policy*, vol. 38, no. 10, pp. 5512–5519, 2010.

[25] C. S. Kaunda, C. Z. Kimambo, and T. K. Nielsen, "Potential of small-scale hydropower for electricity generation in Sub-Saharan Africa," *ISRN Renewable Energy*, vol. 2012, Article ID 132606, 15 pages, 2012.

[26] BHA, *A Guide to UK Mini-Hydro Developments*, BHA, 2005, http://www.british-hydro.org/Useful_Information/A%20Guide-%20to%20UK%20mini-hydro%20development%20v3.pdf.

[27] http://www.microhydropower.net/.

[28] U. G. Wali, "Estimating hydropower potential of an ungauged stream," *International Journal of Emerging Technology and Advanced Engineering*, vol. 3, no. 11, pp. 592–600, 2013.

[29] M. C. Tuna, "Feasibility assessment of hydroelectric power plant in ungauged river basin: a case study," *Arabian Journal for Science and Engineering*, vol. 38, no. 6, pp. 1359–1367, 2013.

[30] Small Hydro power, 2015, http://smallhydro.com.

[31] S. K. Singal, R. P. Saini, and C. S. Raghuvanshi, "Analysis for cost estimation of low head run-of-river small hydropower schemes," *Energy for Sustainable Development*, vol. 14, no. 2, pp. 117–126, 2010.

[32] A. Date and A. Akbarzadeh, "Design and cost analysis of low head simple reaction hydro turbine for remote area power supply," *Renewable Energy*, vol. 34, no. 2, pp. 409–415, 2009.

[33] H. Motwani, S. V. Jain, R. N. Patel et al., "Cost analysis of pump as turbine for pico hydropower plants—a case study," *Procedia Engineering*, vol. 51, pp. 721–726, 2013.

[34] S. Mishra, S. K. Singal, and D. K. Khatod, "Costing of a small hydropower projects," *International Journal of Engineering and Technology*, vol. 4, no. 3, pp. 239–242, 2012.

[35] J. L. Gordon and A. C. Penman, "Quick estimating techniques for small hydro potential," *Journal of Water Power and Dam Construction*, vol. 31, pp. 46–55, 1979.

[36] D. Papantonis, *Small Hydro Power Stations*, Simeon, Athens, Greece, 2001.

[37] G. A. Aggidis, E. Luchinskaya, R. Rothschild, and D. C. Howard, "The costs of small-scale hydro power production: impact on the development of existing potential," *Renewable Energy*, vol. 35, no. 12, pp. 2632–2638, 2010.

[38] Q. F. (Katherine) Zhang, "Small Hydropower Cost Reference Model," May 2015, http://info.ornl.gov/sites/publications/files/pub39663.pdf.

[39] IRENA, *Renewable Energy Technologies: Cost Analysis Series, Hydropower, Volume 1: Power Sector Issue 3/5*, International Renewable Energy Agency, Masdar City, United Arab Emirates, 2012.

[40] IEA-ETSAP, "Technology Brief E12," 2010, http://www.etsap.org/.

[41] SEE hydropower clean water clean energy, http://www.seehydropower.eu/download_tools/details.php?id=1.

[42] Technical/economical evaluation of SHP plants: SMART Mini-Idro tool, April 2015, http://www.seehydropower.eu/.

[43] RETScreen, http://www.retscreen.net/.

[44] HOMERsoftware, http://homerenergy.com/software.html.

[45] J. Nganga, M. Wohlert, and M. Woods, *Powering Africa through Feed-in Tariffs*, Renewable Energy Ventures Kenya, Nairobi, Kenya, 2013, https://www.boell.de/sites/default/files/2013-03-powering-africa_through-feed-in-tariffs.pdf.

[46] J. K. Kaldellis, D. S. Vlachou, and G. Korbakis, "Techno-economic evaluation of small hydro power plants in Greece: a complete sensitivity analysis," *Energy Policy*, vol. 33, no. 15, pp. 1969–1985, 2005.

[47] A. Salmani, S. Sadeghzadeh, and M. R. Naseh, "Optimization and sensitivity analysis of a hybrid system in KISH-IRAN," *International Journal of Emerging Technology and Advanced Engineering*, vol. 4, no. 1, 2014.

Permissions

The contributors of this book come from diverse backgrounds, making this book a truly international effort. This book will bring forth new frontiers with its revolutionizing research information and detailed analysis of the nascent developments around the world.

We would like to thank all the contributing authors for lending their expertise to make the book truly unique. They have played a crucial role in the development of this book. Without their invaluable contributions this book wouldn't have been possible. They have made vital efforts to compile up to date information on the varied aspects of this subject to make this book a valuable addition to the collection of many professionals and students.

This book was conceptualized with the vision of imparting up-to-date information and advanced data in this field. To ensure the same, a matchless editorial board was set up. Every individual on the board went through rigorous rounds of assessment to prove their worth. After which they invested a large part of their time researching and compiling the most relevant data for our readers.

The editorial board has been involved in producing this book since its inception. They have spent rigorous hours researching and exploring the diverse topics which have resulted in the successful publishing of this book. They have passed on their knowledge of decades through this book. To expedite this challenging task, the publisher supported the team at every step. A small team of assistant editors was also appointed to further simplify the editing procedure and attain best results for the readers.

Apart from the editorial board, the designing team has also invested a significant amount of their time in understanding the subject and creating the most relevant covers. They scrutinized every image to scout for the most suitable representation of the subject and create an appropriate cover for the book.

The publishing team has been an ardent support to the editorial, designing and production team. Their endless efforts to recruit the best for this project, has resulted in the accomplishment of this book. They are a veteran in the field of academics and their pool of knowledge is as vast as their experience in printing. Their expertise and guidance has proved useful at every step. Their uncompromising quality standards have made this book an exceptional effort. Their encouragement from time to time has been an inspiration for everyone.

The publisher and the editorial board hope that this book will prove to be a valuable piece of knowledge for researchers, students, practitioners and scholars across the globe.

List of Contributors

Golden Makaka
University of Fort Hare, Private Bag Box X1314, Alice 5700, South Africa

Siddalingappa R. Hotti
Department of Automobile Engineering, PDA College of Engineering, Gulbarga Karnataka 585102, India

Omprakash D. Hebbal
Department of Mechanical Engineering, PDA College of Engineering, Gulbarga Karnataka 585102, India

Haroun A. K. Shahad and Saad K. Wabdan
College of Engineering, University of Babylon, P.O. Box 4, Hilla, Babylon, Iraq

B. Johnson, J. Francis and J. Whitty
School of Computing Engineering and Physical Sciences, University of Central Lancashire, Preston PR1 2HE, UK

J. Howe
School of Computing Engineering and Physical Sciences, University of Central Lancashire, Preston PR1 2HE, UK
Thornton Science Park, University of Chester, Parkgate Road, Chester, Cheshire CH1 4BJ, UK

Neeraj Kanwar, Nikhil Gupta, K. R. Niazi and Anil Swarnkar
Department of Electrical Engineering, Malaviya National Institute of Technology Jaipur, Jaipur 302017, India

Mahir Said, Geoffrey John and Cuthbert Mhilu
Department of Mechanical and Industrial Engineering, University of Dar es Salaam, Dar es Salaam, Tanzania

Samwel Manyele
Department of Chemical and Mining Engineering, University of Dar es Salaam, Dar es Salaam, Tanzania

Vincent Anayochukwu Ani
Department of Electronic Engineering, University of Nigeria (UNN), Nsukka 410001, Nigeria

Shuvankar Podder, Raihan Sayeed Khan and Shah Md Ashraful Alam Mohon
Department of Electrical and Electronic Engineering, Bangladesh University of Engineering and Technology (BUET), Dhaka 1000, Bangladesh

V. N. Ananth Duggirala
Department of EEE, Viswanadha Institute of Technology and Management, Visakhapatnam 531173, India

V. Nagesh Kumar Gundavarapu
Department of EEE, GITAM University, Visakhapatnam, Andhra Pradesh 530045, India

Vincenzo De Marco, Gaetano Florio, and Petronilla Fragiacomo
Department of Mechanical, Energy and Management Engineering, University of Calabria, Arcavacata, Rende, 87036 Cosenza, Italy

Sandra Eriksson
Division for Electricity, Department of Engineering Sciences, Uppsala University, P.O. Box 534, 751 21 Uppsala, Sweden

Rashid Ahammed Ferdaus, Mahir Asif Mohammed, Sanzidur Rahman and Mohammad Abdul Mannan
Faculty of Engineering, American International University-Bangladesh, Road 14, Kemal Ataturk Avenue, Banani, Dhaka 1213, Bangladesh

Sayedus Salehin
Department of Mechanical and Chemical Engineering, Islamic University of Technology (IUT), Organisation of Islamic Cooperation (OIC), Board Bazar, Gazipur 1704, Bangladesh

Sambu Kanteh Sakiliba, Abubakar Sani Hassan and Jianzhong Wu
Institute of Energy, Cardiff University, Queen's Buildings, The Parade, Cardiff CF24 3AA, UK

Edward Saja Sanneh
Ministry of Energy, Banjul, Gambia

Sul Ademi
Institute for Energy and Environment, Department of Electronic & Electrical Engineering, University of Strathclyde, Technology and Innovation Centre, Level 4, 99 George Street, Glasgow G1 1RD, UK

P. K. Halder
Jessore University of Science & Technology, Jessore 7408, Bangladesh

N. Paul
Bangladesh University of Engineering & Technology, Dhaka 1000, Bangladesh

M. R. A. Beg
Rajshahi University of Engineering & Technology, Rajshahi 6204, Bangladesh

Gandhi Habash and Daniel Chapotchkine
Azrieli School of Architecture and Urbanism, Carleton University, Ottawa, ON, Canada K1S 5B6

Peter Fisher, Alec Rancourt and Riadh Habash
School of Electrical Engineering and Computer Science, University of Ottawa, Ottawa, ON, Canada K1N 6N5

Will Norris
DEI & Associates Inc., Waterloo, ON, Canada N2L 4E4

K. Touafek, A. Khelifa, M. Adouane and H. Haloui
Unité de Recherche Appliquée en Energies Renouvelables (URAER), Centre de Développement des Energies Renouvelables (CDER), 47133 Ghardaïa, Algeria

Prakash Kumar and Dheeraj Kumar Palwalia
Department of Electrical Engineering, Rajasthan Technical University Kota, Rajasthan 324010, India

Saravanakumar Rajendran and Debashisha Jena
Department of Electrical Engineering, National Institute of Technology Karnataka, Surathkal, Mangalore 575 025, India

Andres Schmidt
Department of Forest Ecosystems and Society, Oregon State University, Corvallis, OR 97331, USA

Maya Suchaneck
Department of Geography, Ruhr University Bochum, 44780 Bochum, Germany

Rajen Pudur
Electrical Engineering, National Institute of Technology, Yupia, Arunachal Pradesh 791112, India

Sarsing Gao
Department of Electrical Engineering, North Eastern Regional Institute of Science and Technology (NERIST), Nirjuli, Arunachal Pradesh 791 109, India

Zelalem Girma
Electrical and Computer Engineering Department, Arba Minch University, P.O. Box 21, Arba Minch, Ethiopia University of Kassel, Mönchebergstraße 19, 34109 Kassel, Germany

www.ingramcontent.com/pod-product-compliance
Lightning Source LLC
Chambersburg PA
CBHW080634200326
41458CB00013B/4632

* 9 7 8 1 6 3 2 3 9 7 6 7 6 *